A FIELD GUIDE TO

EASTERN FORESTS

THE PETERSON FIELD GUIDE SERIES®

A FIELD GUIDE TO

EASTERN FORESTS

NORTH AMERICA

JOHN KRICHER

Illustrated by

GORDON MORRISON

Photographs by

JOHN KRICHER

SPONSORED BY THE NATIONAL AUDUBON SOCIETY,
THE NATIONAL WILDLIFE FEDERATION, AND
THE ROGER TORY PETERSON INSTITUTE

HOUGHTON MIFFLIN COMPANY
BOSTON NEW YORK 1998

LIBRARY OF CONGRESS CATALOGING-IN-PUBLICATION DATA

Kricher, John C.
A field guide to eastern forests, North America / John C. Kricher ; illustrated by Gordon Morrison ; photographs by John Kricher.
p. cm. — (The Peterson field guide series ; 37)
Sponsored by the National Audubon Society, the National Wildlife Federation, and the Roger Tory Peterson Institute.
Includes index.
ISBN 0-395-92895-8
1. Forest ecology—North America. 2. Forest animals—North America—Identification. 3. Forest plants—North America—Identification. I. Morrison, Gordon. II. National Audubon Society. III. National Wildlife Federation. IV. Roger Tory Peterson Institute. V. Title. VI. Series.
QH102.K75 1998
577.3'0974 — dc21 98-4134

Book design by Anne Chalmers
Typeface: Linotype-Hell Fairfield; Futura Condensed (Adobe)

PRINTED IN THE UNITED STATES OF AMERICA

RMT 10 9 8 7 6 5 4 3 2 1

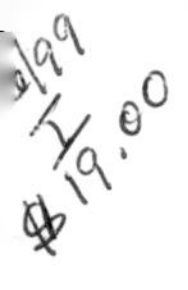

Editor's Note

During the last two or three decades, interest in the natural world has been increasing exponentially; more people are taking to the woods and fields. Some are birdwatchers, others are botanically oriented, many others hunt or fish.

There are also those souls who embrace the environmental ethic—"instant environmentalists," we might call them—who like the idea, but know very little about the interrelationships of nature and what makes the wild world tick. They may recognize fewer than a dozen kinds of birds and even fewer wildflowers or trees, or *no* butterflies except the Monarch and the Cabbage White. The Field Guides were designed to correct this, to make it simpler for them to put names to things, and then go on to learn what animals do (behavior), where they live, and how they interact with each other and with their environment (ecology). This book, number 37 in the Peterson Field Guide Series, is designed to pull things together (or apart). It is one of a new generation of field guides; it takes the ecology of the eastern forests as its theme.

In 1934 my first *Field Guide to the Birds* was published, covering the birds of eastern and central North America. It was designed so that live birds could be readily identified at a distance by their *field marks* without resorting to the "bird in hand" characters that the early ornithologists relied on. During the last half century the binocular and the spotting scope have replaced the shotgun. In like manner the camera has become the modern vehicle for making a botanical collection; no more picking rare flowers, putting them in the vasculum and then the plant press until they are dry enough to fasten to herbarium sheets. In similar manner, butterflies can now be "collected" as transparencies. They need not be caught, put into the killing jar, then pinned through the thorax in a specimen tray.

The Peterson System, as it is now called, is based primarily on patternistic drawings with arrows that pinpoint the key field marks. The rather formal schematic illustrations and the direct comparisons between similar species are the core of the system, a practical method that has gained universal acceptance. This system, which is, in a sense, a pictorial key based on readily noticed visual impressions rather than on technical features, has been extended to other branches of natural history until there are now more than three dozen titles in the Field Guide series. In this new *Field Guide to Eastern Forests*, the Peterson System has been extended to forest types; each type of forest can be identified by its field marks—a unique combination of plant and animal species.

Most readers of this guide to the ecology of our eastern forests by Dr. John Kricher probably have already learned to name many of the trees by using one or both of Dr. George Petrides' books: *A Field Guide to the Trees and Shrubs* (northeastern and central North America) and his more recent *Field Guide to Eastern Trees* (eastern and central North America). Likely as not they also own *A Field Guide to the Birds*, as well as other guides in the series. Do not be concerned, however, that you will need to know the name of everything that grows or flies or crawls in the woods before you can understand and enjoy this guide.

This new approach, developed so skillfully by Dr. Kricher, integrates things so that the woodswalker arrives at a more sophisticated or holistic understanding of the forest and its inhabitants. He explains the diversity and symbiotic relationships that allow them to live together, even though some are rooted to the earth while others, like the birds and butterflies, fly free.

The illustrations of Gordon Morrison artfully support the scholarly text so that we can more quickly put things in order. Most of us are visually oriented and therefore illustrations of this sort are especially important in any field guide.

ROGER TORY PETERSON

Preface

Natural history is an extraordinary subject. The study of nature fascinated Aristotle no less than it did Charles Darwin centuries later. It still fascinates us today. Few subjects so clearly and simultaneously appeal to both the emotions and the intellect. A quiet sojourn in a spring woodland is one of the most relaxing of experiences. The sight of various life forms re-emerging from dormancy to activity provides a splendid sense of continuity. Emotions are recharged as Skunk Cabbage pokes up, Red Maples blossom, and the Spring Peeper chorus breaks what had been winter's silence. Nature is art unbridled by human bias. But, it is far more. In an essay that is now part of an anthology entitled *John Burroughs' America* (New York: Devin-Adair Co., 1967), John Burroughs wrote that:

> The gold of nature does not look like gold at the first glance. It must be smelted and refined in the mind of the observer.

Thus is born the study of natural history. The intellectual stimulation provided by nature was defined by Burroughs:

> How insignificant appear most of the facts which one sees in his walks, in the life of the birds, the flowers, the animals, or in the phases of the landscape or the look of the sky—insignificant until they are put through some mental or emotional process and their true value appears.

The idea for this Field Guide developed as I led numerous public groups on ecology field trips and nature seminars. The inspiration comes directly from John Burroughs's philosophy. Rather than merely naming the plants and animals encountered, I discuss them as they function within their environment. Discovering a mixed flock of foraging bird species begins a discussion of the benefits of flocking, possible competition among species, and anatomical and behavioral feeding specializations. Many questions arise. Although the birds may be feeding in the same tree, are they eating the same foods? Does each species capture food differently from the others? For that matter, why do birds form mixed flocks in the first place? What advantages in terms of survival are obtained by species traveling together? If we stop to inspect a decomposing log, we may uncover a Red-backed Salamander. What makes the log decompose? How do the molecules in the dead log become available for re-use by live plants and animals? A Monarch Butterfly alights on a Common Milkweed. Why is the insect so brilliantly colored? Why did it select the milkweed? Why? Even the shape of a leaf or branching pattern of a tree raises the question, "why?" Is the tree's particular leaf shape or arrangement of leaves on the tree somehow beneficial to it? If so, in what ways? Is it beneficial for some flowers to be blue and others yellow or white? Why do some fruits turn bright red, then deep purple?

The above questions are examples of the sorts of riddles that abound in nature. Indeed, these and many other such questions form the bread and butter of modern ecology. Most of these questions do have at least partial answers. With the proper guide at hand, a field trip at any season can become a very eye-opening experience for one who wants to be more fully attuned to nature's complexities. Late one afternoon, after a day-long nature seminar, one participant approached and, sounding a bit frustrated, said she would like to "take me home with her" so I would be available whenever she needed some natural history observation explained. The idea was flattering but not exactly practical. More commonly, I have been asked repeatedly where the type of information I present can be obtained. The disappointing answer has been that the information was not very readily available but was widely dispersed among ecology texts and wildlife and ecology journals. My purpose in preparing this guide is to bring together and illustrate many of the most interesting facets of natural history that characterize forests and fields throughout eastern North America. Starting with observations made in the field, you will learn how to explain and interpret what you have seen.

Is it possible to know and understand a forest or meadow by merely knowing what species are present? Charles Darwin would have answered no. He illustrated nature's complex interactions using the example of a tangled bank of vegetation: the plants, the insects, the birds, and the mammals all interact in ways not readily evident to the casual observer. Myriads of subtle but fascinating interactions are occurring daily within the tangled bank. Such interactions give definition to ecology, the study of organisms in relation to their environments. Ecology is the modern scientific expression of the study of natural history. Both Thoreau and Burroughs were ecologists. Like Darwin, they understood and valued the notion that to view an environment ecologically, one must realize that the whole is more than the sum of the parts. The science of ecology, which traces many of its roots directly to natural history, attempts to understand how the many different species found together in nature affect one another. The word *interaction* carries a great deal of weight in ecological study. In nature, nothing exists in a vacuum.

The successful student of natural history becomes a storyteller *par excellence.* Consider, for instance, the many colorful mushrooms that appear in late summer and autumn. Many species grow very quickly, often appearing virtually overnight. These mushrooms represent merely the visible reproductive structures of a vast underground network of fungal strands, called mycelia. Some of these fungi grow from the soil into root systems of common tree species such as oaks and pines, but they are not all parasites. Indeed, many supply the trees with vital nutrients, which they take from the soil. The trees in turn supply the fungi with some of the sugary products of their photosynthetic efforts. The trees actually *need* the fungi just as the fungi *need* the trees. Below the soil surface where the mushrooms grow lies a story of finely tuned evolutionary cooperation that is by no means rare in nature (see Chapter 8 for more on this example). The mushrooms and their relationship with trees and soil are part of the "gold of nature." Ecologists have smelted and refined it. It is an interesting and meaningful story. John Burroughs would likely be pleased.

JOHN KRICHER

Acknowledgments

Though I have been privileged to travel widely and gather material firsthand, I have also drawn heavily from published research in writing this book. Particularly helpful have been journals and periodicals such as *Ecology, Ecological Monographs, American Naturalist,* and *Natural History,* as well as others.

I have directly availed myself of the published labors of many researchers: W. G. Abrahamson and M. Gadgil (growth form and reproduction in goldenrods), R. A. Askins (pollination, avian ecology), D. P. Barash (sociobiology, behavioral ecology), G. E. Bard (old field succession), B. V. Barnes (forest ecology), F. G. Barth (insect pollination), F. A. Bazzaz (adaptations of old field herbs, old field succession), L. C. Bliss (alpine ecology), P. T. Boag and P. R. Grant (Galápagos finch evolution), F. H. Bormann (northern hardwood forests), D. J. Borror (insect classification and natural history), D. B. Botkin (disturbance ecology), J. R. Bray (Wisconsin forests), E. L. Braun (forest community types), L. Brower (Monarch unpalatability), M. F. Buell (New Jersey plant ecology), J. E. Cantlon (New Jersey forests), P. A. Colinvaux (general ecology), B. Robichaud (New Jersey plant ecology), J. H. Connell and R. O. Slatyer (succession), R. E. Cook (seed longevity), J. T. Curtis (Wisconsin forests), J. Cypher and D. H. Boucher (beech-maple coexistence), R. Daubenmire (plant ecology), G. Daniel and J. Sullivan (northern forests), H. R. and P. A. Delcourt (Ice Age forests), D. Dindal (soil animal ecology), W. H. Drury and I. C. T. Nisbet (succession), F. C. Evans (bee ecology), M. S. Ficken (avian ecology), D. E. Dussourd and T. Eisner (vein-cutting insects), R. T. T. Forman and students (landscape ecology), S. W. Frost (insect natural history), H. A. Gleason and A. Conquist (plant geography), M. A. Godfrey (southeastern Piedmont forests), R. H. Goodwin (shrublands), F. B. Golley (old field ecology), J. W. Hannon (pollination), J. Harper (plant population ecology), B. Heinrich (Bumble Bee energetics), D. J. Hicks and B. F. Chabot (deciduous forest ecology), R. T. Holmes (avian ecology), A. M. A. Holthuijzen and T. L. Sharik (Eastern Red Cedar ecology), H. S. Horn (tree architecture, forest succession), N. Jorgensen (southern New England forests), C. Keever (old field succession, Pennsylvania forests), S. C. Kendeigh (general ecology), E. D. Ketterson and V. Nolan, Jr. (junco ecology), R. M. Knutson (Skunk Cabbage natural history), J. R. Krebs and N. B. Davies (behavioral ecology), F. T. Kuserk (slime molds), G. H. LaRoi (boreal forests), G. E. Likens (northern hardwood forests), R. L. Lindroth, J. M. Scriber, and M. T. Stephen (plant defense compounds), B. E. Lyon and R. D. Montgomerie (delayed plumage maturation), P. L. Marks (origin of old field plants, ecology of Pin Cherry), C. F. Maycock (Great Lakes boreal forests), R. P. McIntosh (succession, New York forests), D. K. McLain and D. J. Shure (insect herbivory), V. McMillan (dragonfly mating behavior), S. J. McNaughton (plant ecology), C. D. Monk (southern mixed hardwood forests), A. F. Motten (pollination ecology), D. H. Morse (milkweed ecology, avian ecology), J. G. and P. R. Needham (aquatic animal ecology), W. A. Niering (vegetation development), T. A. Nigh and S. G. Pallardy (Missouri forests), W. C. Oechel and W. T. Lawrence (boreal forest ecology), D. Owen (crypsis, mimicry), S. T. A. Pickett (adaptations of old field species), E. Quaterman (southern mixed hardwood forests), T. D. Sargent (underwing moth ecology), W. H. Schlesinger (cypress swamp ecology), V. E. Shelford (forest community types), H. H. Shugart, Jr. (forest succession), D. J. Shure and H. Ragsdale (outcrop ecology, floodplain ecology, old field succession), T. G. Siccama (northern hardwood forests), P. D. Smallwood and W. D. Peters (Gray Squirrels and tannins), S. M. Smith (courtship feeding, Coral Snake mimicry), D. G. Sprugel (Balsam Fir waves), S. H. Spurr (forest ecology and New England forests), E. W. Stiles and students (temperate zone fruit ecology), E. Tramer (plant growth strategies), J. L. Vankat (major forest community types), D. W. Waller (Jewelweed reproductive ecology), J. A. Wallwork (soil animal ecology), P. A. Werner (goldenrod ecology), R. H. Whittaker (Great Smoky Mountain forests, general ecology), M. F. Willson (vertebrate ecology), and E. O. Wilson (proximate and ultimate causation in evolution, sociobiology).

I am very grateful to the following persons for kindly agreeing to review various parts of the manuscript while in preparation: Robert A. Askins, Ralph E. J. Boerner, W. Dean Cocking, Frank T. Kuserk, Edward J. Larow, William A. Niering, Donald J. Shure, Bruce A. Sorrie, Edmund W. Stiles, and Stephen G. Tilley. Wayne R. Petersen provided a valuable critique of the plates. Each reviewer made numerous suggestions that substantially and sometimes dramatically improved the final product. Errors of any sort that perchance have filtered through are strictly my doing.

Donald J. Shure kindly permitted me to reproduce his photographs of granite outcrop communities and southeastern pineland succession.

Ivana Magovcevic, Beth Picard, and Ellen Smith helped me with word processing and I am most grateful to each for her efforts.

I became an ecologist because certain people sparked my interest and shared their insights with extreme generosity. Their imprints are in this book and I regret that some have passed away and will not see it. Murray F. Buell, Florence C. Griscom, Herbert H. Mills, Paul G. Pearson, Phillip R. Pearson, Jr., and John Small were each my ecological mentors. They taught me how to see natural history. I am very much indebted to each of them.

This book was originally an endeavor conceived and shared with Sara E. Bennett. I am most grateful to Sally for her ideas and enthusiasm in the early days of the project.

Harry Foster, my editor at Houghton Mifflin, has been most helpful and supportive from the ideas phase right on up through the final edit. Also, he's one fine editor. I am most grateful for the assistance and skill that Barbara R. Stratton brought to the tedious process of copy editing such a complex manuscript.

Gordon Morrison has made this book a visual treat. His plates have captured exactly the essence of what I sought as I formulated the ideas for the guide. He is a joy to work with and I consider myself extremely lucky that Gordon agreed to take on the project.

My former wife, Linda, as is her way, offered continuous support and encouragement throughout the long period required to write this book. She and I have shared a life of ecology since Rutgers in 1968. Our often-crowded Sunday tours of Hutcheson Memorial Forest, though we did not know it at the time, were really the first drafts of this book. It is to Linda that the book is dedicated.

Contents

A FIELD GUIDE TO EASTERN FORESTS

How to Use This Book

Almost everyone begins to use a field guide by thumbing through the pages and looking at the plates of illustrations. A glance at this book reveals immediately a departure from the usual type of field guide: *A Field Guide to Eastern Forests* is not taxonomically arranged family by family, genus by genus. Instead, emphasis has been placed upon the ways plants and animals interact together. Thus you will discover plates containing many different combinations of species. One plate shows a mixed-species flock of forest birds. Two plates compare various flowers that are pollinated by wind with some that are pollinated by insects. Other plates illustrate animals living on goldenrod and those associated with milkweed. Major forest communities are illustrated in a series of plates showing the most characteristic plants and animals associated with each of the different forest regions of eastern North America. Such diverse combinations of species groups may surprise some readers who are used to finding all the oaks shown together and all the thrushes on the same plate. However, the intention with this guide is to direct the reader toward nature *interpretation.* I try to provide a vehicle for understanding how the complex fabric of nature is woven and held together. This guide is not intended to be comprehensive for field identification. Other, specialized guides in the Peterson Series will serve better for identification purposes. This guide, however, is comprehensive in another context. I have selected numerous examples of species that interact in intriguing and meaningful ways. Using these examples, most of which are widespread in eastern North America, and all of which are readily visible in the field, you will be able to comprehend and better appreciate just how nature works. This is a field guide to ecology.

As Roger Tory Peterson pointed out years ago, there is a logical way to identify birds in the field. The Peterson field-mark system focuses on patterns of similarity among groups of birds and then focuses specifically on characteristics unique to each species. Thus one learns to recognize a woodpecker before learning to identify a Yellow-bellied Sapsucker. The success of the Peterson system derives from its ability to direct users toward various levels of pattern recognition. Is it a grebe or a duck? If a duck, is it a diver or a dabbler? With a sensitivity toward pattern recognition, complexities become less complex, and identifications more reliable. Recognizing field-mark patterns allows one to "see" more efficiently.

Just as pattern recognition can be used to master groups such as birds, insects, and ferns, so may it be used to understand natural history. A woodland may appear insurmountably complex, but it isn't. Patterns abound if one learns how to recognize them and, once recognized, these patterns are the keys to understanding how nature functions.

As an example, consider the pattern of leafing-out seen in an early spring woodland. The woodland itself is structurally organized (see Fig. 1, p. 10). There is a canopy of tall trees (such as various oaks and maples), an understory of shorter trees (such as Flowering Dogwood and Sassafras), a shrub layer, often of viburnums and laurels, and a herbaceous layer of ferns and wildflowers. The herbaceous layer is the first to become green in spring, followed by the shrubs, then the subcanopy. The tall canopy trees tend to leaf out last. Clearly, a temporal pattern as well as a spatial

Landscape Diversity. Old fields and wet meadows intermingle with forests, creating a diverse environmental mosaic that supports a wide variety of plant and animal species.

pattern exists. In this guide I attempt not only to demonstrate these patterns but to interpret them as well. The example just cited is discussed in more detail at the beginning of Chapter 2.

Like animal and plant identifications, natural history interpretation is easier when patterns are clearly recognizable. Using this guide, the reader will discover how to interpret patterns in the field that otherwise might be overlooked. I identify and illustrate characteristics of nature, *ecological field marks*. Though the concept admittedly requires some stretching, the Peterson system can be adapted to a functional, interpretive approach as well as a taxonomic approach. This book does exactly that.

Fields and Forests

Theoretically, one could envision the eastern United States as a vast, potentially unbroken tract of forest. Someone once argued that an ambitious squirrel could travel from New England to Arkansas without ever having to leave the treetops. But even in theory, the idea of an uninterrupted forest primeval is flawed. We know perfectly well that numerous species of plants and animals exist as intrinsic parts of weedy brushy fields. Common Ragweed and Field Sparrows simply do not occur in forests. These organisms are well adapted for their sunny open habitats and would, indeed, be at a serious disadvantage in a shady forest. The very existence of such species argues that fields occur as naturally as do forests throughout eastern North America. Therefore, I include the natural history of fields in this guide. Ecologically, fields are really environments that have been disturbed in some manner. Left alone, most fields throughout eastern North America slowly become forests (see Chapter 4).

As an example, Eastern Red Cedar (*Juniperus virginiana*) grows extensively in open habitats from the Midwest through the Northeast. Other species, mostly pines, replace Red Cedar along southern coastal regions. Red Cedar is a native of eastern North America and requires high levels of sunlight for its rapid growth and to complete its life cycle. Shading by taller tree species eventually kills it. Red Cedar depends on open grassy areas for its survival, many of which are created by occasional disturbances to forests, such as fire or windthrow. The cutting of forests for pasture and agriculture has vastly increased the amount of open habitat suitable for Red Cedar and the tree has clearly prospered as a result. However, it was present as part of the ecology of eastern North America before human settlement, persisting in naturally occurring grassy areas, newly opened areas, and outcrops. When a forest clearing occurs, Red Cedar berries, carried and defecated by birds such as Robins and Northern Mockingbirds, germinate easily. Red Cedar seedlings survive in pastures because neither cattle nor horses appear to enjoy eating them. The presence of abundant Red Cedar in an open field is an ecological field mark. It points to both habitat disturbance and possible past use of the land for pasture. It demonstrates the seed-dispersal powers of birds (see Chapters 4 and 7 for more on these subjects). Fields and forests must be considered together as, ultimately, they form parts of the same temporal tapestry.

AREA COVERED: The area covered by this guide includes eastern North America from approximately the 100th meridian eastward to New England and the Southeast. Most of the guide focuses on the Eastern Deciduous Forest, a vast complex characterized by many species of broad-leaved trees that drop their leaves in winter (see Chapters 2 and 3). This forest is so widespread and complex that different groups of tree species characterize different geographic regions. These groups, or communities, are presented in Chapter 3. Also included are the Boreal Forest, a primarily coniferous evergreen forest of the northern states and Canada; the Southern Mixed Pine-Oak Forest, which is really a unique segment of the Deciduous Forest; and the Subtropical Forest, found only in southern Florida.

GENERAL ORGANIZATION: This guide is divided into eight chapters. The approach taken in each chapter is to present various patterns that you can observe, followed by explanations that discuss why a given pattern may exist as it does. Chapter 2 is an introduction to patterns of forest ecology. It will show you how to look sharply at details of forest natural history and includes a brief questionnaire designed to help you learn how to "see ecologically." Chapter 3 presents each of the major forest types of eastern North America. The plates illustrate selected common *indicator species,* which, in combination, characterize each of the forest communities. With Chapter 3 and the plates to guide you, you should be able to recognize any forest type you encounter and learn what to look for as you travel throughout the region of the guide. Chapter 4 focuses on fields and the changes that occur as many abandoned fields gradually become forests, a process called *old field succession* or *vegetation development*. You will learn about the natural history of common weeds—why some are annuals and others perennials. Some of the most common field plants, ranging from native and alien weeds to shrubs and fast-growing trees, are illustrated on the plates. Using Chapter 4 as your guide, you should be able to interpret landscapes in terms of levels of disturbance and maturity. You should be able to look at a field, a sand dune, or a rocky

Seasonality. Eastern forest ecology is dominated by the changing of the seasons. These photographs were taken at the same place in the same year in spring, summer, fall, and winter.

outcrop, and know what's going on ecologically. Chapter 5 provides background about the process of adaptation and natural selection. With examples, you will learn to think about the complex processes of natural selection, adaptation, and evolution and to understand how these processes shape organisms. Chapters 6–8 take a seasonal approach.

THE SEASONAL APPROACH: Eastern fields and forests, whether in Vermont or in Georgia, are all part of the temperate zone. Summer is a time of growth and winter a time of quiescence. Among the most distinctive patterns of eastern forests are those involving the changes of the seasons. Recognizing this fact, I have organized the latter parts of the guide by season. There are chapters on spring (Chapter 6), summer (Chapter 7), and autumn/winter (Chapter 8). Each chapter begins with a brief introduction outlining the sorts of events associated with that particular time of the year. After reading the introduction, go to the plates at the end of the chapter. These illustrate some of the patterns you should try to identify, the ecological field marks. The plates, however, provide examples only. Although this guide is as comprehensive as possible regarding the events of nature, I do not attempt to include all species. However, the examples selected encompass species with broad ranges throughout the geographic area served by the guide.

GETTING STARTED: Nature is never dull, and you can begin interpretive natural history at any season. Ideally, you should familiarize yourself with the plates and read the first four chapters of the guide at the outset, because these chapters provide a valuable overview and each is applicable to all seasons.

A brief description appears opposite each plate, along with a cross-reference to the page where each topic is discussed in more detail. The text gives a more complete interpretation of the pattern illustrated. Additional illustrations appear throughout the text as supplements to the plates. Where certain patterns may be evident in more than one season, chapters are appropriately cross-referenced. Descriptions are concise enough to be read in the field, while observing the pattern under consideration. Each seasonal chapter is arranged to emphasize plants first, then insects and other invertebrates, followed by birds and other vertebrates.

Various terms are introduced where needed (e.g., ecosystem, succession, metamorphosis, etc.). They are defined when they occur in the text. Unnecessary terminology and jargon have been avoided.

Upon hearing my description of this book while it was in preparation, a friend of mine termed it a "second-generation field guide." He noted that for many people attracted to natural history, identification is a great beginning, but they soon realize that there is more to understanding and enjoying nature than merely naming species. With this guide to patterns and interrelationships at hand, you can venture into the field and discover "nature's true value." Have fun.

Forest Field Marks

Seeking Patterns

The eastern half of North America is clothed in wood. More than 600 species of trees and shrubs produce a diverse forest that extends from the northernmost limits of tree growth in Canada southward to the semitropical climes of Florida. Within this vast area, some trees, like Red Maple, can be found growing in woodlands of every eastern state. Others, like Virginia Live Oak, are much more restricted in distribution. In addition to trees and shrubs, eastern forests include hundreds of species of herbaceous plants, fungi, mosses, liverworts, ferns, and even algae. Animals, too, are diverse. Birds are perhaps the most conspicuous animals seen in forests. Though more secretive than birds, numerous reptiles, amphibians, and mammals will be evident to the observant visitor. Far more abundant are the millions of insects, spiders, mites, and other invertebrates that cohabit a woodland community. All interact in interesting and important ways.

So diverse is any woodlot that no single inventory exists that includes all of the plant, animal, and microscopic species in any one given area. To the naturalist, this complexity is both a blessing and a curse. With so many interacting parts, how can one hope to unravel and understand how forests work? The old adage about being unable "to see the forest for the trees" can seem only too accurate. Nature looks complicated, but, although complex, it has patterns. With a trained eye you can learn to recognize common characteristics and patterns in different types of forests. Although the plants and animals may vary from one forest to another, the processes by which nature works are the same. In this chapter you will learn that you indeed can "see the forest *and* the trees."

Patterns are evident in any forest. In this chapter I have ar-

Fern Layer. Ferns grow rapidly while light is relatively abundant, prior to the canopy leafing out in spring. Species shown is mostly Hayscented Fern.

ranged them in the same order I use when I point them out to groups of beginners. I begin with the obvious, constantly posing the question, "why?" By the time we have taken a brief walk through forest and neighboring field, we have noted and discussed over a dozen broad patterns. You can do the same thing. Taken together, these patterns will tell you a great deal about any forest. You will know something about the forest's present state, its past, and its future. You will develop a feeling for the way the forest is structured, the interaction between forest and the soil, and the roles played by the animal inhabitants of the forest. You will also learn to recognize and understand the dynamic interface between field and forest.

Stratification

OBSERVATIONS: Within forests, green leaves are characteristically arranged in vertical layers, a pattern referred to as *stratification*. The crowns of the tallest trees combine to form a *canopy*, which is usually clearly defined. Beneath the canopy there is often (but not always) an *understory*, consisting of smaller tree species such as Flowering Dogwood or Sassafras. Below the understory there is usually a *shrub layer*, made up of species such as laurels and viburnums. The shrub layer tops an *herb layer* of ferns and wildflowers. This layer is particularly evident in spring, when many wildflower species, the spring ephemerals, are in bloom. Finally, atop the soil itself, is a *litter layer* of decomposing leaves and wood. Fully mature forests typically display clear stratification, while young forests may be less clearly stratified.

Fig. 1. Forest strata.

EXPLANATION: Stratification allows various plant-growth forms to coexist within a given habitat. This is more than just a difference in height. In spring, green ascends from the ground upward. Spring ephemeral wildflowers, such as Jack-in-the-pulpit, Mayapple, and Spring-beauty, leaf out first. Skunk Cabbage is in nearly full leaf and fiddleheads of ferns are opening while buds on canopy trees are still tightly closed. Species of the shrub layer become green next, following the wildflowers and ferns in the upward pattern. Finally, the understory and canopy species spread their leaves. The reason for this orderly progression of greening from the ground up has to do with plant growth and sunlight. Light from the sun is required by all plants as an energy source. By leafing out earlier than trees, herbs and shrubs have more sunlight and can grow faster. Although as spring turns to summer the days will become longer and the temperature warmer, sunlight at the forest floor will steadily decrease due to shading. When the canopy is fully developed, often as little as 1% of the light striking the canopy finds its way to the forest floor. The remaining 99% is reflected or absorbed by the canopy leaves. Stratification is one pattern by which several distinct kinds of plants can share the same space and resources.

Stratification is most evident in moderately moist, or *mesic,* forests (see below). Drier forests display a more open canopy, and the trees are shorter in stature. The understory of dry forests is often comprised of blueberries and huckleberries, members of the heath family. Very wet, swampy forests are also poorly stratified, typically with an open canopy.

OBSERVATIONS: The many species of trees and shrubs that grow throughout eastern North America all fall into a few growth types. For instance, one forest will be characterized by mostly *needle-leaved* trees, such as pines, spruces, firs, and hemlock, while another will be comprised mostly of *broad-leaved* species, such as oaks, maples, hickories, and beech. Still another may contain a mixture of needle-leaved and broad-leaved species. The relative abundance of needle-leaved and broad-leaved trees varies with the climate. Another important characteristic is whether trees and shrubs are *deciduous* or *evergreen.* Deciduous trees drop their leaves in autumn and evergreen species remain in leaf all year around. In North America, most needle-leaved trees are evergreen and most broad-leaved trees are deciduous. The few broad-leaved species which are evergreen tend to occur in the southern states.

The general pattern, from Far North to Deep South is: evergreen needle-leaved trees, deciduous broad-leaved trees, and mixed deciduous and evergreen broad-leaved trees. Of course, many woodlands are mixtures of various growth types.

EXPLANATION: Trees vary in their tolerance to extremes of climate, and are especially influenced by temperature and precipitation. Because climate varies considerably across eastern North America, so do the tree species of the forests.

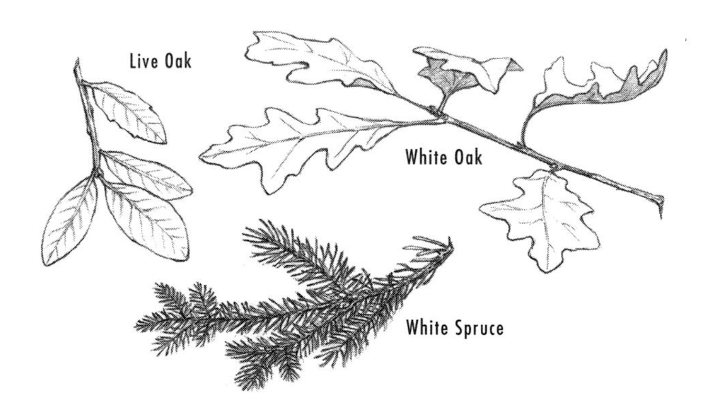

Fig. 2. Deciduous and evergreen leaves.

From the Midwest to the northeastern United States and Canada, average annual precipitation varies from 28 to 60 in. (70–150 cm) and, though fairly evenly distributed among the months of the year, in winter it often falls as snow. Annual precipitation generally decreases both northward and westward. In the southeastern states rainfall averages between 40 and 60 in. annually (100–150 cm), increasing southward, and mostly falling in spring and summer. Precipitation is highest, about 98 in. (250 cm) per year, in the southern Appalachians. The growing season is shortest in the northern states and Canada and longest in the South. For instance, Ontario has a 140-day season, New York City a 175-day season, and North Carolina enjoys a 200-day growing season. Northward the growing season is more condensed, because colder temperatures both reduce the rate at which organisms can function and turn precipitation into snow, which, because of frozen soil, cannot be taken up by the plants until the spring thaw. Needle-leaved species survive best under these conditions because their waxy needles, stems, and roots are filled with resinous chemicals that serve as botanical antifreeze. They can last through a tough winter without damage from freezing, and the needles are in place and able to begin photosynthesis as soon as the ground thaws and liquid water is available. Even the pointed, conical shape of a Balsam Fir or White Spruce serves to minimize snow buildup and is thus adaptive for the cold winter climate. In northern areas, the growing season may be as short as 65 days. In New England the last killing frost of winter may occur as early as mid-March or well into May, varying the length of the annual growing season. Throughout most of the northeastern states the season lasts about 150 days.

The difference between a 65- and a 150-day growing season is an important determinant of tree growth form. Needle-leaved species become dormant over the long winter and conserve their needles. Broad-leaved trees, given the additional 85 days or more of growing season, drop their leaves in winter and have sufficient time to grow new leaves in spring. Broad leaves offer proportionally more exposed surface than needle leaves, an advantage for capturing the sun's rays. For this reason broad-leaved species predominate in areas with sufficiently long growing seasons. From extreme southern New Jersey to the Deep South the growing season exceeds 200 days per year. In most of the South it averages 250 days, and in southern Florida, the growing season essentially lasts all year. Under such conditions many broad-leaved species remain evergreen. These species rarely suffer sufficient stress to induce dormancy.

INDICATOR SPECIES

OBSERVATIONS: In any habitat certain species will be very abundant and conspicuous. In many parts of eastern North America, oaks and hickories comprise over 90% of the tree species in a forest. These are the *indicator species* of a community termed the Oak-Hickory Forest (see Chapter 3). In addition to oaks and hickories, Blue Jays, Wild Turkeys, and Gray Squirrels are indicator species of the Oak-Hickory Forest. Both the birds and the squirrels eat acorns and hickory nuts.

Indicator species will likely be the first species you notice when you survey a habitat. For instance, in the Boreal Forest (see Chapter 3) you will find a great abundance of White Spruce and Balsam Fir and you might encounter North America's largest hoofed mammal, the Moose. The Boreal Forest is often termed the "spruce-moose forest." Some habitat types are almost entirely defined by a single or very few indicator species. Baldcypress Swamps, Mangrove Swamp Forests, and Beech-Maple Forests are examples (see Chapter 3). Other habitats, such as the Appalachian Cove Forest, are comprised of many species, and these species must be considered together to define the forest type. In the Appalachian Cove Forest, two tree species, White Basswood and Carolina Silverbell, though often not numerically abundant, are considered to be indicator species of this forest type because of their uniqueness to the area. Many forests are identified by an entire constellation of species. In Chapter 3, each of the major forest types of eastern North America is discussed and the indicator species named.

EXPLANATION: Indicator species, by their uniqueness or abundance, confer a distinct identity on the habitat. The term *dominant species* is often used interchangeably, but dominant refers only to a large population of a species being present and not necessarily to the effect of the dominant species on other less abundant species. Indeed, some indicator species (like White Basswood in the Cove Forest, for example) may have small populations but still be highly indicative of a certain habitat. Another example is the presence of insectivorous plants in bogs. Though not obvious and often far from abundant, insectivorous plants only grow under the highly acidic conditions found most commonly in bogs (see Chapter 5).

Plants depend on both climate and soil (see below) for their survival. Unlike animals, a plant cannot escape an unsuitable area by merely running, hopping, or flying away. Therefore, plant distribution is tightly linked to climate and soil type, both of which vary considerably across eastern North America. Bald-

Mountain Laurel. This evergreen member of the shrub layer grows throughout eastern North America. Flowers have anthers arranged in arching spokes; spokes release when contacted by an insect, an adaptation for pollination.

cypress could not survive the northern winters but thrives in hot humid southern swamplands. When we recognize a species as being an indicator of a certain habitat, what we are saying is that this species is biologically adapted to survive under the conditions imposed upon it in that area. Each species has certain tolerance limits beyond which it cannot persist. American Holly is abundant throughout the Southeast, but it becomes progressively rarer to the north as the climate begins to exceed its tolerance limits.

Not all species serve as indicators of habitats. Red Maple occurs in virtually every forest type in the East and is therefore rarely a suitable indicator of any particular habitat (the exception being Red Maple Swamp Forests of the Northeast). White Oak is another species that is too widely distributed to be a good indicator, though it is a prominent member of the Oak-Hickory Forest. Only in combination with other oaks and several hickory species is it a useful indicator.

Plants and animals that do serve as useful indicators of their habitats are described in the later chapters and on their accompanying plates. Below is a brief list of some of the most common and widespread forest plant species which, because of their tolerance of a variety of soil and moisture conditions, are generally not, by themselves, indicators of particular habitats. However, several of these plants, such as Black Willow, White Oak, Red Maple, and American Beech, *are* indicators of specific forest communities when in combination with certain other species. It is the *combination* that is important, not just the individual species.

Black Willow
Hophornbeam
American Hornbeam
American Beech
White Oak
Pignut Hickory
American Elm
Tuliptree
Sassafras
Witch-hazel
Black Cherry
Red Maple
Flowering Dogwood
Red-osier Dogwood
Smooth Sumac
Mountain Laurel
Buttonbush
Red Mulberry

Animals as Indicator Species

As mentioned above, animals, because of their mobility, tend to be less suitable as indicator species than plants. Many of the birds and mammals of eastern North America range very widely in virtually all forest types. A list of some of these non-indicator species is given here. Other animals, however, do indicate specific habitat types because their ecological requirements restrict them to a single habitat. An example is the Brown-headed Nuthatch, found only in Southern Pine-Oak Forests. Animals may also be *partial* indicator species. The Prothonotary Warbler, for example, is found only in swamps and along southern rivers. The swamps may be composed of various tree species but must be swamps. Animals useful as indicator species are listed along with plants in Chapter 3.

Eastern Screech Owl. This species occurs in both red (shown here) and gray color phases and is a common resident in most forests and woodlots throughout all eastern states.

BIRDS — PERMANENT RESIDENTS (PL. 3)

Red-tailed Hawk
Ruffed Grouse
Wild Turkey
Great Horned Owl
Eastern Screech Owl
Common Flicker
Pileated Woodpecker
Downy Woodpecker
Hairy Woodpecker
American Crow
Blue Jay
White-breasted Nuthatch
Brown Creeper
Northern Cardinal
Song Sparrow

BIRDS — SUMMER RESIDENTS (PL. 3)

Broad-winged Hawk
Whip-poor-will
Ruby-throated Hummingbird
Great Crested Flycatcher
Eastern Wood-pewee
Eastern Phoebe
Barn Swallow
House Wren
Gray Catbird
American Robin
Wood Thrush
Red-eyed Vireo
Black-and-white Warbler
Ovenbird
Common Grackle
Northern Oriole
Rose-breasted Grosbeak
Indigo Bunting
Rufous-sided Towhee

Pileated Woodpecker diggings. Oval holes are characteristic only of Pileated Woodpeckers, a widespread forest species most abundant in the southern states.

Red Fox. Often seen during the daylight hours, the Red Fox ranges widely throughout eastern forests and often hunts in open fields.

MAMMALS (PL. 2)

- Virginia Opossum (expanding range northward)
- Big Brown Bat
- Eastern Mole
- Eastern Chipmunk
- Northern and Southern flying squirrels
- Woodchuck
- Deer Mouse
- White-footed Mouse
- Eastern Cottontail
- Striped Skunk
- Raccoon
- Short-tailed Shrew
- Beaver
- Gray Squirrel (very abundant in oaks)
- Long-tailed Weasel
- Mink
- River Otter
- Red Fox
- Gray Fox
- White-tailed Deer
- Black Bear
- Bobcat

Many of the most common mammal species are wide ranging, not confined to any particular forest type. As such, they are not indicator species, but they are frequently encountered. The following section will introduce you to some of them.

BLACK BEAR

Once persecuted into near extinction throughout much of its former range, eastern populations of Black Bear are rebounding. Always relatively abundant in such places as the Great Smoky Mountains and parts of Florida, the Black Bear is becoming increasingly abundant throughout much of New England and the north-central states as well as the Southeast. The species remains common throughout much of Canada and Alaska. Black Bears will live in any kind of eastern forest, including swamps. They are classified as carnivores but, in reality, will eat a wide range of fruits, roots, and other vegetable items as well as meat. Fish are also a common part of a bear's diet. The bears are scavengers; many people experience their first Black Bear sighting as they watch the animal picking over garbage at a dump. Eastern populations of Black Bears are, indeed, black, while western populations vary from black to brown to cinnamon. Northern animals winter in dens, entering into deep sleep. However, they are not true hibernators. Young, usually twins or triplets, are born in January and February. An infant Black Bear that will grow to perhaps 590 pounds by adulthood (males are larger and heavier than females) weighs but half a pound at birth. Black Bears rarely harm humans but should be regarded as potentially dangerous and thus not disturbed or approached, especially sows with cubs.

BIG BROWN BAT

One of the commonest bats of eastern forests, often observed at dusk on summer nights as it flies erratically about in search of insect prey, captured in the air. Bats, the only true flying mammals, are also among the only animals to have evolved a prey detection system based on sonar, in which the animal emits high-frequency sounds (generally inaudible to the human ear) that bounce against objects, both moving and stationary, and that are reflected back to the bat to guide its flight, a reason why bats typically have very large pinnae (external ear flaps). Big Brown Bats are well named, as they have wingspreads of about 10–13 in., making them among the largest of the eastern bats. Eastern populations are typically dark brown. This species is widespread, occurring from northern Mexico throughout the United States and well into Canada, including the Far West. Highly nocturnal, bats sleep by day, and this species roosts in caves, hollow trees, and often buildings, with small groups tightly clustered together as they dangle upside down. Some bat species migrate to avoid winter while others hibernate. Big Brown Bats are known to do either, depending upon whether they are northern or more southern populations. The other widespread species abundant in the East is the Little Brown Myotis, or Little Brown Bat. It is smaller and flies with more rapid wingbeats than the Big Brown Bat.

STRIPED SKUNK

Common and widespread, occurring in all 50 states, northern Mexico, and all Canadian provinces. Found in all forest types as well as forest edge, old fields, and prairie. The bold and easily recognized black and white coloration, visible both day and night, acts as a form of warning coloration (see Chapter 7), as skunks defend themselves effectively through noxious scent. Members of the weasel family (in which all species have powerful musk glands), skunks are particularly potent. Two scent glands located on either side of the anus are capable of squirting a substance called methylmercaptan, from 10–15 ft., directly at any potential predator. Skunks normally warn before spraying, usually by stamping their feet. Consequently, skunks are rarely attacked, except by Great Horned Owls, which may be insensitive to the normally overpowering odor. Skunk odor may be at least partly removed by treatment with tomato juice or household ammonia. Mostly active at night, skunks are omnivores, eating a wide range of plant and animal foods. Females give birth in spring, and young normally accompany the mother through midsummer. Active year-round. The only other skunk species found east of the Mississippi, the Spotted Skunk, is found mostly in the Midwest and southern states, where it occurs with the Striped Skunk. Striped Skunks grow to a length of about 30 in., males larger than females, and may weigh up to 14 pounds.

WHITE-TAILED DEER

The most widespread deer in the western hemisphere, its range extending from mid-Canada throughout the United States, and on into Mexico and Central and South America. Abundant throughout much of the East. White-tailed Deer are named for the wide, flattened tail that is held erect as the animal bounds away, flashing a bright white underside. Overall, this species is russet tan in summer and grayish in winter. Only males, called bucks, have antlers. A large buck may weigh nearly 400 pounds and stand nearly 4 ft. tall at the shoulder. Bucks are substantially larger than females, called does, which weigh only up to 250 pounds and stand 3 ft. at the shoulder. Fawns are light tan with white spotting. Best described as a generalist, White-tails occupy a diverse array of habitats and eat a wide variety of plant foods. You may encounter White-tails in open fields, in pastures and agricultural areas, or in open woodlands and dense forests. They are essentially browsers, feeding on leaves of many deciduous trees as well as numerous kinds of conifer needles. They are also fond of various nuts, including acorns and beech nuts. Around farmland, White-tails favor cornfields. Because of elimination or reduction of predators over most of its range, as well as rigorous hunting restrictions, White-tailed Deer are abundant in many areas. Local populations sometimes become "too successful" and overgraze, leading to higher disease rates and malnutrition in the herd.

During much of the year, bucks and does remain in separate

herds, though they gather together in "deer yards" in winter. A buck may mate with several does, or only one. A healthy doe will normally give birth to two fawns, occasionally to three. Fawns initially have virtually no scent, an aid in avoiding detection by predators such as Coyotes, and a fawn will tend to crouch and freeze rather than run when a potential predator is near. Look for bark rubbed smooth by White-tailed bucks vigorously scraping their antlers. When antlers grow they are initially covered with fine hair called velvet. Deer scrape off the velvet before the onset of the rutting season.

BEAVER

The Beaver, which weighs an average of 50 pounds (but sometimes reaches 100 pounds), is the largest rodent in North America and the second largest in the world, exceeded only by the South American Capybara, which reaches 120 pounds. Beaver are widespread throughout North America, including much of the East. They are aquatic animals, well known for their engineering activities that involve creating impoundments through the ambitious building of dams. Beaver make profound habitat alterations, creating habitat for many kinds of plants and animals. They often use aspens in their dam and lodge building, and they are fond of eating aspen bark as well. Active all year, Beaver collect food for winter and store it in their lodge. Mostly nocturnal, Beaver are best seen at dawn or dusk, when they are active. They are wary and, when a Beaver detects danger, it is apt to slap the water with its tail, a warning to other Beaver (its relatives) to take cover. Beaver are unmistakable: large size; naked, scaly, flat tail. Their hind feet are webbed, an obvious adaptation for efficient swimming, and their nostrils, ears, and eyes are protected by membranes when they are swimming beneath the water's surface. Beaver can remain underwater for up to 15 minutes. They waterproof their thick fur with an oil (castoreum) taken from glands near the anus and applied with comblike nails on the hind toes. Beavers live in family units and are thought to pair for life. Young are born in late spring.

RACCOON

Easily identified by the bushy ringed tail, pointed nose, and black facial mask. Common and increasing throughout many suburban areas in the East. Occurs throughout the U.S. except for parts of the Rocky Mountains and Great Basin Desert. Found in all forest types, woodland edges, old fields. Skilled at climbing trees. Raccoons are members of the family Procyonidae, along with the Ringtail and Coati, both of which are confined to the Southwest and tropical America. Omnivores, they eat a wide variety of fruits, nuts, and various animal prey ranging from birds' eggs to crayfish. They tend to occur near streams and are known to dip some of their food in water before consuming it. Raccoons may weigh al most 50 pounds and be nearly a yard in length. They are typically rather bold, tipping garbage cans over. It should be remembered that they are very powerful and potentially fierce animals and should not be cornered. They tend to be active year-round, sometimes denning up for part of the northern winter, but never hibernating. Generally active at night, they are frequent victims of automobiles. Young, normally a litter of four, are born in spring.

RED FOX

Once significantly reduced in population by fur trappers, the Red Fox and its relative the Gray Fox are expanding their populations in many areas. Occurs throughout North America (except the Great Basin Desert and parts of the Southwest) ranging as far north as Alaska. Found in all forest types, woodland edge, old fields. Like wolves and Coyotes, members of the canine family. Red Fox is easily recognized by its overall rusty red coloration, white undersides and chin, and white tip on the bushy tail. The similar-sized Gray Fox is gray above, reddish below, with a black tip on its tail. Red Fox can weigh up to 14 pounds and be 40 in. in body length with a 17-in. tail. Primarily nocturnal, though often seen in daylight, foxes are omnivores, taking many kinds of plant food (fruits, grasses, nuts, corn) as well as animals such as various birds, squirrels, Woodchucks, and crayfish. Also will take large insects. Active year-round, they mate in winter, with litters of 4–8 pups born in spring. Mother maintains a den while the young are dependent. Red Foxes occur in several pelt colorations: black phase, silver phase, and mixed phases, as well as the normal red phase.

EASTERN COTTONTAIL

Generally abundant throughout most of the East, occurring as far west as the Dakotas and parts of west Texas and Arizona. Found in most forest types except Boreal Forest. Also common in old fields, thickets. Unmistakable, with large white "powderpuff" tail. Typically hop. The similar New England Cottontail has a rusty nape with black between its ears, field marks lacking in the Eastern Cottontail. Active year-round, often seen foraging on snow. Vegetarian: eats a wide variety of vegetable matter. Females may produce 3–4 litters per year and each litter normally has 4–5 young. High mortality rate, as they are preyed upon by numerous creatures ranging from hawks and owls to weasels, Raccoons, and foxes. Adults rarely live more than one year. Though many people assume rabbits to be rodents, they are not. Rodents have a single pair of upper incisors, while rabbits and hares have two pair, one of several skeletal characteristics putting them into a different family, the Leporidae.

WHITE-FOOTED MOUSE

Typically the most abundant mammal in forests, though somewhat cautious and nocturnal, so not often seen. A small (6–8 in.) mouse with large ears, large rounded eyes, and a long (3.5 in.) bicolored tail. Coat is rusty brown to grayish, belly always white. Occurs in all forest types throughout most of the East, ranging to the Dakotas, Montana, and northern Colorado as well as all of Texas and much of Arizona, but absent from the Southeast and Florida, as well as most of Canada. Replaced in the Southeast by the similar Cotton and Old Field mice, and in the North and West by the Deer Mouse. Omnivorous, eating seeds, fruits, and a variety of small animals, mostly insects. Good tree climbers, they are active in winter, roosting in nests and surviving on seed caches. Heavily preyed upon by many avian and mammalian predators, as well as various snake species.

Many species of birds can be found throughout most eastern forests; thus, being so widespread, they are not indicative of any particular forest type. Some are permanent residents, and some are present only during the summer breeding season.

RED-TAILED HAWK

(25 in.) A large, bulky, soaring hawk with a white chest, a distinctive "belly band," and rufous on the upper tail. The Red-tail is typically seen soaring in lazy circles on warm thermal air currents with its tail widely spread. It often perches on utility poles. Feeds heavily on mammals such as rabbits and various rodents. Makes a large, often conspicuous nest. Permanent resident occurring throughout the West as well.

BROAD-WINGED HAWK

(19 in.) Small soaring hawk with barred tail, reddish streaks across breast. During spring, courting pairs are often conspicuous as they soar above the forest calling loudly. Very generalized predators, Broad-wings devour small mammals, birds, reptiles, and amphibians, often captured along the forest edge. Long distance migrants, Broad-wings migrate via the Isthmus of Panama to winter in South America. Large numbers of migrating Broad-wings (a formation called a kettle) can be seen soaring together during September and October.

GREAT HORNED OWL

(25 in.) A large nocturnal owl with distinct feather tufts above each eye. White throat (bib), orange disk around each eye, heavily barred breast. Makes a resounding, deep hoot. Great Horneds occur in all types of forests and are common residents in deserts as well, found throughout North America and well into South America. In the East they begin courtship and mating in winter, nesting either in an abandoned hawk, crow, or heron nest or evicting the "rightful owners" and taking over the nest (Great Horneds are known for their aggressiveness). They feed on mammals such as rabbits, squirrels, and skunks, but will also take snakes and some birds, including hawks. House cats often fall victim to Great Horned Owls.

AMERICAN CROW

(21 in.) An abundant, unmistakable permanent resident of forests and farmlands throughout most of North America. Gregarious, uniformly black with a purple sheen, makes a distinctive *caw!* Crows are among the most adaptable of birds and are believed to be of unusually high intelligence. Though probably woodland inhabitants originally, crows have adapted well to agricultural, suburban, and urban ecosystems. They are omnivorous, eating a very wide range of foods. Routinely prey on nests and nestlings of other bird species. In winter, crows often congregate in a large roosting colony called a "rookery," after the European bird called the Rook, a kind of crow. Permanent residents throughout most of the U.S. Canadian populations migrate south during winter.

COMMON FLICKER

(12–14 in.) A brown woodpecker with a spotted breast, often seen on the ground, shows a conspicuous white rump patch in flight, yellow under wings and tail. Flickers are often seen on the ground, where they devour ants. They adapt well to open country and are common residents in wooded suburbs. Migrates in the Far North, but permanent resident throughout most of the U.S. In the West, flickers have gray heads and red under wings and tail. Call is a loud *whicka!*

RUFFED GROUSE

(18 in.) A large chickenlike bird with a wide gray or rusty fan-shaped tail, with a black band at the tip. Often flies up suddenly, with a loud whirr of wings. In spring, males attract females by using their wings to drum atop a hollow log, creating a low-pitched rumbling sound that carries impressively. Permanent resident in all forest types throughout northern states, Canada,

and the Appalachians. Absent from Deep South. In some areas, Ruffed Grouse populations are cyclic, varying regularly in abundance, for reasons not well understood (see page 451).

GREAT CRESTED FLYCATCHER

(10 in.) This flycatcher with a grayish face, bright yellow belly, and rusty tail often perches high in the treetops. Feeds by flying from perch, snatching insect in midair. Eats a wide variety of insect prey and will take fruit as well. Voice a loud *wheep!*, often heard from the dense treetop canopy. Found in any type of deciduous or mixed forest in the East, but absent from Boreal Forests. Nests in tree cavities, where it often lines its nest with snakeskin. Winters in tropics, from Mexico to northern South America, and some birds linger in Florida.

RED-EYED VIREO

(6 in.) A greenish, slender bird with a white stripe above its red eyes (a difficult field mark to see) and a gray crown. Usually found in the canopy but may descend to the understory to feed. Searches upper branches for insects, which it captures by gleaning, snatching prey from branches and leaves. Wide-ranging and probably among the most abundant of summer resident woodland birds, adapting well to suburban woodlands. Song a monotonous warble, given incessantly throughout the day. Winters in South America, generally in Amazonia.

WOOD THRUSH

(9 in.) An upright thrush with a rusty head and heavily spotted breast. Usually seen in the understory or on the ground, where it forages, robinlike, for arthropods and worms. Like all thrushes, Wood Thrushes also eat many kinds of fruits. Song is flutelike, highly melodious. Brown-headed Cowbirds have caused a decline in population among Wood Thrushes in many areas (see page 386). In some woodlots, Wood Thrushes actually raise more cowbirds than thrushes! Winters in tropics, mostly in Central America, where its forested habitat has been severely reduced, another potential threat to Wood Thrush conservation.

OVENBIRD

(5 in.) A ground-dwelling wood warbler that walks methodically on branches and the forest floor. Olive above, with a white eye-ring and orange crown. Feeds on a wide variety of arthropods and worms, as well as various fruits. Song a loud, whistled, repeated *teacher!* Common name derives from the unusual nest (see Plate 47). Winters in tropics, with populations in the Caribbean and Central America.

OBSERVATIONS: The number of species in a habitat reflects its *species richness* or diversity. One of the best ways to begin to know a forest or any other habitat is to take inventory of the species of plants and birds. Both groups of organisms are conspicuous and both show interesting trends in species richness as you move north or south (see below). Species richness of trees varies considerably among forest types. Some forests, like the northern Boreal Forest or the coastal Mixed Pine-Oak Forests, have low species diversities. Large tracts of Boreal Forest consist essentially of only two species, White Spruce and Balsam Fir. In contrast, the Appalachian Cove Forest may have more than 25 tree species in a single acre.

EXPLANATION: Four forces of nature determine if a species is present in a habitat. The first is distribution and dispersal. Can the propagules (seeds, juveniles, etc.) of a species find their way to the habitat? Many European weeds that abound in fields throughout North America would not be here if the land had not been colonized. Seeds of these species were brought from the Old World on the boots and in the hay bales of settlers. Once here, they thrived (see Chapter 4 for more on this topic). The second force determining species richness is physiological tolerance limits. Can the species survive in the habitat? Many southern species of plants succumb to the rigors of a northern winter because they lack genes that would provide them with the physiological ability to tolerate frost. It is fairly obvious but nonetheless important to keep in mind that a species cannot persist in an area unless all of its physiological requirements are met. Some species are simply more finicky than others in this regard. For a species to survive in a habitat it must be able to deal with biotic forces, such as competition from other species with similar needs, as well as predation, parasitism, and disease. A blight that arose during the early part of this century has essentially eliminated the American Chestnut as a major component of eastern forests where, prior to the invasion of the blight, it thrived (see Oak-Hickory Forest, Chapter 3). Finally, a species must have a certain measure of good luck with the unpredictable forces of weather, climate change, and other natural events outside of its control. Nature is often unpredictable, and the laws of chance influence species richness.

SPECIES-RICHNESS PATTERNS

Many species of birds depart from their nesting grounds in the late summer and fall and migrate to sunnier areas. The logic behind this phenomenon seems obvious: birds migrate to escape the rigors of winter, when climate is severe and food supply short.

Though migration is risky (see Chapter 5), it may be less risky than attempting to endure prolonged cold, snow, and ice storms. By its very occurrence, migration tells us that harsh climates exact large physiological costs. Some organisms can pay these costs. Downy Woodpeckers remain during winter's worst cold and Red Pines normally survive winter easily. Many wildflowers survive below ground during winter. But other organisms are physiologically unable to pay the cost. They must leave or they must find a more hospitable climate to begin with.

Warm climates have more species of plants, birds, mammals, reptiles, amphibians, and insects and other arthropods than do cold climates. As you travel from south to north you encounter fewer species. This may be because it is difficult for many species to evolve the physiological adaptations necessary to range into colder, biologically less favorable areas (see Chapter 5). But warm climates are not ideal either. The forces of biological competition, predation, and parasitism may be intensified in areas abounding with species. One reason migration of birds from the tropics to the temperate zone may have begun is because northern areas probably harbor both more insect food and fewer competitors and predators during the breeding season, when migrants must gather food not only for themselves but also for young. Though flying north from the tropics to breed in the temperate zone has risks, the abundant insect food, coupled with fewer predators and competitors, makes the venture advantageous (see Chapter 5).

Soil Characteristics

OBSERVATIONS: The soil is the substrate upon which all forest life depends. In a literal sense it is the foundation of the forest. Soils are affected by regional variations in climate and vegetation. The field marks of a soil are its color, texture (is it sandy or clayey?), and wetness. Roadcuts are excellent places to see *soil profiles,* a vertical layering of texture and color that characterizes many soils. The following soil types are found in eastern North America:

TRUE SPODOSOL soils are very gray in color, with a thin black band of highly decomposed organic material called *mor humus* (see Chapter 8) on the surface. These soils used to be called "podzols," from the Russian word meaning "ash earth." If you dig down through the gray layer, you will find that the color changes to dark brown or sometimes reddish. True spodosol soil is found in northern coniferous forests throughout Canada and North America. Atop the soil is a litter layer of decomposing needles, often so thick that the ground is soft to walk upon.

GRAY-BROWN SPODOSOL (sometimes called **ALFISOL**) is the common soil type found beneath eastern broad-leaved deciduous forests. The color is browner than the gray of true spodosols and the litter layer of dropped leaves is not as thick as in coniferous forests. The soil is less acidic, forming a humus called *mull* (see Chapter 8) that hosts many more earthworms than are found in true spodosols. This soil type occurs throughout the Northeast and mid-Atlantic states.

YELLOW SOIL is found on the Coastal Plain from North Carolina through eastern Texas. Yellowish on the surface and grayish yellow below, this soil type is very sandy compared with other eastern soils. Pine and Pine-Oak forests are the common forest types growing on yellow soil. The litter layer is often thin.

RED SOIL, or **OXISOL,** is found throughout the southeastern states and can range in color from orange-red to bright salmon. Beneath the red is a deeper layer of yellow-brown soil. Southern Hardwood Forests, Southern Oak-Hickory Forests, and some Mixed Pine-Oak Forests are found on this soil type. As on yellow soil, the litter layer tends to be thin.

PRAIRIE SOIL, or **CHERNOZEM,** is colored deep brown at the surface and lighter brown below. It occurs on what was originally tallgrass prairie in Illinois, Missouri, Iowa, Wisconsin, and Minnesota. Many of the original prairie areas have been converted to agriculture and forests have begun to grow on some areas of prairie soil. True prairie is rare in these states today.

EXPLANATION: Soil is a product of the combined effects of geology, rainfall, temperature, and vegetation. Geology, the bedrock and its continuous erosion, provides the basic elemental material. Soil particles come in three size categories, the mixture of which gives the soil texture. Sand grains are the largest soil particles, followed by intermediate-sized *silt* particles and microscopic-sized *clay* particles. Good healthy soils are a mixture of sand, silt, and clay and are called *loams.* Loam soil feels a bit "gritty" but not sandy or slimy like clay soils. Coastal soils, because they are geologically quite young, are largely sandy. Clay soils are commonest in the South, where rainfall is highest. *Gravel* is characteristic of glaciated soils in northern areas. Gravel consists of large rounded stones and rocks within the soil.

Soils contain mixtures of minerals such as potassium, sodium, iron, magnesium, and calcium. These elements react chemically with water from rain and snowfall—reactions affected by temperature—and form the compounds that give regional soils their colors. Because of climatic differences, specifically the higher rainfall of the East, soils there differ from midwestern soils. Soils in the South differ from northern soils because warmer temperatures influence chemical reactions in the soil. When water moves through the soil it acts to remove minerals from the surface and deposit them at various depths. This process is termed *leaching*. The different colors and layering patterns, called *soil horizons*, that characterize the various soil types are produced by leaching, as it differs among climatic regions.

Precipitation also adds to the acidity of the soil. Water is composed of both hydrogen and oxygen atoms (H_2O) and, as it moves through the soil, it removes minerals (attached to clay particles) but leaves behind atoms of hydrogen. The more hydrogen atoms there are, the more acidic the soil. Slightly acidic soils are normal in regions of high precipitation.

When leaves decompose they add minerals to the soil. Leaves decompose at different rates depending upon region and species (see Chapter 8). Decomposition slows during the long cold northern winter and is quicker in the South, where the climate is warmer. Conifer needles decompose more slowly than leaves from deciduous trees, so litter accumulates more thickly beneath conifers. Conifer needles and many deciduous leaves also add to the acidity of the soil through their decomposition.

Sandy soils, with their large sand grains, are well aerated, as oxygen has no difficulty penetrating among the particles. However, water also moves very quickly through sand, making sandy soils prone to drought: their water-retention capacity is very poor. In contrast, clay soils become very tightly packed because the microscopic clay particles tend to stick together. Air, water, and roots have difficulty moving through clay soils. Therefore, clay soils are often poorly aerated and water-logged. Loam combines the best properties of soil texture.

Soil is not merely a chemical factory. A complex community of microbial plants and tiny animals inhabits the soil. These are the real laborers of the decomposition and recycling process. This soil community is described in Chapter 8.

Soil Moisture and Habitat

OBSERVATIONS: It is generally not difficult to tell whether a forest is lush or dry. There are three basic moisture regimes and any forest will fall within one of the three.

MESIC FORESTS occur on well-drained soil, usually a loam. These forests typically have a high species richness (diversity) of trees and other plants and generally appear lush. Most of the Eastern Deciduous Forests can be described as mesic.

XERIC FORESTS occur on sandy or otherwise poor soils that are usually overly dry. Forests of the Coastal Plain are xeric and xeric

Sweetgum Swamp. Hydric environments, a form of extreme environment for trees, are often inhabited by relatively few species. In this case, Sweetgum is the only canopy species.

habitats occur sporadically throughout the East. In northern regions, south-facing mountain slopes are often xeric both because soils are thin on mountain slopes and because warmer temperature (the sun shines longer on a south-facing than on a north-facing slope) adds to the potential for water stress.

HYDRIC FORESTS occur in low-lying areas that retain water. Commonly called swamps, these forests are constantly moist and the soil is usually water-logged. Floodplain forests are also hydric for part of the year but are more mesic during the seasons when flooding does not occur.

Relative moisture level is a critical characteristic for plants. Eastern habitats range from very xeric (dry) to very hydric (soggy), and most forest tracts in the area include examples of mesic, xeric, and hydric sites. For example, within the generally xeric southern Coastal Plain, the pine forests are interrupted by mesic hardwood tracts and hydric Baldcypress Swamps. In the region of the mesic spruce-fir Boreal Forest, hydric areas support bogs and xeric areas support Jack Pine Forests. Any region will have a range of forest types based on variations in drainage patterns.

EXPLANATION: A plant adapted to withstand a xeric habitat is usually not well suited to survive in a hydric habitat and vice versa. Only one species, Red Maple, seems equally at home in both hydric and xeric habitats. Red Maples abound in northeastern swamps, but are also commonly found among the oaks on dry, south-facing slopes. As you have probably guessed, mesic habitats are the least extreme, least stressful of the three habitat types, and more species of plants are found in mesic habitats than in either of the two extremes. Floodplains have fewer tree species than well-drained forests, and swamps may consist essentially of a single species. Coastal Plain forests have a lower species richness than bottomland forests.

A look at the various oaks and hickories will demonstrate that different species are associated with different moisture regimes. Most oaks are found in mesic (well-drained) habitats, although some oaks tolerate xeric (dry) conditions relatively well. Species such as Post Oak, Blackjack Oak, Bear Oak, Chinkapin Oak, and Chestnut Oak are all indicator species of xeric habitats. These species thrive on poor, sandy, dry soils. In contrast, Swamp White Oak, Overcup Oak, Swamp Chestnut Oak, Shingle Oak, and Cherrybark Oak (a variety of the Southern Red Oak) are all indicator species of a hydric habitat, commonly called a swamp.

Most hickory species favor mesic-type environments. Among the hickories, the following distribution of species occurs:

SPECIES	MOISTURE PREFERENCE
Water Hickory	hydric
Swamp (Nutmeg) Hickory	hydric
Shellbark Hickory	hydric-mesic
Pecan	mesic-hydric
Shagbark Hickory	mesic-hydric
Black Walnut	mesic
Mockernut Hickory	mesic
Butternut	mesic-xeric
Bitternut Hickory	mesic-xeric
Pignut Hickory	mesic-xeric
Sand Hickory	xeric
Black Hickory	xeric

Evidence of Fire

OBSERVATIONS: Most people are surprised to learn that occasional fire is not only a natural event but often healthy for a forest. Many species of plants are dependent upon intermittent fires for both survival and reproduction (see p. 71). Though generally infrequent in any given location, lightning-set fires do occur and the evidence of fire is often apparent. The most obvious evidence is burned and charred tree trunks, often with resprouted branches (see Fig. 3). A cross-section of an old tree trunk with many years of growth rings may often show fire scars among the rings. Many species, such as Jack Pine and Longleaf Pine, are indicator species of fire and their life cycles are intimately adapted to periodic fire (see Chapter 3). Fires may occur anywhere but are most common in Pine-Oak Forests and Southern Mixed Pine-Oak Forests (see Chapter 3).

Effects of Fire. This Coastal Pine Forest is subject to periodic natural fires, which tend to suppress the growth of oaks, allowing the pines to regrow and persist. Photo shows young Pitch Pines, Blackjack Oaks, and huckleberries.

Plant Population Patterns

OBSERVATIONS: Plants are distributed in one of three spatial patterns (see Fig. 4):

REGULAR POPULATIONS are seen in orchards and tree plantations. Each individual plant is equally distant from others of its species. Regular populations are generally rare in nature.

CLUMPED POPULATIONS are very common. As the name suggests, a given individual of a species is grouped closely with others of its species in a clump. Many tree species, such as American Beech and aspens, grow in tight clumps. In addition, shrubs such as the huckleberries and herbaceous species such as Starflower, Canada Lily, Bunchberry, many violet species, Ground pines (*Lycopod-*

Fig. 3. Charred tree stump with pine seedling germinating.

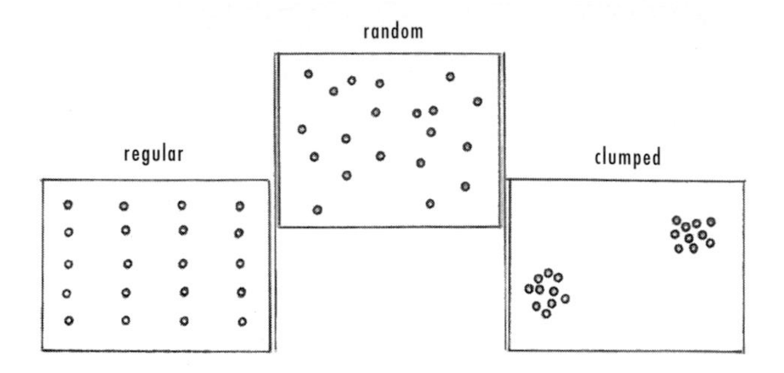

Fig. 4. Distribution patterns.

ium), and Bracken Fern are easily observed to be clumped populations.

Random populations are those in which the individuals seem to have grown by chance, with no relation to each other's presence. Random populations are not always easy to discern. Many populations that appear random at first glance, upon careful measurement, are actually clumped.

EXPLANATION: In nature most plants tend to grow in clumped populations. Plants reproduce from seeds, and most seeds drop near their parent plants. In addition to reproducing sexually by forming seeds, many species reproduce by underground horizontal stems, termed *rhizomes* or *stolons,* which literally interconnect plants that, above ground, appear to be separate individuals. In reality, they are all part of the same plant and are genetically identical! This form of plant growth, termed asexual reproduction, is similar to coral-reef growth in that a colonizing individual can proliferate quickly to efficiently exploit the resources it finds. The name for a group of plants that are genetically identical and all part of the same underground rhizome system is *genet.* Each above-ground individual is called a *ramet.* Genets are clones; examples range from clumps of American Beech, to aspens, sumacs, and many wildflowers, such as Mayapple.

Regular populations, though geometrically neat, are uncommon in nature. Plants cannot move as individuals, and even spacing is an improbable chance event. In the known cases of regular plant distribution, evidence exists that each plant is antagonistic toward others of its species, somewhat in the way animals are territorial (Chapter 6). Many desert shrubs seem to leach chemicals into the soil surrounding them, inhibiting the growth of other in-

Mayapple. An abundant early spring wildflower, Mayapple demonstrates vegetative (asexual) reproduction. Photo shows a genet, or clone, of Mayapple. Each above-ground ramet is connected to the others by an underground stem.

dividuals. Few eastern forest species are antagonistic, but Black Walnut inhibits many species from growing near it by producing a chemical that makes the soil unsuitable for the successful growth of other species (including other Black Walnuts).

Random populations occur when seeds are dispersed by wind or animals (see Chapter 7). Many seeds are randomly distributed, though many others land near a parent plant. Birds may distribute seeds randomly or in a clumped pattern. For instance, an American Robin flying over a field drops a seed randomly when it defecates. But a Robin sitting on a preferred perch, day after day, may drop many seeds in the same location, forming a clump. Also, species that grow asexually by rhizomes may begin as randomly spaced individuals but each individual becomes a genet and thus the final product is highly clumped.

A Forest's Age

OBSERVATIONS: It is not possible to exactly determine a forest's age by mere inspection, but it is possible to make an educated guess as to the overall level of "maturity" of the forest. Old forests contain a well-defined canopy and understory, and dead tree snags will be scattered throughout. Sites of overturned, decomposed stumps will appear as mounds scattered over the forest floor, giving the forest an irregular, rolling topography. Trees will tend to have broad circumferences and be spaced rather far apart. Mature forests usually have a thick litter layer and many fallen and decomposing tree trunks. Most of the sunlight is intercepted by the canopy and understory, and thus mature forests are very highly shaded, some being almost dark. Young or "immature" forests can be identified by the overall high density of trees of small circumference, giving a crowded, spindly appearance. These trees will be of varying heights, and the understory, if present at all, will tend to be poorly defined. Light is often abundant but patchy on the forest floor, and thick shrub and herb layers are the result. In a young forest the litter layer will be thin. Many immature forests have resulted from regrowth following a recent disturbance. You may encounter areas of large fallen dead trees with small living trees replacing those killed by fire, hurricane, or other forces. Many tree species such as Red Maple can successfully resprout if the root systems are undamaged. The resprouted trees often have multiple trunks. The field marks of disturbance are discussed in more detail in Chapter 4.

EXPLANATION: The term *virgin forest* refers to a forest that has never been cut or otherwise affected by human activity. There are only a few small tracts of virgin forest remaining in the eastern United States. Virtually all our forests have regrown from abandoned farmland during the past hundred years or so. In addition, disturbances from the forces of weather or insects are ever present. These forces can destroy a mature forest but in its place will grow a young *successional* or *second-growth forest.* A forest may look serene and unchanging, but each forest has a history and a future and what you observe is but a moment in its lifetime.

Forest Gaps

OBSERVATIONS: A *gap* is a localized forest opening created either by a small-scale disturbance, such as the death of a large canopy tree, or a larger disturbance, such as a hurricane or fire. Windthrow and lightning are common causes of small-scale gaps. Light floods into gaps, creating excellent conditions for the growth of subcanopy and shrub species as well as vines, such as Poison-ivy and Virginia Creeper. Gaps often develop a dense, almost jungle-like understory. Some gaps may be large, covering as much as an acre or more. These larger gaps are created by such forces as hurricanes or fire. Large gaps support species found in successional areas (see Chapter 4 and p. 35).

EXPLANATION: Gaps produce patchiness in a forest. Close inspection of most forests will reveal a mosaic of gaps of differing sizes and ages. They are caused by unpredictable disturbances, such as *windthrow* of an old canopy tree. They provide conditions for the rapid growth and reproduction of species which, under denser shade, would grow and reproduce far more slowly. Ranging in size from very small to quite large in area, gaps help provide a refuge for the many species of plants that grow best in habitats with high light intensity. Gaps eventually close as those species characteristic of the canopy return and grow to their full statures. However, as older gaps close, new ones are created by naturally occurring events. Most forests will consist of numerous gaps of varying ages. Gaps add much to the diversity of forests and provide conditions for species that do not thrive under low light.

Predicting a Forest's Future

OBSERVATIONS: Compare the species of trees that comprise the canopy with those that are growing as tiny seedlings and small saplings on the forest floor. If the canopy consists of primarily oak species, do you find oaks as seedlings and saplings? If you do, then it is safe to conclude that the forest is self-reproducing and will continue to be an oak forest. If you find, however, that the species present as seedlings and saplings differ substantially from those established in the canopy, this indicates that the species composition of the forest may be changing. These changes are slow, taking the equivalent of several human lifetimes.

EXPLANATION: Forest development is very slow because the lifetimes of trees are long. A forest may appear quite stable when in fact its species composition is changing. Some species, such as Sugar Maple, are very shade tolerant as seedlings and saplings. These tiny trees can persist for many years in the shrub and herb layers while growing very slowly. Eventually, by exploiting gaps, they take their places in the canopy, replacing other species. In forests ranging from New Jersey to Missouri, Sugar Maple is currently becoming more abundant at the expense of oak species. Forests change because their component tree species have different rates of growth and different tolerance limits.

Ecotones

OBSERVATIONS: Forests do not exist in a vacuum. Most forests are surrounded by different types of habitats, such as varying-aged fields, croplands, or residential areas. The border between two types of habitats, such as where a forest meets a field, is called an *ecotone* (see Fig. 5). Ecotones are composed of a mixture of species from the neighboring habitats but are, in their own way, unique. An ecotone can be recognized by its location as a border area, by its mixture of species, some of which differ from those in the interior forest, and by its less stratified structure. Ecotones are often described as "brushy" or "shrubby." Ecotone species are often the same as those found in disturbed areas (see Chapter 3). The species richness of ecotones is often higher than that of ei-

Fig. 5. Ecotone between forest and old field.

ther of the bordering habitats because the ecotone contains species from both habitats. Ecotones may be broad; a forest may intergrade very gradually with another habitat. An ecotone may also be abrupt, with a sharp border, so that you can literally take a single step from one habitat into another very different habitat. Broad forest ecotones are more interesting areas than abrupt ecotones because of the slow transition that can be seen between non-forest and forest.

EXPLANATION: Within a forest ecotone two constellations of species, one from a forest and one from a field or other type of habitat, are mixing and interacting. Competition among plant species is occurring for light, soil moisture, soil nutrients, and space. The ecotone really represents a transitional area as one habitat becomes dominant at the expense of another. Throughout eastern North America, forests expand their borders unless checked by human activity such as cutting. If you are standing in a forest-field ecotone and facing the forest, the forest is slowly moving in your direction. In Chapter 4, you will learn why.

Old Fields

OBSERVATIONS: Ecotones and forest gaps are examples of change, of forest dynamics. Throughout eastern North America there are open fields of weedy species, shrubs, and sun-loving, fast-growing trees. These open areas, called *old fields,* are often extensive in area. Brushy fields contain annual, biennial, and perennial herbaceous plants, mixed among grasses and woody species such as Red Cedar, Poison-ivy, or young Tuliptree. There is an abundance of herbaceous species, the so-called weeds, as well as grasses and woody species. Some of the grasses, the bunch grasses, are in dense tufts. Thick patches of shrubs, especially members of the heath family such as blueberries or huckleberries, may be present. Old fields are a distinctive type of landscape and form the habitat of many species not found in forests. Usually, however, young forest trees, initially dense and spindly, indicate the eventual conversion of field to forest.

EXPLANATION: Old fields are undergoing a process called *old field succession* or *vegetation development*. An open area, perhaps initially exposed to the bare soil, is filling with species, invaded first by pioneer species (often "weeds") and later by longer-lived, slower-growing species. The initial invaders are exploiting the open environment, growing and reproducing quickly. These are termed *shade-intolerant* species because they require high amounts of sunlight and cannot grow when shaded. Common herbaceous weed species such as the ragweeds, asters, and goldenrods, as well as vines such as Poison-ivy, and trees such as Eastern Red Cedar, Gray Birch, and Loblolly Pine are all classed as intolerant. Some species, such as Northern Red Oak, Red Maple, and White Oak, are neither highly intolerant of shade, nor particularly tolerant. These species usually attain numerical dominance after the intolerant pioneer species have died out, and may persist indefinitely. Some species, such as Sugar Maple, Flowering Dogwood, and American Beech, survive very well under shade and can re-

Old Field. Abandoned agricultural or pasturelands revert eventually to natural ecosystems, invaded by a series of different grasses, herbs, shrubs, and trees.

place less tolerant species. Sugar Maple can survive as seedlings and saplings under deep shade and sprout up when gaps permit entrance of sunlight. These *shade-tolerant* species will assume canopy status. Once a canopy of tolerant species forms, the seedlings of intolerant species, which require high light intensities, fail to survive. The topic of ecological succession is considered in detail in Chapter 4.

Animals of the Forest

Plants, though prominent and obvious, are far from the only life forms of a forest. Animals are abundant and diverse, and most important forest processes involve animals as well as plants. The following are the major groups of animals that can be observed in a forest. In the chapters that follow, most examples will discuss animals from these groups as well as plants.

Insects and Other Arthropods

Spring and summer are good seasons to observe insects, spiders, and other arthropods (joint-legged animals with external skeletons of a material called chitin). Insects abound at all levels of the forest and many canopy species are quite difficult to observe because they normally do not leave the canopy. Many, such as most butterflies and many beetles, are brilliantly colored and obvious but at least as many are highly cryptic and hard to see. Green leaves harbor green insects (katydids, leafhoppers) that resemble leaves. Beetles, spiders, and moths can look remarkably like bark. The forest litter hosts harmless herbivorous millipedes, highly

Net-winged Beetle. Insects such as this species are essential pollinators of numerous plant species.

predaceous centipedes and spiders, and hosts of beetles. Flowers attract insects such as bees, flies, and butterflies, and depend on these insects to pollinate them (see p. 273). The pollinators attract predatory insects such as ambush bugs, as well as the ever-present spiders. Many "bees" are really flies that closely resemble bees, deriving some protection from predators by appearing similar to stinging bees. Occasionally a forest will be obviously under the attack of insects such as Gypsy Moths or Spruce Budworms, and many trees will be defoliated by their larvae. In Chapter 7, many of the fascinating adaptations of insects will be presented.

Birds

Birds, though present at any time of year, are most diverse and abundant during the spring, summer, and fall, when migration and breeding are occurring. Temperate forests are comprised of two major bird communities: the permanent residents, which reside all year around, and the summer visitors, who join the residents during the breeding season, migrating north in spring and south in fall (see Pl. 3). To these are added winter visitors, many of which vary in numbers from one year to the next. Birds feed heavily on insects during the spring and summer breeding season but begin switching to a fruit diet as autumn and migration time approaches. Fruiting plant species in forests seem timed to develop fruits just as migrants are passing through; thus the migrant birds serve as seed dispersers (see Chapter 7).

Most bird species are diurnal (active during the day) and easy to observe. Many important patterns of animal behavior, such as territoriality, flocking, courtship, mating, nesting, and foraging, can be seen by looking at birds. For this reason many of the examples in the chapters which follow deal with birds.

Mammals

Most mammals are generally nocturnal and therefore not nearly as easy to observe as birds. However, with persistence and patience, you will see many species (see Pl. 2) and come to understand many facets of their natural history. The most obvious forest mammals tend to be the nut-eaters, the squirrels and chipmunks. However, the most abundant mammals of the forest are the mice, such as the Deer Mouse and White-footed Mouse. Though abundant in number, these tiny rodents, which form an important food source for owls, weasels, and other predators, are only occasionally glimpsed. White-tailed Deer can be very abundant in many forests and occasionally overbrowse a forest in winter. Along edges between forests and fields live Cottontail Rabbits and Woodchucks, and old fields host Meadow Voles, husky little

Meadow Vole. These small mammals often abound in old fields, experiencing cyclic population fluctuations (page 450) and providing an important food source for predatory animals.

rodents that undergo cyclic oscillations in abundance (see Chapter 8). Predatory mammals, such as weasels, raccoons, and foxes, are present in low numbers and are basically nocturnal.

Reptiles and Amphibians

Reptiles have a body covering of dry scales, claws on their toes, and, unlike birds and mammals, are not capable of producing sufficient heat to maintain a constant high body temperature. For this reason they are often described as "cold-blooded," although the term ectothermic is more accurate. Snakes, lizards, turtles, and crocodilians are all reptiles, and each of these groups becomes less abundant from the warm to the cold climes. Forest reptiles inhabit the forest floor, and are often hidden among the leaf litter or under logs and rocks.

Amphibians differ from reptiles in that they lack scales and claws and must remain moist. Like reptiles, they are ectothermic and often feel slimy and cold. Frogs, toads, and salamanders are amphibians. Most amphibians are found in moist areas, not only because their skins require moisture, but because they must reproduce in water. Amphibian eggs usually hatch into larvae, which live in water before undergoing their metamorphosis into adults. In spite of their requirement for virtually constant moisture, many amphibian species live in forests beneath logs and leaf litter.

The art of searching for reptiles and amphibians involves the ability to recognize likely hiding places and carefully turn over rocks and logs.

The Forest Food Chain and Ecological Pyramid

OBSERVATIONS: Forests are green. The organisms that, by their very mass, dominate the habitat are the trees. Indeed, that's why we call it a forest. The combined weight of the living organisms in a habitat is called *biomass.* Obviously, the biomass of the plants in a forest far exceeds the biomass of the animals. Though there may be literally millions of caterpillars, flies, cicadas, aphids, birds, and squirrels living in the forest, their combined weights are far less than the combined weights of the plants of the canopy, subcanopy, shrubs, and herbs. Moreover, if you survey the animals with care, you will find that animals that feed on plants, the *herbivores,* have a higher biomass than those animals that feed on flesh, the *carnivores.*

The biomass of vegetation can be envisioned as the base of an ecological pyramid, forming the forest food chain (see Fig. 6). The next level up is the herbivores, a much smaller level in biomass, though populated by millions of caterpillars, aphids, bugs, mice, squirrels, and Woodchucks. Dependent on the herbivores are the primary carnivores, things like spiders, mantids, and insectivorous birds. Above them are the secondary carnivores, such as weasels and skunks, and, above these, is a tiny level called the top carnivores, the owls, hawks, foxes, and other large carnivores. The position on the ecological pyramid corresponds to the distance the organism is from the ultimate energy source, the

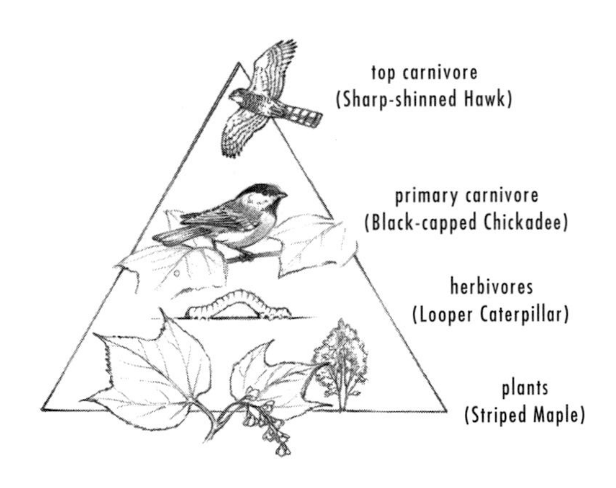

Fig. 6. Ecological pyramid.

Sharp-shinned Hawk. This top carnivore species feeds essentially entirely on small birds, such as chickadees.

sun. The largest and rarest animals are the top carnivores. It requires many mice, grouse, squirrels, skunks, and other animals to sustain a pair of Great Horned Owls. Thus the top carnivores are by far the least numerous animals in the food chain. Many animals occupy more than one level on the ecological food pyramid. Raccoons eat both plant and animal material. So do chickadees. These animals are *omnivores.*

With a bit of observation, you will soon be able to discern ecological food chains. For instance, in the Northern Hardwood Forest, the leaves of the Striped Maple are commonly eaten by the caterpillar of the Green Looper moth, a herbivore. The common Black-capped Chickadee can be seen flitting about the branches, picking off loopers. The chickadee may fall prey to a Sharp-shinned Hawk or a Long-tailed Weasel. Energy from the sun, captured by the Striped Maple, has moved from the maple, through the looper, through the chickadee, to a hawk or weasel. When the hawk or weasel dies, the energy in its body will be utilized by fungi and bacteria, as well as by many soil animals, the *decomposers.* This is but one of thousands of examples of how energy passes from one organism to another through a food chain. Any habitat contains numerous interconnected food chains, a food web.

EXPLANATION: In nature, virtually everything seems to serve as food for something else. Many of the adaptations discussed in this guide involve how predators successfully capture prey and how prey are equipped to avoid capture. Obviously, nature is organized in part by the rule of "eat or be eaten."

Where does the energy come from? The answer is clear: from

the sun. Ultimately, all food energy that passes through any living body, be it plant or animal, originated 93 million miles from earth in the thermonuclear furnace of the sun. The sun's energy travels to earth as electromagnetic radiation, much of which is in the wavelengths we call visible light.

Green plants have the chemical skills to convert sunlight into food through a process called photosynthesis. In doing so, plants make it possible for all other forms of life to exist. We humans cannot get fat simply by lying out under the sun. Plants can and do. The green plants form the first link in the ecological food chain, and are the base of the ecological pyramid of energy. They are the *producers*.

Animals are *consumers*. Many animals, from katydids to cottontails, eat only plants. These herbivores have digestive systems adapted to breaking up cellulose and lignin plus other complex structural chemicals of plants. Herbivores typically have longer digestive systems, which help them digest tough plant fiber. Ruminants, such as deer, have five-chambered stomachs in which dense populations of bacteria aid in breaking down plant material. Ruminants regurgitate partially digested plants and continue grinding the material when they chew their cud. Rabbits pass fecal pellets that are reingested for a second trip through the gut, to further digest the material. Termites, one of the few kinds of animals that eat wood, digest the wood with the help of protozoans living in their gut.

Katydids are eaten by shrews and cottontails by weasels. These flesh-eaters, the carnivores, are always at least three links away from the sun on the ecological food chain. Some predators, like the large hawks and owls, are routinely five links from the sun. They are called top carnivores because nothing eats them.

In our example above, the Striped Maple was fed upon by the Green Looper, which was eaten by a Black-capped Chickadee, which fell prey to a Long-tailed Weasel. The weasel may be the victim of a Great Horned Owl, making the food chain five links long. The sun's energy has been transferred through space to earth, captured by the Striped Maple. The maple uses some of the energy to stay alive, so less energy is available to the looper than the maple began with. Likewise, the looper uses energy for its needs, so less energy is available to the chickadee. Chickadees are larger than loopers, so more energy is required to make a chickadee than to make a looper, a fact of life that even further reduces the number of chickadees that can be supported by a looper population. Finally, because there are fewer loopers than leaves, chickadees must hunt for loopers, using energy to ultimately gain energy. What all this means is that in order to support just one chickadee, there must be many hundreds of loopers, or other suitable food sources. There simply isn't enough energy available for chickadees to outnumber loopers, any more than there is for loopers to outnumber Striped Maple leaves. The same principle explains why there are fewer Long-tailed Weasels than chickadees, and fewer Great Horned Owls than weasels. The weasel got only the energy that was left over after it passed through the plant, looper, and chickadee. The owl got even less.

Plant and Animal Identification

Though the purpose of this guide is specifically *not* identification, you will need to do some basic sorting of the plants and animals in order to make the guide more useful. Identification is fun and challenging, and you will most likely be using this guide in conjunction with others that stress identification. The other guides of the Peterson Series are specifically designed to help you accurately identify everything from birds through beetles. I include here only the most basic steps in identifying the major groups you will encounter.

Plants

One convenient thing about plants is that they don't run or fly away. By spending a few moments scrutinizing the different parts of a tree, a shrub, or a flower, it is usually possible to identify it. Here are the fundamental characteristics that will aid you in identifying plants.

Trees come in two basic kinds, broad-leaved and needle-leaved. Broad-leaved trees have wide flat leaves. Oaks and maples are typical examples. The leaf may be simple or compound. A simple leaf consists of a single blade on a leaf stalk called the petiole. A compound leaf consists of several leaflets on a single petiole. Beech leaves are simple but hickory leaves are compound. In some trees, the leaves grow opposite each other on a twig. Maples and viburnums are common examples. On most trees and shrubs, however, leaves alternate. Leaves may be oval, heart-shaped, or elongate. Some trees have leaves with small, sawlike teeth around the edge. Leaves may have lobes, which may be rounded or pointed. Some leaves feel dry and leathery, some papery. Many broad-leaved trees are deciduous, dropping leaves in autumn and growing new leaves in spring. Some broad-leaved shrubs and trees, particularly in the southern states, are evergreen, however.

Needle-leaved trees include the pines, spruces, firs, hemlocks, larches, junipers, and cypresses. With very few exceptions (the American Larch, or Tamarack, being one), all are evergreen. Nee-

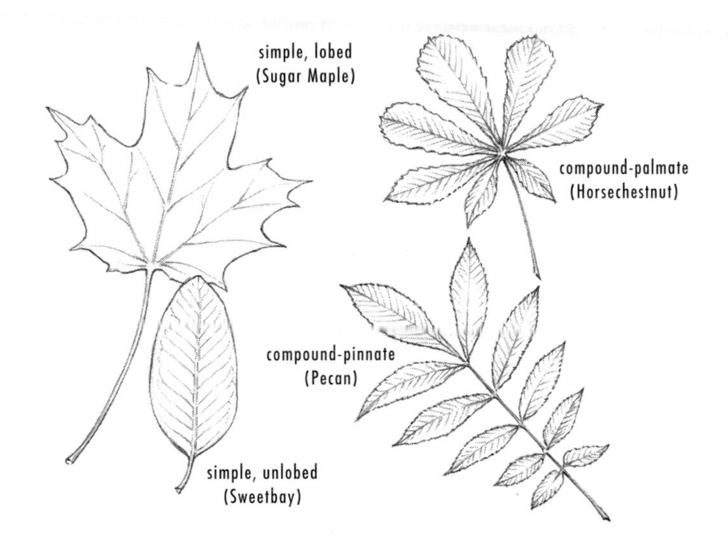

Fig. 7. Simple and compound leaves.

dles may be stiff or soft, long or short. They usually grow in clumps called bundles, and the number of needles in a bundle helps identify the tree.

Needle-leaved trees are conifers, which means that they do not have flowers but produce seeds in cones. Cones are initially green but turn brown. They may feel prickly or smooth. Pine and spruce cones hang down, but fir cones stand upright on the branch.

Bark color and texture are often useful for identifying trees. Bark may be scaly, furrowed, ridged, or smooth. It may adhere tightly or peel off in strips, scales, or plates. It may be gray, reddish, or some other color. Often bark changes in color and texture as the tree ages. Young trees tend to have smooth textured bark compared with older trees, which usually have furrowed bark.

Broad-leaved trees, shrubs, and wildflowers reproduce by means of flowers. When flowers are present, they are very useful in identifying the plant. Many trees bear small flowers that are hard to see and are not, therefore, very useful as field marks. This can be overcome, in some cases, by using binoculars. Some trees, however, such as Tuliptree and Redbud, are easily identified by their flowers. Even flowers in the hand can pose identification problems, and a hand lens is often very useful, not only for identi-

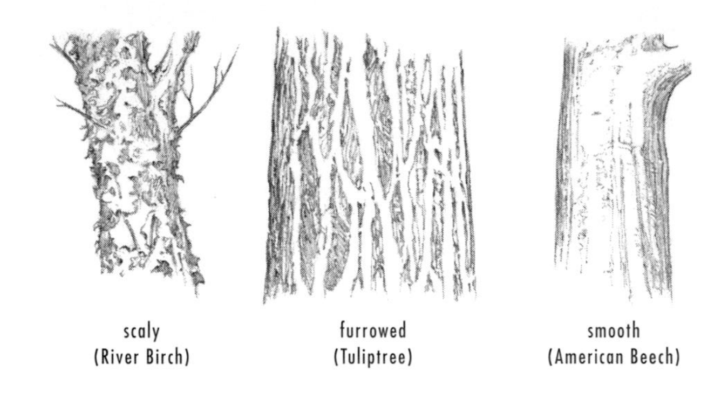

Fig. 8. Representative bark types.

fication, but also for seeing the various structures of the flower. Petal color is important in identifying flowers. In many species bracts are important as well. Bracts are modified leaves that often resemble petals and grow from the flower base. In Flowering Dogwood, the bracts are more prominent than the true flowers at the center.

For identifying wildflowers, color and shape of the flower are usually important field marks. Many wildflowers, especially those in forests, bloom in spring and early summer. Many old field wildflowers bloom in mid- and late summer. Flower characteristics are discussed at greater length in Chapter 6.

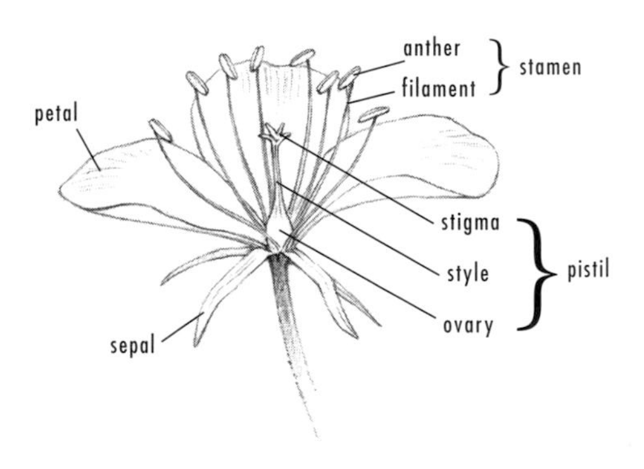

Fig. 9. Flower parts.

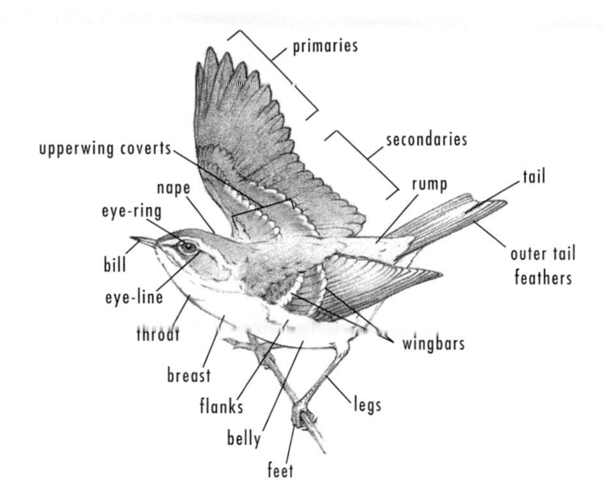

Fig. 10. Parts of a bird.

Birds

Birds are generally conspicuous. They are often brightly colored, vocal, and active during the daylight hours. To identify birds, look for field marks such as overall size and shape. Is the bird robin-sized, sparrow-sized, or crow-sized? Note the bill size and shape. Is the bill chunky and thick, like that of a grosbeak, or slender, like that of an oriole? Note the color pattern. Does the bird have wingbars, wing patches, a white rump, or white outer tail feathers? Does it have an eye-ring? Is its breast streaked or spotted? When the bird flies, does it fly straight or undulate? When on the ground, does it hop or walk? Do you find it in flocks or is it solitary? What is its habitat? Do you find it inside a forest, along a forest edge, or in an open field? For success in bird identification, binoculars are essential.

Mammals

Mammals generally lack the bright colors of birds, but they are by no means dull. Their coat colors, often very subtle in tone, range from pure black or white to many shades of brown, gray, and rufous. Important field marks include overall size and shape, characteristics of the tail (ringed, bushy, naked, short, or long), and the markings on the face. Many mammal species are either crepuscular (active at dawn and dusk) or nocturnal, and many are fairly secretive. Seeing them poses a challenge; many are glimpsed when illuminated by headlights as they cross highways.

Reptiles and Amphibians

Reptiles are represented in our area by the snakes, lizards, turtles, and the American Alligator. Amphibians are represented by the salamanders, frogs, and toads. Both reptiles and amphibians are often very colorful but can be very well camouflaged in their environments. Many take refuge beneath rocks and logs, and thus you must search them out. Amphibians, as the name implies, tend to favor wet areas such as vernal ponds, streams, swamps, marshes, and wet meadows. Many, however, can be found in moist situations in mesic forests. Amphibian watching is most successful during the spring breeding period when frogs and toads are calling. You must be prepared to go out at night, preferably with a head lamp, and get wet in order to see a diversity of salamanders, frogs, and toads. To identify salamanders, note the color pattern, relative thickness of the hind legs compared with the front legs, and the length of the tail. To identify anurans (frogs and toads), look first at the overall size, shape, and color. Note if the animal is a true frog or a treefrog by looking for toes with wide flattened tips, the suction cups of treefrogs. Is the animal a toad, by virtue of having dry "warty" skin? Note the voices. Anurans are easily identified by voice.

Reptiles can be searched for during the daylight hours, by carefully turning over rocks and other debris likely to shelter a resting serpent. (Stand *behind* the rock if you do this.) *Be careful* when looking for snakes—some are poisonous. To identify snakes, note color and blotching pattern, and shape of the head (slender,

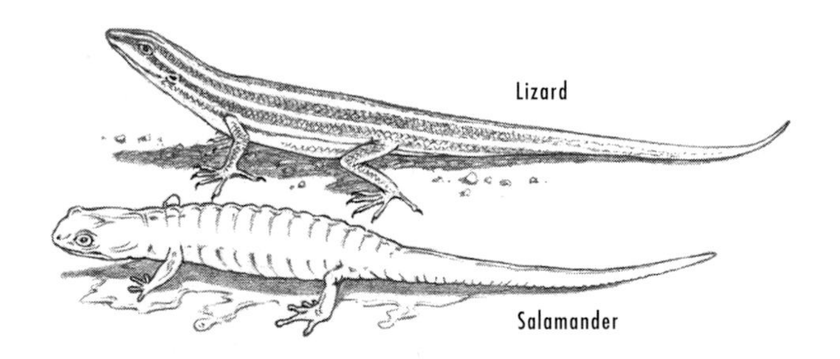

Fig. 11. Salamander and lizard compared.

widened, diamond-shaped?). Lizards are not highly diverse in our area and, like the snakes, they are represented by more species in the southern states. Lizards usually sit motionless but scurry quickly when discovered. Note their color, patterning, and general degree of "spininess." Turtles are slow and easy to identify. Several species, like Eastern Box Turtle and Wood Turtle, are primarily forest dwellers, but many more species are pond turtles.

Insects and Other Invertebrates

There are myriads of species of insects and the ones included in this guide merely skim the surface. Since insects are generally quite small, a hand lens is often very useful for identification. The best way to begin is to recognize the basic major insect orders—is the insect a grasshopper, a beetle, a dragonfly, a sucking bug, or a bee? Butterflies and moths have brightly and complexly patterned wing coloration, aiding greatly in identification. Insect watching is important in using this guide. Settle on a patch of wildflowers and note what visits the flower, what is crawling on the stem, and what is perambulating about the leaf litter. Spotting insects, many of which are very well camouflaged against their background environments, is a skill you may have to learn. Look closely in flowers, on bark, and among leaves. You'll be amazed at what you find there.

Spiders, centipedes, millipedes, and other invertebrates are also touched upon in this guide. Like insects, these small invertebrates require close observation and a hand lens is valuable in looking at them. Many are found under decomposing logs and among the forest litter layer. Don't be afraid to take a closer look at the array of soil-litter animals. Most don't bite and all have interesting natural histories.

Fig. 12. Jumping spider.

Getting Started: A Forest Field-mark Questionnaire

This chapter has introduced you to the field marks of forests. As a primer to guide you in "forest watching," I include this questionnaire, which summarizes the field marks discussed in this chapter. You can use it to check out your own local forests.

STRATIFICATION: Is the forest clearly stratified into canopy, understory, shrub, and herbaceous layers (see Fig. 1)? Is the forest unstratified, with no clear separation between canopy and understory?

TYPES OF TREES: Is the forest comprised of mostly needle-leaved or broad-leaved tree species? Needle-leaved trees dominate to the north and along coastal plains. If mostly broad-leaved, how many species are evergreen and how many deciduous? Broad-leaved evergreens are by far most common in the South.

INDICATOR SPECIES: Which species of trees, shrubs, and herbaceous plants are most abundant and thus most likely to be suitable indicator species? Which birds and other animals are most common and obvious? Once you have established the identity of the indicator species, check Chapter 3 to learn which of the eastern forest types your forest represents.

SPECIES RICHNESS: As you walk 500 yards in a straight line, does the forest seem to consist of many species or just a few among the various plant strata and animal groups? The most species-rich forests occur in the Appalachian bottomlands and, in general, in the southern states.

SOIL: What color is the soil? Grayish soils are typical in northern coniferous forests and reddish soils predominate in the South. Does the soil feel sandy or claylike? Sandy soils are most common on coastal plains. Is the litter layer thick with many undecomposed and partially decomposed leaves or is it thin?

SOIL MOISTURE: Does the soil feel moist or dry? Does the forest appear lush, or are most of the leaves thick and waxy? Does the forest appear to be mesic, xeric, or hydric? Any region will have examples of all three forest moisture regimes. Which indicator species (or combination) characterizes each regime in your area?

EVIDENCE OF FIRE: Do you see any indication that the forest has recently been exposed to fire? Are there charred stumps or fire scars on the tree trunks? Coastal Plain forests typically show evidence of fire, but interior forests may as well.

PLANT POPULATION PATTERNS: Do you observe plant populations that seem highly clumped? Look not only at trees but also at shrubs and herbaceous species. Do any species appear to be regularly distributed? Do any appear random?

THE FOREST'S AGE: Are the canopy trees individually large and widely spaced or are the trunks thin and crowded? Are there large dead snags and fallen trees scattered throughout the forest? Is the forest floor highly shaded or does light penetrate the canopy well? How clearly developed is the understory? Is the litter thick on the forest floor?

FOREST GAPS: Do you see evidence of local disturbance such as a recently fallen tree? Do you observe more dense understory and vine growth in the gap area? How much patchiness is there overall in the forest? Do you see any species in the gap areas that are not present in the undisturbed areas?

PREDICTING THE FOREST'S FUTURE: Are the seedling and sapling tree species the same as those of the canopy? Do you think the forest will continue as it is, or are the seedling and sapling tree species different enough from those of the canopy that the forest will change in time?

ECOTONES: In areas where the forest borders other habitats, is the transition from forest gradual and broad or abrupt? Do you see a mixture of forest and non-forest plant species in the border area?

OLD FIELDS: What differences are there between old fields and forests? What is the relative amount of open area and shade in various fields in your area? How much cover is taken by grasses and weedy herbs? Are there clumps of shrubs? What size are the trees? Do you see old fields that show evidence of gradually changing into forests?

ANIMALS: What are the most obvious insects, birds, mammals, and reptiles and amphibians? Which animals are easiest to see? Which are most abundant?

FOOD CHAINS AND THE PYRAMID OF ENERGY: Can you confirm that producers far outweigh consumers? Do you see more herbivores than carnivores? What food chains are evident in your forest area? Can you trace the pathway of energy from the sun through the different energy levels of the ecological pyramid (see Fig. 6)?

SEASONALITY: What season of the year is it? Refer now to the appropriate chapter on the season you are observing. You should also read Chapters 3 and 4 to learn about specific forests, old fields, and other unique habitats.

TREES — NON-INDICATOR SPECIES (P. 15)

These six species are among the most widespread in the East, thus none of these species by itself serves as an indicator of any particular forest type. However, in combination with other species, some can be useful indicators.

AMERICAN ELM

Known for its tall, straight trunk and elliptical, symmetrical crown. Double-toothed, alternate, light green leaves (yellow in fall), asymmetrical at base. Small clusters of whitish flowers (see Pl. 35) in spring. Seeds flat, encased in a notched wing. Bark deeply furrowed.

WITCH-HAZEL

Alternate, widely notched leaves (yellow in fall), asymmetrical at base. Small, 4-petaled, yellow flowers appear in fall and winter. Grows as a small tree (up to 30 ft. tall) or shrub.

SASSAFRAS

Abundant understory species with aromatic, alternate, untoothed leaves having three shapes: unlobed, single-lobed, or double-lobed. Leaves turn red, then yellow in fall. Bark deeply ridged. Small, yellowish green flowers in spring (see Pl. 36). Fruits small blue berries (see Pl. 49).

RED MAPLE

Opposite leaves with 3 prominent lobes and 2 small basal lobes. Leaves brilliant crimson in fall. Bark smooth gray in young trees, furrowed in older trees. Tiny red flowers in spring, before leaves open. Seeds with reddish wings.

FLOWERING DOGWOOD

Understory tree, often indicative of high-calcium soils. Leaves opposite, oval, with prominent veins. Leaves deep red in fall. Flowers tiny, red and yellow, surrounded by 4 large white bracts (see Pl. 36). Berries red (see Pl. 49). Bark scaly, gray.

AMERICAN HORNBEAM

Small, shrublike understory tree with alternate, double-saw-toothed leaves (orange-red in fall). Leaf has long tip. Fruits nut-like, greenish, in green scales hanging from branch. Bark very smooth, gray, streaked, suggesting muscle fibers.

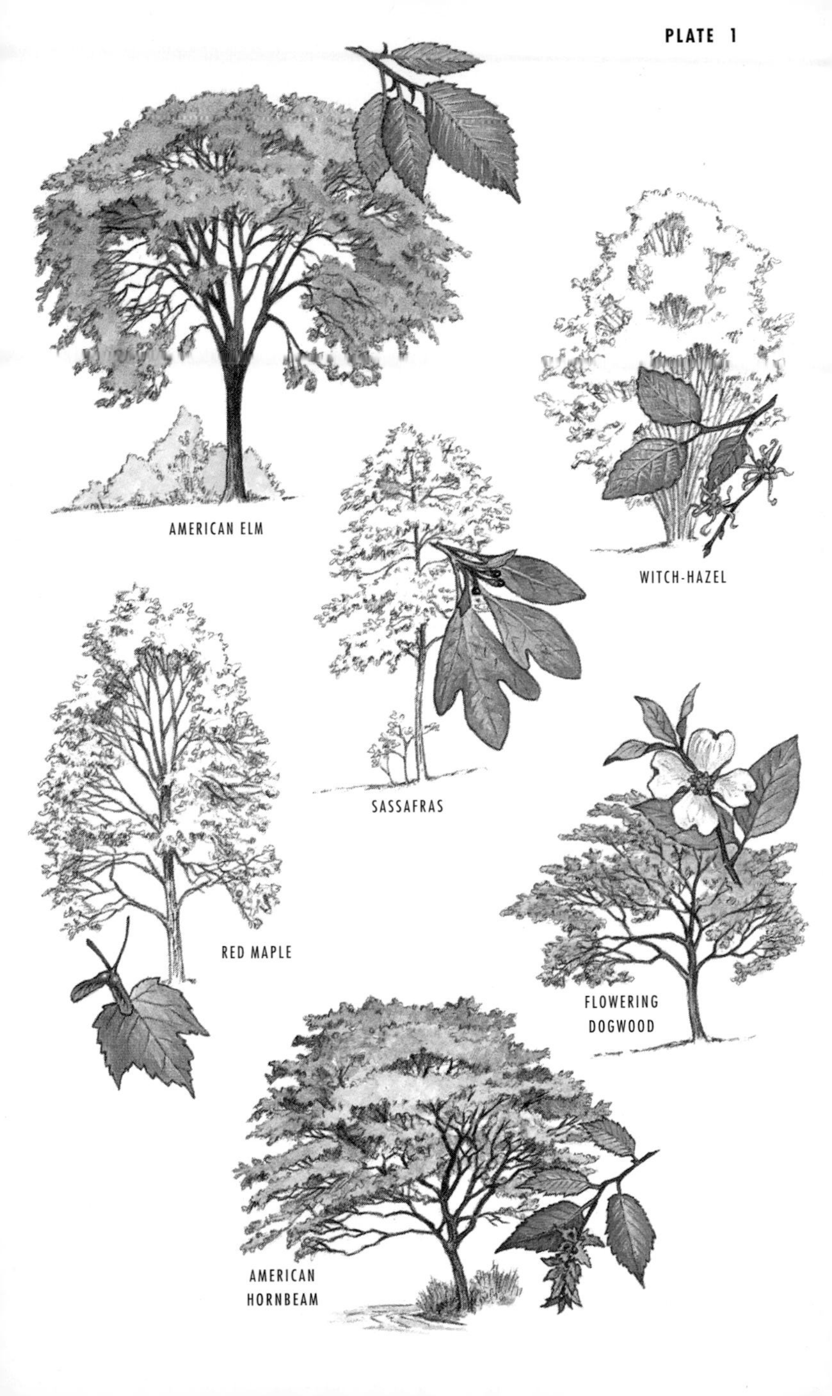
AMERICAN ELM
WITCH-HAZEL
SASSAFRAS
RED MAPLE
FLOWERING DOGWOOD
AMERICAN HORNBEAM

MAMMALS — NON-INDICATOR SPECIES (P. 17)

BLACK BEAR

The only eastern bear. Usually black, but a cinnamon phase occurs also.

BIG BROWN BAT

Our largest eastern bat; wingspan 10–13 in. Uniformly brown. Flies with steady wingbeats. Roosts in caves, hollow trees.

STRIPED SKUNK

Cat-sized; black with white stripes along body and bushy, black and white tail. White face stripe. Nocturnal.

WHITE-TAILED DEER

Large, light brownish tan with white under tail, visible as animal flees. Only males have antlers. Forms herds in winter, family groups in summer.

BEAVER

Largest American rodent; body up to 36 in. long; tail naked, flat, scaly, up to 12 in. long. Reduced in population over much of its range. Prefers lakes and streams with birches, aspens.

RACCOON

Large, robust, with a black mask on a pointed face and a bushy tail with black rings. Common in wooded suburbs, forests.

RED FOX

Doglike but with a thick bushy tail. Very reddish overall, but with white tail tip and belly. **Gray Fox** is smaller but has a gray upper body and tail, with no white tail tip.

EASTERN COTTONTAIL

Buffy brown with large white "cotton" tail, easily visible when rabbit is running. Ears small, for a rabbit.

WHITE-FOOTED MOUSE

Rufous upper body; white below, including feet. Large black eyes, long tail, large ears. **Deer Mouse** is very similar but tends to be more brownish gray.

BIG BROWN BAT
BLACK BEAR
STRIPED SKUNK
WHITE-TAILED DEER
BEAVER
RED FOX
RACCOON
EASTERN COTTONTAIL
WHITE-FOOTED MOUSE

FOREST BIRDS — NON-INDICATOR SPECIES (P. 16)

RED-TAILED HAWK

A large (to 25 in.), bulky, soaring hawk with a white chest and rufous tail. Tail often widely spread when soaring; rufous above, pale below, without banding.

GREAT HORNED OWL

A large (to 25 in.) nocturnal owl with distinct tufts above its eyes. Breast heavily barred, rufous disk around each eye, white throat bib. Voice a deep *hoot.*

AMERICAN CROW

Large (to 21 in.), abundant, permanent resident of forests and farmlands. Gregarious. Uniformly black with a purple sheen.

COMMON FLICKER

Medium-sized (12–14 in.) brown woodpecker with a white rump, very conspicuous in flight. Often on ground, eating ants. Males have a black "mustache." Permanent resident except in North. Voice a loud, repeating *whicka.* See also Pl. 47.

RUFFED GROUSE

A chickenlike forest bird with a wide, gray or rufous, fan-shaped tail, with a black band at tip.

BROAD-WINGED HAWK

Medium-sized (to 19 in.) soaring hawk with pale underwings and a banded tail. Black bands on tail distinct in flight.

GREAT CRESTED FLYCATCHER

A robin-sized tyrant flycatcher that commonly perches in crowns of shade trees. Grayish face, yellow breast and belly, rufous tail. Voice a loud *wheep!* Migrant.

RED-EYED VIREO

A sparrow-sized, greenish bird with a white stripe above its red eyes and a gray crown. Resident in forest canopy. Song a monotonous repeated warble. Migrant.

WOOD THRUSH

Smaller than a Robin but shaped similarly. Hops on forest floor. Rufous head, heavily spotted breast. Song flute-like, highly melodious. Migrant.

OVENBIRD

A sparrow-sized, ground-dwelling wood warbler that walks. Olive above, with a white eye-ring and orange crown; breast streaked with black. Song a loud, whistled *teacher, teacher, teacher.* Migrant. See also Pl. 47.

RED-TAILED HAWK
GREAT HORNED OWL
COMMON FLICKER
male
AMERICAN CROW
RUFFED GROUSE
GREAT CRESTED FLYCATCHER
BROAD-WINGED HAWK
RED-EYED VIREO
WOOD THRUSH
OVENBIRD

Eastern Forest Communities

COMMUNITIES OF PLANTS AND ANIMALS

A forest is an assemblage of plants and animals coexisting and interacting. Taken together, these plants and animals constitute an ecological community. Because eastern North America is such a vast area, the factors that determine the composition of forest communities vary considerably (see following section), and so do the forests. No two forests are identical, but it is nonetheless possible to recognize major regional forest communities. These forests occupy large ranges (see map on p. 60) and are associated with the climate and soils of the particular region. In this chapter, each of these major forest communities is described and information is given on their natural histories. It is possible to recognize a forest type either by identifying the indicator species or by noting which forest community prevails in the area where you are observing.

What Determines a Community?

CLIMATE: Climate is the most important influence on the distribution of plants and animals. The amount of sunlight received, the temperature, amount of precipitation, and exposure to wind are all responsible for the overall pattern of forest distribution. If the climate were totally uniform throughout eastern North America, there would be far less variability among forests. Sunlight and temperature interact to determine length of the growing season (see Chapter 2), a major determinant of the species composition.

Local effects of topography are also important. If a forest is located on a north-facing slope it will be exposed to cooler and often moister conditions than if it is facing south. The result is that a mountain may have very different forest communities on its north-facing and south-facing slopes. North-facing slopes typically have "northern affinity species" such as Beech and Sugar Maple. South-facing slopes have "southern affinity species" such as oaks and hickories. Slope effect is most pronounced in northern latitudes.

GEOLOGY AND SOIL: The nature of the substrate also has a very strong influence on the biological community. Soil acidity, texture, and drainage are important characteristics. Geological characteristics and soil type vary throughout eastern North America and combine with climate in affecting forests (see Chapter 2). For example, some species, such as Eastern Red Cedar and Northern White-cedar, are common only on soils rich in limestone. Other species, like Jack Pine, prosper only on sandy soils too poor for many other species. A forest may be xeric (see Chapter 2) because it is located in a hot, dry climate or because it sits on sandy, highly drained soil that cannot retain moisture. Though not as important a determinant as climate, soil and geological characteristics cannot be neglected in interpreting forest types.

FIRE: You will probably be surprised at how many forests (as well as prairie and savanna) are strongly under the influence of occasional, naturally occurring fires. Fire burns leaf litter and brush, helping release minerals to the soil where they can be recycled to the vegetation. Many plant species, such as Pitch Pine, Longleaf Pine, and Jack Pine, depend on fire for their reproduction (see pp. 71 and 92). Fire also destroys species that would otherwise compete with fire-adapted species.

BIOTIC INTERACTIONS: Much of what goes on in nature involves interactions. Animals influence plants by grazing on them, by pollinating them, and by spreading their seeds. Plants compete among themselves for light, water, and minerals. But cooperation of a sort also occurs. Many tree species depend on fungi growing in their root systems to aid in the intake of minerals from the soil. Trees such as the Black Locust contain bacteria in their roots that take nitrogen from the atmosphere and convert it to a form useful to the locust tree (p. 185). Birds often forage together in mixed flocks. Some members of the flock may compete against others for food, but more food may be discovered because there are more birds searching. Together, the flock may stand a better chance of detecting predators than if the birds foraged individually. Interactions among species will be a major focus in Chapters 5–8.

HUMAN INFLUENCE: Human activities have tremendously altered and influenced forests. Over 80% of the forests in the New England states were cut and the land converted to agriculture following European settlement. Prior to the coming of the Europeans, Native Americans set fires that cleared underbrush in forests. Eastern

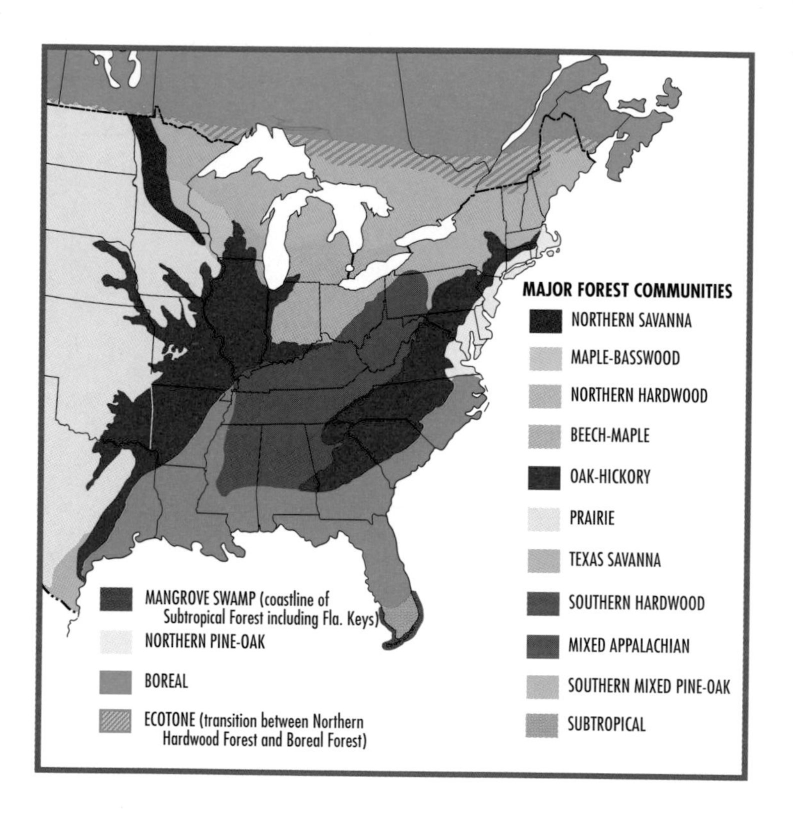

MAJOR FOREST COMMUNITIES
NORTHERN SAVANNA
MAPLE-BASSWOOD
NORTHERN HARDWOOD
BEECH-MAPLE
OAK-HICKORY
PRAIRIE
TEXAS SAVANNA
SOUTHERN HARDWOOD
MIXED APPALACHIAN
SOUTHERN MIXED PINE-OAK
SUBTROPICAL
MANGROVE SWAMP (coastline of Subtropical Forest including Fla. Keys)
NORTHERN PINE-OAK
BOREAL
ECOTONE (transition between Northern Hardwood Forest and Boreal Forest)

forests are now becoming re-established following the shift of agriculture to the Midwest in the latter part of the 19th century. Human-made pollution also influences forests. Eastern forests are currently being subjected to acid rain, the effects of which are still unclear.

EASTERN FOREST COMMUNITIES

Each forest community is described in the following format:

INDICATOR PLANTS

- **TREES**
- **SHRUBS**
- **HERBACEOUS SPECIES AND GRASSES**

INDICATOR ANIMALS

- **BIRDS**
- **MAMMALS**
- **OTHER ANIMALS**

DESCRIPTION

- **SIMILAR FOREST COMMUNITIES**
- **RANGE**
- **REMARKS**

A WORD ON INDICATOR PLANTS

The indicator species listed for each forest occur commonly enough that you should not miss them (though you may have to search a bit). Rare species, though some may be good indicators, are generally omitted because they usually require too much searching.

Combinations of indicator species are the best way to identify a major forest community. Sugar Maple, when in abundance with Beech in the Midwest, indicates the Beech-Maple Forest community. When Sugar Maple is in combination with American Basswood in Minnesota and Wisconsin it indicates the Maple-Basswood Forest. Finally, when Sugar Maple occurs with an abundance of Yellow Birch, Eastern Hemlock, Beech, and White Pine in New England, it indicates the Northern Hardwood Forest.

A WORD ON INDICATOR ANIMALS

Most large animals range throughout eastern North America and so do not make very good indicator species. However, as with plants, certain combinations of animals, especially birds, do indicate a particular forest community. Remember, it is the *combination* of species that is important.

Some uncommon or local species are included because they are unique. For instance, the Red-cockaded Woodpecker is a local resident throughout southern pine forests (see Pl. 12). This species is difficult to find but worth the effort. It is very dependent on pine forests and is therefore an excellent, though unfortunately rare, indicator species.

A Word on Range

Over large geographic regions, because forest species composition is so much a function of climate and soil, different forest communities can be pinpointed accurately by range. If you are in coastal South Carolina, you are in the range of the Southern Mixed Pine-Oak Forest. If you are in eastern Pennsylvania you will probably see Oak-Hickory Forest. However, in certain places like the Appalachian and Allegheny Mountains, the effects of altitude and slope create different climates over very short distances. The result is a mixture of forest communities over an extremely small range. Similar variation in forest type occurs locally throughout eastern North America. Virtually all types of the forest communities found in the East can be identified within Great Smoky Mountains National Park (see p. 97).

Boreal Forest (Pl. 4)

Indicator Plants

TREES: White Spruce, Black Spruce, Balsam Fir, Paper Birch, Bigtooth Aspen, Quaking Aspen, Balsam Poplar, Tamarack (American Larch), Eastern Hemlock, White Pine, Red Pine, Jack Pine. Red Spruce and Fraser Fir are part of the Appalachian Extension of the Boreal Forest, replacing White Spruce and Balsam Fir in the Appalachians.

SHRUBS: Mountain Maple, Mountain-ash, Green Alder, Bearberry, Mountain-holly, Lowbush Blueberry, Elderberry, Thimbleberry, Sheep Laurel (Lambkill).

HERBACEOUS SPECIES: Red Baneberry, Nodding Trillium, Bunchberry, Canada Mayflower, Creeping Wintergreen, Goldthread, Common Wood-sorrel, Clintonia (Corn-lily), Twinflower, Starflower, Large-leaved Aster, Northern White Violet, Shinleaf, One-sided Pyrola. Fireweed and Orange Hawkweed are abundant in disturbed open areas. Sphagnum Moss is abundant on wet sites.

Indicator Animals

BIRDS: Gray Jay, Common Raven, Boreal Chickadee, Spruce Grouse, Red Crossbill, White-winged Crossbill, Pine Grosbeak, Evening Grosbeak, Pine Siskin, Red-breasted Nuthatch, Winter Wren,

Ruby-crowned Kinglet, Golden-crowned Kinglet, Black-backed Woodpecker, Northern Three-toed Woodpecker, Tennessee Warbler, Yellow-rumped Warbler, Blackpoll Warbler, Cape May Warbler, Bay-breasted Warbler, Black-throated Green Warbler, Mourning Warbler, Blackburnian Warbler, Magnolia Warbler, Wilson's Warbler, Swainson's Thrush, Gray-cheeked Thrush, White-throated Sparrow.

MAMMALS: Moose, Red Squirrel, Beaver, Porcupine, Snowshoe Hare, Lynx, Marten, Fisher.

Description

The Boreal Forest is an evergreen, needle-leaved forest of "Christmas trees," with birch, poplar, and aspen growing on disturbed sites. Spruces have cones that dangle below the branches, while firs have upright cones on the branches. Tree bark is covered by pale lichens. The forest interior is often so dark that it lacks understory trees or shrubs. The litter is soft and thick with the accumulation of fallen decomposing needles. Low spreading shrubs such as Bearberry carpet the forest floor, along with herbs such as Bunchberry. Winters are cold and prolonged with much snow cover. The soil is highly acidic gray spodosol (see p. 26). Soils are young and nutrient-poor because they cover an area very recently glaciated. Bogs commonly occur (see p. 67). Fires occur frequently in many boreal areas, especially Jack Pine Forests (see p. 70). The Appalachian Extension of the Boreal Forest (see p. 97) is dominated exclusively by Red Spruce and Fraser Fir.

SIMILAR FOREST COMMUNITIES: Where the Boreal Forest meets the Northern Hardwood Forest considerable overlap of species occurs, producing forests which are mixtures between the two communities. Jack Pine Forests (see p. 70) occur on very poor soils in boreal areas. Boreal Bogs (see p. 67) occur on hydric (swampy) sites.

RANGE: Extreme northern New England, New York, northern Michigan, Wisconsin, and Minnesota north through Canada to treeline. One arm of this forest, the Appalachian Extension, follows high elevations along the Appalachian Mountains from Maine to Georgia. Red Spruce replaces White Spruce and Fraser Fir replaces Balsam Fir along the Appalachian Extension.

REMARKS: The Boreal Forest is essentially an unbroken tract of conifers that extends across northern North America through northern Europe and Asia. It is often called the "spruce-moose" forest. The forest has a low species richness of trees, with many tracts comprised almost entirely of White Spruce and Balsam Fir. These species are harvested for lumber and pulpwood and are commercially grown for Christmas trees.

The towering Moose, the world's largest hoofed animal, roams the Boreal Forest and Bog. A full-grown bull Moose may top 6 ft. at the shoulder. Moose are solitary, feeding mostly on aquatic plants. The rutting season, when males are highly aggressive in pursuit of females, is fall. Calves are born in spring.

Other common mammals include three rodent species, the Red Squirrel, the Porcupine, and the Beaver. Red Squirrels are ubiquitous inhabitants of needle-leaved trees, where they feed on conifer seeds extracted from the cones. Like Gray Squirrels (p. 84), they cache seeds as insurance against winter shortages. Porcupines feed mostly on bark. Basically nocturnal and rather sluggish, the well-protected Porcupine is most often seen atop a pine, aspen, or poplar. Beaver (see Pl. 2) are found in many forests but are quite common in boreal areas. They can exert a very profound influence on the landscape by damming streams and creating ponds. Dusk is the best time to look for Beavers, as they are essentially nocturnal.

The Snowshoe Hare and Lynx both show anatomical adaptations to survive in boreal areas. The hare changes color with the seasons (hence it is also called the Varying Hare). Snowshoe Hare typically are buffy brown in summer and white in winter, thus their pelage provides good camouflage in both seasons. In winter, hares grow dense fur pads on their paws, a characteristic shared by the Lynx, an uncommon boreal cat. The Lynx has very wide paws, an effective adaptation for running over snow. Lynx prey heavily on Snowshoe Hare.

Campers in Boreal Forests are visited by the Gray Jay, a tame and inquisitive bird. Other boreal birds, like the Spruce Grouse and Pine Grosbeak (see Pl. 52), are also prone to ignore humans and permit close approach. Uniquely adapted to feed on

Moose. Found wherever there is Boreal Forest, the long-legged and long-snouted Moose is well adapted to feed on vegetation in northern boggy lakes.

conifer seeds are the two crossbill species (see Pls. 4 and 52). These birds have crossed mandibles, which enable them to probe into cones and extract seeds. Some boreal birds, like the Pine and Evening grosbeaks and the crossbills, occasionally invade more southern latitudes, an event called an irruption (see p. 445 and Pl. 52).

The spring migration brings waves of warblers, like the colorful Blackburnian. Warblers are insectivorous and prosper on the abundance of budworms (moth larvae) and other insects that emerge during the growing season. Studies have shown that various warbler species tend to specialize both their feeding location and behavior within a spruce or fir tree. Some, like the Cape May Warbler, spend most of their time feeding high in the tree, on the outer branches. Others, like the Bay-breasted, feed mostly in the middle of the tree, often near the trunk. Some search methodically for insects, others flit actively from branch to branch. In this manner, competition among the warbler species is minimized. Each has nearly exclusive access to one area within the tree.

Most boreal tree species are evergreen with needle-like leaves. The conical shape of these trees adapts them well to shedding snow in winter, preventing branch damage. Needles also represent an adaptation to the short growing season and prolonged winter. Frozen soils prevent water from reaching leaves in winter, but the growing season makes it uneconomical for a tree to drop and regrow new leaves annually. Needles remain on the tree all winter, and the thick clusters of short needles on branches of spruces and firs break up the flow of wind across the branches, thus reducing wind chill and evaporation. Studies have shown that the arrangement of needles on these trees tends to minimize heat loss by convection, thus trees stay warmer.

Fig. 13. Porcupine.

Coated with a thick waxy cuticle that minimizes water loss, and filled with a resinous, sugary "antifreeze" that prevents frost damage, needles endure the winter drought by becoming dormant, but they can quickly begin photosynthesis after the spring thaw. The fragrant odor of Balsam Fir resin is a pleasant byproduct of its winter adaptations. Balsam Fir needles also contain chemicals that discourage insects from eating them. Some of these chemicals mimic insect growth hormones and interfere with metamorphosis from larva to adult. Some insects, however, like the Spruce Budworm (see below), can cause substantial damage.

The thick litter layer of needles is indicative of the low rate of decomposition in the Boreal Forest. The long, cold winter greatly slows the activity of bacteria, fungi, and animal decomposers. The acidity of the soil also retards decomposition (see Boreal Bog, p. 67). The thick litter layer, with resin in the needles, can burn well, and thus substantial fires occur from time to time.

Balsam Fir produces an excellent seed crop every 2–4 years, and the resin in the seeds helps discourage seed predation by crossbills and Red Squirrels. The seeds are small, approximately 60,000 per pound, and they fall from early August through September. Germination does not occur until the following spring, from late May through early July. Seedlings persist in dense shade but grow quickly if "released" by a light gap caused by treefall or other disturbance. Balsam Fir is subject to damage by fire and Spruce Budworms. The budworms kill the overstory, but this releases shaded understory seedlings and saplings, and the damaged stand can be regenerated. The Spruce Budworm cycle for any given area runs between 60 and 80 years.

Red Spruce. The shapes of boreal conifers help them to quickly shed heavy snow as the boughs bend with the wind.

Balsam Fir is subject to another unique cycle, called *waves.* On slopes facing prevailing winds, the most exposed line of Balsam Firs often dies from exposure. Downslope trees die first, but their loss exposes increasingly upslope trees, which die next; thus a wave of dying trees moves upward. Regeneration follows. There can be as many as 98,000 young trees per hectare (2½ acres) 20–30 years after the passage of a wave. This stage, called the juvenile or pole stage of regeneration, is followed by a thinning due to competition among the young trees. The forest enters the mature stage after 40–60 years. Another killing wave begins after 75–80 years and the cycle repeats.

At higher elevations along the Appalachian Mountain chain, climate favors boreal species, and so the Boreal Forest reaches southward into southern states. The Appalachian Extension of the Boreal Forest hosts many bird species otherwise characteristic of northern latitudes, including the Northern Junco, Golden-crowned Kinglet, Red-breasted Nuthatch, Brown Creeper, Winter Wren, and Blackburnian, Black-throated Green, and Canada warblers. Fraser Fir replaces Balsam Fir, and Red Spruce, rather than White Spruce, characterizes the Appalachian Extension.

Boreal Bog (Pl. 5)

Indicator Plants

TREES: Black Spruce, Tamarack (American Larch), Northern White-cedar. Old, filled-in bogs are typically "invaded" by Balsam Fir, Paper Birch, Balsam Poplar, and Black Ash.

SHRUBS: Labrador-tea, Leatherleaf, Bog Rosemary, Rhodora, Pale Laurel, Poison-sumac, Winterberry (Northern Holly).

HERBACEOUS SPECIES: Sphagnum Moss, Pitcher-plant, sundews, Cotton-grass, cranberries, Creeping Snowberry. Three-leaved False Solomon's-seal, various orchids, various sedges.

Indicator Animals

BIRDS: Olive-sided Flycatcher, Nashville Warbler, Lincoln's Sparrow, Rusty Blackbird, Yellow-bellied Flycatcher.

MAMMALS: Porcupine, Moose (only in northern areas).

Description

Boreal Bogs are moist, peat-filled areas with an abundance of Sphagnum Moss. Bogs are so moist that they are often quite unstable, producing the effect of quaking when walked upon. *Be careful*—it is possible to break through and become injured or entrapped. Bogs are also recognized by an abundance of Black Spruce and Tamarack (American Larch), usually growing most densely around the edge of the bog. Toward the bog's center grow various shrubs, all of which bear thick waxy leaves. Cranberry may trail along the ground, which will be otherwise covered by various sedge species and a moist, soft mat of Sphagnum Moss, among which can be found scattered insectivorous plants, bog orchids, and Cottongrass. Areas of open water may occur.

SIMILAR FOREST COMMUNITIES: Boreal Bogs share many species with the Boreal Forest and Northern Hardwood Forest. Many bogs are transient. Over hundreds to thousands of years (see below) these bogs will be replaced by the forest community prevailing in the region. The reverse is also true: some areas are becoming more boglike. Therefore, it is common to discover habitats with mixtures of Boreal-Forest and Bog Species.

RANGE: Bogs are scattered extensively throughout Canada and the northern United States as far south as New Jersey and Ohio. They are most abundant in the ranges of the Northern Hardwood and Boreal forests.

REMARKS: Boreal Bogs are utter treats for the naturalist. The spires of Black Spruces serve as perches for the vocal and robust Olive-sided Flycatcher. Rusty Blackbirds (which make a sound like the squeak of rusted hinges), along with Lincoln's Sparrows, skulk in the thick Labrador-tea. Nashville and Mourning warblers sing from dense clumps of Leatherleaf. Inspection of the sphagnum mat will reveal the presence of such insectivorous plants as Sundew and Pitcher-plant. Rare orchids are often found on the mat.

Boreal Bogs trace their histories back to the retreat of the glaciers beginning about 20,000 years ago. When the last glacier melted it left behind numerous depressions of various sizes which filled with fresh water to become lakes, some of which were formed in such a way that no natural drainage was possible. It is these lakes, isolated from tributaries, that became bogs as sediment from surrounding areas washed in and accumulated. The combination of soil type (spodosol—see p. 26), lack of drainage, and presence of Sphagnum Moss makes these bogs very acidic. Sphagnum extracts minerals from the water by exchanging mineral atoms for atoms of hydrogen. These released hydrogen atoms add to the acidity of the bog.

The high acid concentration found in bogs results in much-reduced decomposition rates because the decomposer bacteria do not survive well in highly acidic conditions. Things simply don't rot in bogs. Bog waters are usually dark brown, similar to the color of strong tea. This color is caused by the slow leaching of chemicals called tannins from leaves that drop into the bog. Because decomposition is so slow, available nitrogen is in short supply, a condition favoring insectivorous plants such as Pitcher-

A Filling Bog. Only a small section of open water remains as this bog has become almost completely covered by vegetation, increasing the build-up of peat.

plant and Sundew, because these odd plants are adapted to capture insects to obtain nitrogen (see p. 257).

Some Boreal Bogs show a clear pattern of zonation. If viewed from the air, such bogs somewhat resemble a target with concentric zones of vegetation surrounding a central "bull's-eye" of open water. The outermost ring in the target is comprised of non-bog tree species because the soil farthest from open water is richer and more loam-like. Next comes a ring of bog tree species, the spruces and larch. These are followed by a dense ring of shrubs, followed by a zone of Sphagnum Moss, sedge, and associated herbs. A small zone of open water may or may not occur, depending on how old the bog is. Visualized through thousands of years of time, the outermost rings move progressively inward, eclipsing and replacing the inner rings until the bog is completely filled and no bog species remain.

In reality it is uncommon for a bog to fill so precisely. In some areas bogs can increase in size, rather than becoming forest habitat. Sphagnum Moss and peat sometimes accumulate sufficiently to actually raise the water table, favoring bog conditions and resulting in an increase in the bog's area.

Because a bog may require thousands of years to fill, bog sediments become the archives of past vegetation patterns in the region. Pollen blown from plants that grow near the bog-lake accumulate among the sediment particles, and because the high acidity prevents rotting, the pollen produced thousands of years ago can still be identified today. Ecologists extract sediment cores from bogs and reconstruct the vegetation history of the area by identifying the kinds of pollen present. From such work it is possible to learn the rate at which various species returned northward following glacial retreat.

Tamarack, or American Larch, is unusual among the conifers because it is deciduous. In autumn its needles turn bright yellow and drop from the tree. New needles grow back in spring. Tamarack is found in both pure and mixed stands and, though abundant in bogs, it can also tolerate drier soils. It is frequently mixed with Balsam Fir, Black Spruce, and White Spruce, but its high tolerance of acid and soggy conditions makes it very common in Boreal Bogs. Because Tamarack's widely spaced feathery branches allow much light to shine through, many shrubs tend to grow under or near Tamarack stands. These include alders, Red-osier Dogwood, Bearberry, huckleberries, Sweet Gale, blueberries, and cranberries. Tamaracks must age a bit for optimal cone production. The best cone crops occur when a tree is between 50 and 150 years old. A single mature tree can have 20,000 cones and produce approximately 300,000 good seeds in a good year. One enemy of Tamarack is the Porcupine. It strips the outer bark to feed on the inner bark, injuring the tree, sometimes fatally. Lack of light is another enemy. Tamarack is easily shaded out by other species but persists in Boreal Bogs, where the wet acidic soil prevents other species from invading.

Black Spruce also abounds in bog areas. More shade tolerant than Tamarack, Black Spruce is also fire tolerant. Cones are most abundant atop the tree and are therefore less apt to be injured by ground fires, making Black Spruce a better colonizer following fire than Tamarack. Both Black Spruce and Tamarack have shallow root systems and are subject to windthrow. Black Spruce can grow to a height of 80 ft. but is usually much smaller in sphagnum-rich areas. In nutrient-poor bogs a tree 100 years old may be less than 10 ft. tall.

Jack Pine Forest

Indicator Plants

TREES: Jack Pine (abundant), Red Pine, Red Maple, aspens, Paper Birch, Black Spruce (occasional).

SHRUBS: Green Alder, Mountain Maple, Wintergreen, Trailing Arbutus.

HERBACEOUS SPECIES: Barren Strawberry, plus others common to the Boreal Forest (see p. 62).

Indicator Animals

One bird species, the Kirtland's Warbler, nests *exclusively* in low Jack Pine Forests. It is very locally distributed on the Michigan lower peninsula.

DESCRIPTION

Jack Pines are small to medium-sized (up to 80 ft. tall but usually shorter) pines that dominate on poor soils, giving the name "pine barrens" to the area. Jack Pine Forests tend to be quite open and, in some places, form savannas (see p. 120), often with an abundance of grasses and/or shrubs. At the western edge of the range, in northwestern Wisconsin, Jack Pines are stunted and scattered among the mixed grasses and shrubs of the prairie.

SIMILAR FOREST COMMUNITIES: Jack Pine Forests bear a similarity to coastal plain Northern Pine-Oak Forests (p. 90), but Jack Pine rather than Pitch Pine is the dominant species. Jack Pines associate more with typically boreal species. Also see Northern Savanna (p. 120).

RANGE: Great Lakes states, especially northern areas of Michigan, Minnesota, and Wisconsin, extending northward well into eastern and western Canada.

REMARKS: Jack Pines, which have paired cones that curl inward toward the tip of the branch, are among the most fire-dependent of any pine species. Jack Pine Forests owe their existence to periodic fire. Without it, these trees would be replaced by other boreal species. Balsam Fir and White Spruce as well as many hardwoods sprout as seedlings beneath Jack Pines and, because Jack Pine is entirely intolerant of shade, would eventually replace it if it were not for fires. Jack Pine Forests burn, on the average, every 125–180 years, more frequently in some areas.

Jack Pine cones do not open without heat of at least 116 degrees F, and the ash from the fire provides an ideal mineral-rich bed for the seeds. Cones may remain tightly closed on trees for 25 years, not dropping off, but holding viable seeds until the heat of a fire pops the cones open, releasing the seeds. Not all Jack Pine populations have tightly closed cones, however. Some have cones that open annually even in the absence of fire.

Fig. 14. Kirtland's Warbler on a Jack Pine branch.

The very rare Kirtland's Warbler nests only in Jack Pine Forests on the Lower Peninsula of Michigan. The bird has very specific ecological requirements, nesting only in stands no more than 20 years old with thick ground cover. Stands of at least 80 acres are necessary and there must be a high density of pine branches near the ground. Management of these Jack Pine stands through selective regular burning is essential to maintain the Kirtland's Warbler's nesting sites.

EASTERN DECIDUOUS FOREST COMMUNITIES

The Eastern Deciduous Forest is a complex of forest communities dominated by trees that drop leaves in winter (deciduous trees). It ranges from northern New England to southern Florida and extends westward until it is gradually replaced by Prairie Grasslands from western Minnesota, Iowa, and North Dakota south to Texas. Throughout the Eastern Deciduous Forest mature trees range in height from 60 to more than 100 feet, and the forest is usually layered with distinct canopy, understory, shrub, and herbaceous layers (see Fig. 1, p. 10).

Over 500 species of trees and shrubs occur in the Eastern Deciduous Forest, and many different forest communities divide this huge forest into smaller units.

Northern Hardwood Forest (Pl. 6)

Indicator Plants

TREES: Yellow Birch, Sugar Maple, American Beech, Eastern Hemlock, White Pine, Red Pine, Northern Red Oak, Gray Birch, Paper Birch, Pin Cherry, Balsam Poplar, American Mountain-ash, Mountain Maple, Red Spruce.

UNDERSTORY AND SHRUBS: Striped Maple, Hobblebush, Nannyberry, Mountain Laurel, Canada Honeysuckle.

HERBACEOUS SPECIES: Painted Trillium, Goldthread, Hairy Beardtongue, Common Wood-sorrel, Kidney-leaved Violet, Pink Lady's-slipper, Wood Lily, Spotted Wintergreen, Wild Bergamot, Wild Sarsaparilla. Many Boreal and Oak-Hickory herbaceous species occur also.

Indicator Animals

BIRDS: White-throated Sparrow, Northern Junco, Purple Finch, Northern Waterthrush, Mourning Warbler, Canada Warbler, American Redstart, Black-throated Blue Warbler, Blue-headed Vireo, Hermit Thrush, Black-capped Chickadee, Brown Creeper, Golden-crowned Kinglet, Yellow-bellied Sapsucker, Cedar Waxwing.

MAMMALS: Red-backed Vole, Snowshoe (Varying) Hare, Red Squirrel, Porcupine, White-tailed Deer (abundant).

Description

The Northern Hardwood Forest is the northernmost deciduous forest community. It has many species in common with the Boreal Forest to the north and the Oak-Hickory Forest to the south and is often, therefore, called the "transition forest." Nonetheless, the Northern Hardwood Forest is distinctly dominated by three deciduous trees, Yellow Birch, Sugar Maple, and American Beech. The typical forest interior has a well-developed understory of Striped Maple and Hobblebush and many wildflower and fern species are present. Two conifers, Eastern Hemlock and White Pine, often grow abundantly among the broad-leaved species. Eastern Hemlock is found in moist cool areas such as ravines and north-facing mountainsides and White Pine grows on exposed and/or disturbed sites. White Pine, Pin or Fire Cherry, and Gray Birch are important species on disturbed sites (see Chapter 4).

SIMILAR FOREST COMMUNITIES: See Boreal Forest, Beech-Maple Forest, Maple-Basswood Forest, and Oak-Hickory Forest.

RANGE: Eastern Canada and northern New England and New York west through Canada to northern Michigan, Wisconsin, and Minnesota.

REMARKS: The Northern Hardwood Forest is among the most picturesque of the Eastern Deciduous Forests. In autumn it is a mosaic of colors, with its orange-red Sugar Maples, tan Beech leaves, and yellow birch leaves among the dark green Hemlocks and light green White Pines. Sugar Maples are commonly tapped in spring for their sap, which is boiled to extract the ingredients of maple syrup. Also in spring many wildflower species are in bloom, carpeting the forest floor along with well over a dozen fern species. White-throated Sparrows whistle *Old Sam Peabody, Peabody, Peabody,* and Northern Juncos dart through the understory, flashing their white outer tail feathers. The Yellow-bellied Sapsucker taps methodically on a Yellow Birch trunk as warblers such as the Black-throated Blue forage actively for insects. Also skulking among the foliage is the Solitary Vireo, larger and with a thicker bill than the warblers with which it shares the summer's insect harvest. At dusk the flute-like songs of the Hermit and Swainson's thrushes are heard.

Yellow Birch is present in Northern Hardwood Forests, both on disturbed and mature forest sites. This species is prolific at seed production—during a year of a good seed crop (every 1–2 years), there may be more than a million seeds per acre, all dispersed by

Pink Lady's-slipper. This common orchid is found in the largely acidic soils that characterize the Northern Hardwood Forest.

wind. Yellow Birch seeds do poorly in thick litter. Instead, they thrive best in gaps caused by windthrow, logging, or fire. Yellow Birch seeds do very well in cracks in boulders and on rotting logs. Consequently, the root systems of many Yellow Birches are stilted, indicating that the tree began by growing over a log or stump. In good light, Yellow Birch can grow quickly, attaining a height of 10 ft. in 6 years. A mature tree may grow to 100 ft., be nearly a yard in diameter, and live for over 200 years. Yellow Birch is the most shade tolerant of the birches and remains indefinitely as a canopy species.

The Eastern Hemlock, with its short, flattened needles and tiny brown cones, grows in dense stands in cool, humid areas throughout the Northern Hardwood Forest. Golden-crowned Kinglets are common nesters in hemlock stands so dense that the microclimate is noticeably cooler beneath them, and the forest floor is very densely shaded. In areas subjected to recent fire, Eastern Hemlock often grows in mixed stands with White Pine. However, hemlock seedlings, unlike White Pines, grow quite well in shade as well as open areas, and therefore Eastern Hemlock comes to be a very persistent species in mature forests. Hemlock seeds are generally dispersed by wind; the cones open only when dry and remain closed when wet. A mature hemlock can produce cones for more than 450 years, and a healthy tree will produce its first cone crops at about 20 years of age. Densely shaded trees do not produce cones, indicating that competition for light is crucial in the life cycle of an Eastern Hemlock. Like most trees of the northern forests, the Eastern Hemlock does not produce abundant seed crops every year, but rather every 2 to 3 years. Seedlings survive best on cool, moist sites with soil enriched by decomposing litter. Seedlings require constant moisture, and their rate of growth is slow. So well adapted to low levels of light are the seedlings that exposure to full sunlight actually inhibits their growth! Seedlings survive under low light for many years—a seedling may be 40–60 years old and only be 6 ft. in height! When the canopy opens a bit, seedlings grow more quickly and assume their place among the canopy dwellers. The oldest hemlock recorded was 988 years old, and the tallest on record was 160 ft. Because they lack a strong tap root and do not reproduce well by shoots, Eastern Hemlocks are killed by windthrows and their need for moisture makes them sensitive to droughts. They are, however, well adapted for winter cold. They have been measured as resistant to −112 degrees F in the cold winter months.

White Pine, Gray Birch, and Pin or Fire Cherry are all important successional species, common on open disturbed sites. They are discussed in Chapter 4.

New England Alpine Community

Indicator Plants

TREES: Balsam Fir (stunted and prostrate—see below), Black Spruce, Mountain Birch.

SHRUBS: Shrubs are dwarfed and can be confused with wildflowers. Look closely! Dwarf Bilberry, Mossplant, Alpine Bearberry, Snowberry, Mountain Heath, Alpine Azalea, Alpine Willow (actually a tree that grows like a shrub).

HERBACEOUS SPECIES: Alpine Bistort, Mountain Wood-sorrel, Diapensia, Alpine Cress, Alpine Speedwell, Alpine Violet, Mountain Aster, Moss Campion, Mountain Heath, Alpine Willow-herb, Alpine Azalea, Alpine Goldenrod, Mountain Goldenrod, Mountain Avens.

Description

The Northern Hardwood Forest is interrupted in New England by the Presidential Range of the White Mountains. The combination of northern latitude plus high altitude on the mountainside results in a distinct treeline, beyond which trees are replaced by prostrate shrubs, wildflowers, and lichens. Treeline occurs at 5200 ft. on south-facing slopes but at 4800 ft. on north-facing slopes. Treeline, the point beyond which no trees grow, is generally sharply defined. Trees from both the Northern Hardwood Forest and Boreal Forest intermingle below treeline, with northern hardwoods dropping out as altitude increases. Boreal species remain but become increasingly stunted at higher altitudes. Trees, particularly Balsam Fir, become shrublike and spread prostrate over the ground (see **REMARKS** below), as woody shrubs, which may resemble wildflowers. Above treeline, the habitat is called *alpine tundra,* and it is an open windswept habitat where wildflowers abound, along with lichens and mosses.

SIMILAR PLANT COMMUNITIES: None. This community shares some species with Northern Hardwood and Boreal forests, but approximately 75 species of herbaceous plants are unique to alpine tundra.

RANGE: The Presidential Range of the White Mountains, especially Mt. Washington, in northern New England.

REMARKS: Moving approximately 400 ft. up a mountain is ecologically much like moving 100 miles north. By climbing a mountain, you visit several major habitats associated primarily with different latitudes. Nowhere in eastern North America is this more evident than on a peak such as Mt. Washington in the Presidential Range of the White Mountains.

As you climb Mt. Washington you gradually leave the Northern Hardwood Forest and pass through Boreal Forest (see p. 62). Balsam Fir, Red Spruce, and Tamarack (American Larch) move in as American Beech, Sugar Maple, and Yellow Birch drop out. Higher still, you begin to see the effects of the chilling, drying wind on trees. Fewer species are present and those that are begin to look different. Flag trees—Balsam Firs with branches only on their protected leeward sides—give way to flattened, shrublike Balsam Firs and Black Spruces that spread nearly prostrate over the ground. This odd growth form, called *krummholz* (see Fig. 15), after the German word for "twisted wood," is the result of severe winter winds killing any branches that grow above the protected covering of the snow.

Finally, not even krummholz survives and the ground is covered only by dwarf shrubs and many species of herbaceous plants, a habitat called alpine tundra. Tundra begins beyond treeline at higher latitudes in northern Canada, but is also found atop high

Fig. 15. Moss Campion and krummholz.

Mt. Washington, New Hampshire. Alpine tundra species characterize this mountaintop, upon which is recorded the most severe weather in New England.

mountains in the temperate zone. Besides the dwarf shrubs and wildflowers, small moss mats grow in rock crevices and lichens abound over the surfaces of granite rocks. Note that the names of many of the commonest tundra species begin with "alpine" or "mountain." Approximately 75 species are found only on the alpine tundra, and not at lower altitudes.

The wildflowers may look delicate but they are hardy—only trampling by humans really does them in. Most tiny plants, like Moss Campion, are perennials. A single plant may be over 100 years old. Leaves of alpine plants are tightly clustered, thick, and waxy, and are often covered with fine hairs, all adaptations to retain moisture and minimize evaporative water loss.

Snow is the savior of Mt. Washington's plant life. More than 70 in. of precipitation fall annually, most of it as snow, forming a winter blanket that protects the ground-hugging plants at a "cozy" 32 degrees F when the air temperature and wind chill above the snow are usually well below zero.

Beech-Maple Forest (Pl. 7)

Indicator Plants

TREES: American Beech, Sugar Maple, Ohio Buckeye, White Ash, Tuliptree, White Oak, Eastern Hemlock (northeastern section), Flowering Dogwood, Witch-hazel.

SHRUBS: No shrubs are true indicators but some of the most abundant species are listed here: Spicebush, Mapleleaf Viburnum, Canada Honeysuckle, Northern Bush-honeysuckle, and Red Elderberry.

HERBACEOUS SPECIES: No wildflower assemblage is unique to this forest type, but some of the most common species are listed here: Wild

Geranium, Solomon's-seal, Squirrel-corn (Wild Bleeding-heart), Canada Mayflower, and Wild Sarsaparilla. Many other species occur. Spinulose Shield Fern is also common.

Indicator Animals

None. No bird or mammal species is uniquely abundant in this forest compared with other neighboring forests. Bird and mammal communities will be essentially a mixture of those of the Northern Hardwoods and Oak-Hickory communities. The Cerulean Warbler, though it occurs elsewhere, is common here.

Description

The combined abundance of mature American Beech and Sugar Maple, which combined can comprise up to 90% of the trees, defines this forest. No species are unique to the Beech-Maple Forest but the soil, recently glaciated, supports the two principal tree species in abundance. A sharp boundary exists between this forest and the Mixed Appalachian Forest (see below) that coincides with the terminal moraine, the southernmost boundary of the Wisconsin Glacier.

SIMILAR FOREST COMMUNITIES: This forest has species in common with the Oak-Hickory, Northern Hardwood, and Maple-Basswood forests, as well as the Mixed Appalachian Forest. Rely on range and the abundance of Beech and Sugar Maple in identifying it.

RANGE: Extreme southwestern New York west through most of Ohio, northern Illinois, and southern Michigan.

REMARKS: In many areas American Beech tends to dominate the canopy and Sugar Maple the understory. This, however, can change (see below). An upland forest may contain up to 70% Beech in the canopy, while Sugar Maple is more dominant along slopes.

American Beech can grow to a height of 80 ft. and live for more than 360 years. Its smooth gray bark and papery, toothed leaves make it one of the easiest trees to identify. Beech is highly shade tolerant and is usually persistent in mature forests throughout its range. Beech is an excellent stump-sprout tree and also sends up many stems via root suckers. Typically, a mature Beech will be surrounded by young sprouts that have all arisen from the roots. Any given tree will usually have 7–12 sucker stems, but some may have many more.

Beech nuts ripen in the fall. Two to three nuts are found per bur and the seeds are generally heavy. Beech is a masting tree (see p. 441), meaning that in any given part of its range most trees produce large seed crops in synchrony every 2–3 years. In staggering numbers, the now-extinct Passenger Pigeon once fed heavily on Beech mast. Beech grows very slowly and produces no seeds until approximately age 40.

Sugar Maple. This species, shown in its fall color, is numerically dominant in several broadleaf forest types.

Beech bark is thin, and Beech trees therefore succumb easily to fire. They are also injured by winter frost, and damaged bark permits entry of sap-rotting fungi.

Throughout the range of the Beech-Maple Forest, Beech and Sugar Maple are equally dominant. Why doesn't one replace the other? The answer, shown by experiment, is that seedlings of Beech grow best under a canopy of Sugar Maple, and vice versa. Therefore, Beech seedlings under Beech do not fare as well as seedlings under Sugar Maple, but the same holds true for Sugar Maple seedlings. The abundance of each species therefore tends to oscillate, with the canopy-dominant species being slowly replaced by the understory-dominant species, only to see the process repeat itself. This odd relationship is called *frequency-dependent selection*. The more abundant a species becomes in the canopy, the worse its propagules will fare in the understory.

Sugar Maple is one of the finest fall trees, with its blazing orange-red leaves. Recent research on Sugar Maples shows that they are gap species, surviving well in deep shade but growing very slowly. Some show suppressed growth for up to 150 years! When a disturbance occurs, permitting more light to penetrate the canopy and understory, Sugar Maple accelerates its growth. The typical pattern for a Sugar Maple life cycle is seedling suppression, release (allowing for some upward growth of saplings), more suppression due to shading, followed by more release due to disturbance. Finally, the tree makes it to the canopy, where an individual may live for 100–200 years. Research has shown that Sugar Maple saplings are more resistant to catastrophic wind damage than most other species, including aspen. This enhances the long-term persistence of Sugar Maple.

Indicator Plants

TREES: Sugar Maple, American Basswood, Northern Red Oak, American Elm, Slippery Elm, Butternut, Flowering Dogwood.

SHRUBS: Mapleleaf Viburnum, Fox Grape, Alternate-leaf Dogwood, Hairy Honeysuckle, Wicopy, Bearberry.

HERBACEOUS SPECIES: The spring flora is very rich in this forest and commonly includes Bittersweet, Large-flowered Bellwort, Bloodroot, Sweet Cicely, Wild Ginger, Dutchman's-breeches, Hepatica, White Trillium, and Ginseng.

Indicator Animals

None. All birds and mammals are shared by other forest communities elsewhere.

Description

Two species, Sugar Maple and American Basswood, are numerically dominant in this forest, which is otherwise a mixture of the Northern Hardwood and Oak-Hickory communities. Two elm species are very common, as is Northern Red Oak. This forest occurs on mesic (moist) sites and is replaced by the Oak-Hickory Forest on more xeric (dry) sites.

SIMILAR FOREST COMMUNITIES: The Maple-Basswood Forest has many species in common with the Oak-Hickory Forest, Northern Hardwood Forest, and Beech-Maple Forest. It is the least distinct of the major forest communities of the Eastern Deciduous Forest. Rely on a combined abundance of Sugar Maple and American Basswood to identify it.

RANGE: This forest ranges from central Minnesota south through Wisconsin and northeastern Iowa. It covers the smallest area of any Eastern Deciduous Forest community.

REMARKS: American Basswood, also called Linden, is recognized by its deep furrowed bark and large, alternate, toothed leaves with long petioles (leaf stalks). Flowers are yellow-white clusters hanging from long leaflike bracts. A mature tree will stand 60–80 ft. tall. American Basswood, though abundant in this forest type, enjoys a much wider range, occurring from Maine west to the eastern Dakotas and south to Arkansas. Throughout its range it prefers loam soils on mesic sites. Soil is enhanced by Basswood, more so than by other trees; as its leaves decompose, they add such important minerals as calcium, magnesium, nitrogen, potassium, and phosphorus to the upper soil. Basswood will begin producing viable seeds at age 15 and good seed crops are produced almost every year. Seedlings are often killed by browsing rabbits. Basswood, like Beech, is an extraordinarily good stump-sprouter. It also is highly shade tolerant, and thus its seedlings survive in the shady understory and grow to become canopy members.

Oak-Hickory Forest (Pl. 8)

Indicator Plants

TREES: Northern Red Oak and Southern Red Oak, Black Oak, Scarlet Oak, White Oak, Chestnut Oak, Shumard Oak (South), and other oaks, Pignut Hickory, Mockernut Hickory, Bitternut Hickory, American Chestnut (understory only, but formerly a canopy tree), Flowering Dogwood, Sassafras, Hophornbeam, Hackberry, Green Hawthorn.

HIGHLY MESIC SITES: Tuliptree, American Elm, Sweetgum, and Shagbark Hickory. Red Maple and Sourwood are important constant elements throughout Ohio, Kentucky, and Tennessee. Sugar Maple is an invading tree on many sites.

DISTURBED AREAS: Black Locust, Gray Birch, Eastern Red Cedar, Quaking Aspen, Bigtooth Aspen, Pitch Pine (North), Slash Pine (South), White Pine (North), and Bear Oak (see Pine Oak Forest below).

SHRUBS: Mountain Laurel, Highbush Blueberry, Lowbush Blueberry, Early Lowbush Blueberry, Mapleleaf Viburnum, Deerberry.

HERBACEOUS SPECIES: Wintergreen, Spotted Pipsissewa, Wild Sarsaparilla, Violet Wood-sorrel, Pink Lady's-slipper, Rue-anemone, May-apple, Jack-in-the-pulpit, False Solomon's-seal, Trout-lily, Sessile Bellwort, and many others.

Indicator Animals

BIRDS: Blue Jay, Wild Turkey, Scarlet Tanager (North), Summer Tanager (South), Rose-breasted Grosbeak.

MAMMALS: Gray Squirrel (abundant), Fox Squirrel, Northern Flying Squirrel, Southern Flying Squirrel, Eastern Chipmunk.

Description

The Oak-Hickory Forest is a forest of nut-producing trees. Acorns and hickory nuts are usually obvious among the leaf litter on the forest floor. Roving flocks of Blue Jays moving noisily through the forest and the ubiquitous Gray Squirrels are easy to find throughout the year. Fox Squirrels are also common within their range, as are Eastern Chipmunks. In summer, Scarlet and/or Summer tanagers and Rose-breasted Grosbeaks sing from the canopy, while Ovenbirds sing *teacher, teacher, teacher* from the forest floor. The understory is usually well developed, often with Flowering Dogwood, especially picturesque when blooming in early spring.

Other understory species commonly include Sassafras and Hophornbeam. The shrub layer is distinct, mostly dominated by species characteristic of acidic soils: blueberries, huckleberries, and laurels. Some shrubs are evergreen. Many wildflower species occur, and the abundance of different species varies throughout the large range of the Oak-Hickory Forest.

SIMILAR FOREST COMMUNITIES: The Oak-Hickory Forest intermingles with virtually all other forest types. Some oaks, like White Oak and Red Oak, are common in forests other than Oak-Hickory. An abundance of *both* oak and hickory species identifies this forest. Wetter sites typically have an abundance of American Elm, Tuliptree, and Sweetgum in addition to oaks and hickories, and Sugar Maple is becoming much more common as a canopy species in many areas (see **REMARKS**). The species assemblages in northern areas differ somewhat from those in the South.

RANGE: The Oak-Hickory Forest has the largest range of any of the Eastern Deciduous Forest communities. It forms a giant horseshoe from eastern Massachusetts south through Ohio, Pennsylvania, New Jersey, Virginia, the Carolinas, and on through central Georgia west through Arkansas and eastern Texas, and north to south-central Michigan.

REMARKS: The Oak-Hickory Forest used to be considered two separate but related forest communities. To the south and west was the true Oak-Hickory Forest. To the north and east was the Oak-Chestnut Forest. Until the present century American Chestnut ranged abundantly as a dominant tree in the Oak-Chestnut Forest from New England through Pennsylvania and Ohio to the Appalachian Mountains. However, in 1906 a fungus native to China invaded Chestnut trees in the New York City area. Utterly unresistant, American Chestnuts rapidly succumbed, and the blight spread quickly throughout the entire range of the species. All of the large individuals were killed, but only above ground. Though no longer the tall giants of the past, Chestnuts continue to survive. Roots are blight resistant and continue to send up shoots so that the American Chestnut remains common in the understory of the Oak-Hickory Forest. Usually, by the time the tree is 20 ft. tall or so, the blight catches up with it and it dies back. The loss of the American Chestnut as an important canopy species altered the forest but did not devastate it. Other species, such as the oaks and hickories, have increased in abundance and replaced the Chestnut as canopy species. Today, American Chestnut survives as essentially an understory species.

White Oak enjoys one of the widest ranges of any tree species of eastern North America. Absent only from northern Maine, New York, Michigan, and Wisconsin (where it is replaced by Boreal and Northern Hardwood forests), and in Florida, White Oak can be found virtually anywhere else within the range of this guide. It is widely tolerant of different soil types, moisture levels, and exposure. It is a masting species, like Beech: White Oaks in a given region produce huge numbers of acorns every 4–10 years, and an individual tree may go several years with virtually no acorn production. Local changes in yearly abundance of Gray Squirrels and Blue Jays are clearly tied to acorn abundance. In abundant years, a single tree may produce 2000–7000 acorns. Acorns have a high germination capacity but do fall prey to squirrels and jays. Blue Jays may aid in the spread of White Oaks, however, since they have a tendency to plant acorns away from the parent plant. Gray Squirrels show a preference for burying Red or Black Oak acorns (see below) and thus aid in the dispersal of those species.

White Oak is a good stump-sprouter and recovers well after cutting or mild exposure to fire. White Oak is not as tolerant of shading as are other species such as Sugar Maple (see below) or Eastern Hemlock and tends to be most abundant on dry, more open sites. Gypsy Moths do great damage to White Oaks and to Oak-Hickory Forests in general.

White Oak can become quite a stately tree, reaching 150 ft. in height (though 100 ft. is much more common) and 4 ft. in diameter, living to an age of 600 years. Other common oaks are Northern Red Oak, Black Oak, and Scarlet Oak. Unlike the White Oak, which has leaves with rounded lobes, the leaves of these species have sharp-pointed lobes. Northern Red Oak and Black Oak hybridize over much of their range, and can be confusing to identify. Northern Red Oak is less shade tolerant than White Oak but

Mixed Oak Forest. Oaks are numerically dominant throughout much of the eastern states. This forest, shown in fall, has a blueberry understory.

more tolerant than either Black or Scarlet Oak. Consequently, both Black and Scarlet Oak are most abundant in more open woodlands. Also, Black Oak seedlings are rather intolerant of shade and thus grow best in open woods. Black Oak often occurs in pure stands on drier, more exposed sites, where Northern Red Oak and White Oak are at a disadvantage. Scarlet Oak, named for its brilliant fall color, grows on many soil types throughout eastern North America. Like the other oaks, it produces bumper acorn crops only occasionally—usually every 3–4 years.

Chestnut Oak is named for the resemblance of its leaf to that of American Chestnut. It requires rather open woodlots to become established. Like most oaks it is a prolific sprouter; in one study in the southern Appalachians, 75% of Chestnut Oak reproduction was by sprouts rather than seeds.

The acorns of oaks are a staple of the diets of Wild Turkeys, Blue Jays, Gray and Fox squirrels, and Eastern Chipmunk. Until this century, hosts of Passenger Pigeons and Carolina Parakeets joined these species to feast on the mast harvest. Today, only Blue Jays and squirrels feed very heavily on acorns while they are still on the tree. Wild Turkeys, replenished in population in many areas by wildlife management techniques, feed on them once they are on the ground.

Studies with Gray Squirrels show that all acorns are not equal from the squirrels' point of view. Acorns from the Black and Red oaks have both a higher fat content and a higher tannin content than those from White Oaks. This would seem to pose a dilemma for squirrels. Acorns with high fat contents are good—their calories help the squirrels get through the winter. However, high tannin intake can interfere with protein digestion (see p. 260). Should a squirrel eat high-fat acorns for the added energy, risking the side effects of high tannin concentration, or eat acorns with low tannin but also with less energy? Experimental work has shown that squirrels do eat both types of acorns, but are wary of those with high tannin levels. Rather than eat these high-tannin acorns, the squirrels bury them. There is, however, no evidence to indicate that tannins are in any way harmful to squirrels. Squirrels may be using tannin level as a cue signaling them to bury Red or Black Oak acorns and eat White Oak acorns. Squirrels cannot easily digest sprouted acorns, and White Oak acorns sprout early, whereas acorns from the Red and Black oaks sprout late. White Oak acorns, from a squirrel's point of view, are "perishable," but Red and Black Oak acorns are "storable." A squirrel will eat White Oak acorns but will bury a Red or Black Oak acorn whole, using it for food after the White Oak acorns have been depleted. So, squirrels eat both types of acorns, but at different times.

Hickories are easily recognized by their distinctive compound leaves. Pignut Hickory is the most wide ranging of the group, inhabiting dry slopes and ridges as well as mesic sites. It reproduces through prolific stump sprouts as well as nuts. Seed crops tend to be large every two years. Mockernut Hickory also grows on exposed ridges and hillsides as well as rich mesic soils. It produces abundant seed crops every 2–3 years and the nuts form an important food for Gray, Fox, and Red squirrels, Eastern Chipmunks, and Raccoons.

In many northern areas traditionally identified as Oak-Hickory Forest, Sugar Maple is invading. This remarkably persistent, slow-growing, shade-tolerant tree (see **REMARKS** under Beech-Maple Forest, above) grows slowly in the understory. Sugar Maple is sensitive to severe exposure, fire, and grazing pressure, but where these disturbance factors have been controlled, Sugar Maple is slowly encroaching on the oaks and hickories. It therefore appears that in the northern range, periodic disturbance is necessary for the persistence of Oak-Hickory Forest.

Eastern Red Cedar, Black Locust, and aspens are important successional species on disturbed sites. They are discussed in Chapter 4. Pitch Pine, Slash Pine, and Bear Oak are part of the Northern Pine-Oak Forest, discussed later in this chapter.

Northern Riverine (Floodplain) Forest (Pl. 9)

Indicator Plants

TREES: Eastern Cottonwood, Black Willow, American Elm, Slippery Elm, Eastern Sycamore, Speckled Alder, Green Ash, Black Ash, Red Maple, Silver Maple, Shagbark Hickory, Box-elder, River Birch, Basswood, Swamp White Oak, Pin Oak, Balsam Poplar (North).

SHRUBS: American Elder, Moonseed, River Grape, Poison-ivy, Trumpet-creeper, Peppervine, Buttonbush.

HERBACEOUS SPECIES: Green Dragon, Sweetflag, Ostrich Fern, Wood Nettle, American Black Currant, Stinging Nettle, Jewelweed, Turtlehead.

Indicator Animals

BIRDS: Belted Kingfisher, Bank Swallow, Spotted Sandpiper, Green-backed Heron, Wood Duck, Yellow-throated Vireo, Yellow Warbler, Blue-gray Gnatcatcher.

MAMMALS: Mink, River Otter (both uncommonly observed but relatively numerous).

Description

Riverine forests, as the name implies, occupy moist sites along rivers and floodplains. Spring flooding is an annual occurrence. Riverine forests are not normally sharply separated from neighboring upland forests and are frequently subject to invasion by upland species such as oaks and hickories. The abundance of light along river banks permits an abundance of vines to festoon the branches. Herb cover may be minimal, especially in areas subject to frequent flooding. Dominant trees differ geographically: cottonwoods and ashes characterize the midwestern regions, while Eastern Sycamore and Red Maple are most abundant in northeastern areas. Silver Maple grows tallest and most abundantly in the Ohio River valley. Willows and elms abound everywhere in riverine forests.

SIMILAR FOREST COMMUNITIES: Swamp forests (see below) share many species with riverine communities. There is a mixture of Northern Riverine with Southern Riverine and Baldcypress Swamp Forest species along rivers in the southern states. Where rivers have been diverted or dammed, non-riverine species tend to replace the riverine species. These species may be apparent in the understory while normal riverine species are still prevalent.

RANGE: This forest is found throughout the Eastern Deciduous Forest except in the Deep South, where it is replaced by Baldcypress Forest or Southern Riverine Forest. The species composition is regionally variable (see above). Cottonwoods and ashes in particular follow major river systems deep into the prairie region of the Midwest. As eastern forest gives way to grassland in the midwestern region, ribbons of riverine forest, termed *gallery forests,* hug the moisture-laden soil, providing the westernmost extensions of the eastern forests.

REMARKS: A quiet canoe ride along a riverine forest provides an ideal way of seeing birds. Green-backed Heron and Wood Duck fly up as you approach, the latter calling a sharp *wooo-eeek* as it flies swiftly through the trees. A Spotted Sandpiper teeters on rocks, bobbing its tail up and down as it probes for insects. Where the river has eroded a bank, Belted Kingfishers and Bank Swallows excavate their nests. Oriole nests hang from branches above the river. Yellow-throated Vireos, Great Crested Flycatchers, Yellow Warblers, and other birds sing from the canopy. River Otters and Mink, aquatic members of the weasel family, though present, are not easy to find and provide a challenge for the naturalist.

Rivers are dynamic bodies; the energy contained in a flooding river has a profound effect on shaping the landscape. Over time, rivers twist and turn, depositing soil to form *point bars* where riverine trees and shrubs invade (see p. 199). Should a point bar remain stable for a long period of time, riverine species will be replaced by more upland species as the bar eventually stabilizes and builds up above water level. However, bars are often flooded, resetting the stage for riverine species to invade and thrive.

Floodplains are very fertile areas because nutrient-laden silt washed from surrounding areas is deposited annually during spring floods. However, floodplains are intrinsically unstable due to the natural dynamics of rivers. Riverine tree species are able to grow rapidly and colonize newly created silt bars.

Perhaps the most beautiful tree of the floodplain forest is the Eastern Sycamore. This tree is by no means confined to northern floodplains but also occurs abundantly throughout the South and into the Midwest. Also called American Plane-tree, the Sycamore is easily recognized by its maple-like leaves and whitish peeling bark. Seeds are in drooping, ball-like fruits (often called "buttonballs"), and seeds are very small—200,000 per pound. The tiny seeds are dispersed by water, germinating on mudflats and point bars. Sycamore grows quickly, reaching heights of 70 ft. by the age of 17 years. It is a successful competitor against other floodplain species such as cottonwoods and willows and is often the dominant floodplain species. It is a good colonizer and persists in the mature canopy. It even invades old fields as an early successional tree (see p. 194).

Elms also characterize Northern Riverine Forests. The American Elm, with its easily recognized wide-spreading crown, ranges throughout eastern North America and well into the midwestern states. Dutch elm disease has caused this species to suffer an unfortunate decline in much of its range. American Elm grows well on mesic soils as well as along floodplains, but it is often outcompeted by Sugar Maples and Beech on these sites. Some mature American Elms grow to 125 ft., with diameters of 5 ft. Elms are excellent stump-sprouters and grow back if cut. Leaves are high in calcium and potassium and their decomposition helps renew soil fertility. Slippery Elm is common along many floodplains as well as mesic woodlands. Both wind and water are important agents of seed dispersal for this species.

Two maple species, Silver Maple and Box-elder (Ashleaf Maple), are important floodplain colonizers. Silver Maple ranges well into the midwest, following major river valleys. It is a true alluvial species, and it is a poor competitor except when growing along river banks. Box-elder is the only maple with compound leaves.

Cottonwoods are important successional trees and are discussed in Chapter 4.

Northern Swamp Forest (Pl. 10)

Indicator Plants

TREES: Red Maple (abundant), Atlantic White-cedar, Northern White-cedar, Black Tupelo (Black Gum), Sweetgum, Speckled Alder, Black Ash, Swamp White Oak, Cherrybark Oak, Willow Oak, American Elm, American Holly, Eastern Hemlock, Balsam Fir.

SHRUBS: Sweet Pepperbush, Spicebush, Swamp White Azalea, Great Rhododendron, Highbush Blueberry, Smooth Arrowwood.

HERBACEOUS SPECIES: Skunk Cabbage, Jewelweed, Goldthread, Marsh-marigold, Swamp Saxifrage, Marsh Blue Violet, Turtlehead, Jack-in-the-pulpit, Cardinal-flower, Sphagnum Moss.

Indicator Animals

BIRDS: Yellow Warbler, Louisiana Waterthrush, Alder Flycatcher, Blue-gray Gnatcatcher, Wood Duck, Green-backed Heron, Barred Owl.

Description

The abundance of Red Maple and the presence of other water-loving species identifies this forest community. Red Maple is sometimes virtually the only tree species present, especially in northern regions, but other species, such as Speckled Alder and both the Atlantic and Northern white-cedars, can become dominant as well. Other trees, such as Willow Oak, Swamp White Oak, and Cherrybark Oak, become part of the southern component. The understory is usually thick with shrub cover, but cedar-dominated swamps can be dark enough to retard shrub growth. Skunk Cabbage is the most characteristic herbaceous species, but other species such as Jewelweed, Cardinal-flower, Jack-in-the-pulpit, and Marsh-marigold are common.

In coastal areas many Northern Swamp Forests are mixtures of Red Maple and Atlantic White-cedar. Inland, Red Maple mixes with Northern White-cedar. Red Maple competes well against both cedar species because of its superior shade tolerance, though it is less water tolerant than the cedars. In very boglike areas with a high water table and much standing water, cedars tend to dominate.

SIMILAR FOREST COMMUNITIES: See Northern Riverine Forest and Southern Mixed Hardwood Swamp Forest. In northern areas, there is some overlap between this forest and Boreal Bog.

RANGE: This forest is found throughout northern North America from New England through New York, New Jersey, Pennsylvania, Ohio, and the midwestern states.

REMARKS: The Northern Swamp Forests host some of the most brilliant

Skunk Cabbage. An indicator, by its abundance, of the Northern Swamp Forest, this species also occurs in hydric areas in many other forest types. It is one of the first wildflowers to open in spring.

wildflowers and one of the most odorous species. The earliest to bloom is Skunk Cabbage, which often begins poking through the mud when there is still snow on the ground. Named for its rank odor, Skunk Cabbage belongs to a tropical plant family called the arums, which includes such familiar house plants as philodendrons and *Dieffenbachia*. It contains calcium oxalate, a caustic chemical which makes the mouth burn and probably helps protect the plant from plant-eating insects. From mid-March throughout the summer its broad leaves show almost no sign of insect damage. Bees and flies pollinate Skunk Cabbage. The unusual flower complex consists of many tiny flowers grouped in a rounded podlike cluster called a spadix, which is topped by an umbrella-like leafy structure, the spathe. Jack-in-the-pulpit bears a similar floral structure. Surprisingly, Skunk Cabbage flowers are warm. They actually melt their way through ice and snow, reaching temperatures of up to 72 degrees F. Because of the warmth of the flower, Honey Bees can pollinate Skunk Cabbage even when the air temperature is just a few degrees above freezing. The flower uses oxygen at prodigious rates in order to maintain its high temperature. When air temperatures approach freezing, the flower burns oxygen at a rate similar to the high respiration of a hummingbird. Why the high flower temperature? The warmth undoubtedly helps both pollen and eggs to grow faster and mature earlier than in other wildflowers.

In early spring the Skunk Cabbage is joined by the bright yellow Marsh-marigold, also called Cowslip, which adds color to the swamp. In summer the deep red Cardinal-flower (Pl. 40) attracts Ruby-throated Hummingbirds, and Turtlehead, with its two-lipped (resembling a turtle's head) white flowers, pokes up among the wide leaves of Skunk Cabbage.

Sweet Pepperbush, which blooms in midsummer, has fragrant flowers located on a spike poking above the shrub. The bright white flowers are pollinated by many bee and fly species. Flowers open from the bottom of the spike upward, so that the "oldest" are on the bottom and the "youngest" on the top. This has important consequences for pollination. As the flowers open, only the pollen-containing stamens are initially exposed. Any insect visiting the flower will pick up pollen in the process. Because the female part of the flower is not yet mature, the flower is functionally male. However, the stamens soon bend outward and the pistil grows upward, beyond the stamens, converting the flower to a functional female. The oldest flowers are at the base of the spike; those at the base are females and those at the top are males. Insects typically move *up* a flower stalk, so that a bee carrying pollen from a given Sweet Pepperbush will fly to the base of a flower stalk of another, pollinate the female flowers, and move upward, picking up new pollen before it departs to visit another plant. This is an effective system of cross pollination, and it occurs in many other plant species.

Speckled Alder, a shrublike tree, often grows in very dense stands. Like the legume species discussed in Chapter 4, it has the ability to take oxygen from the atmosphere and fix it in chemical form. This results in enhanced soil fertility.

Northern Pine-Oak Forest (Pl. 11)

Indicator Plants

TREES: Pitch Pine, Virginia Pine, Bear Oak, Blackjack Oak, Chinkapin Oak, Scarlet Oak, Post Oak, Black Oak, Eastern Red Cedar.

SHRUBS: Bearberry, huckleberries, Inkberry, Broom Crowberry, Lowbush Blueberry, Sheep Laurel (Lambkill), Wild-raisin.

HERBACEOUS SPECIES AND VINES: Blazing-star, Butterflyweed, Pinesap, Poverty Grass, Rough Hawkweed, Wild Lupine, Wintergreen, Little Bluestem grass.

Indicator Animals

BIRDS: Pine Warbler, Prairie Warbler, Rufous-sided Towhee, Chipping Sparrow, Common Flicker, Brown Thrasher, Northern Bobwhite, Whip-poor-will, Great Horned Owl, Mourning Dove, Eastern Bluebird (uncommon).

MAMMALS: Gray Squirrel and Eastern Chipmunk (both abundant).

DESCRIPTION

This is the predominant forest community on dry sandy soils along the northern Coastal Plain (Cape Cod and the southern New England coast, south through New Jersey to the Carolinas). The pines appear "scrubby" and often grow in dense single-species stands. Oaks may grow both upright or shrublike and often make a thick spreading shrub layer beneath the pines. Huckleberries are usually abundant, along with other heaths, and add to the dense shrub layer. Evidence of fire is usually apparent. In older, fire-protected tracts Red, Black, and White oaks gradually replace the pines. Atlantic White-cedar Swamps (see White-cedar Swamp Forest, p. 95) occur on mesic (moist) sites.

SIMILAR FOREST COMMUNITIES: See Oak-Hickory Forest, Southern Mixed Pine-Oak Forest, and White-cedar Swamp Forest. Also see boreal Jack Pine Forest.

RANGE: Coastal New England from Massachusetts (including all of Cape Cod, Nantucket, and Martha's Vineyard) south through Long Island, New Jersey, and eastern North Carolina. A similar forest type is found in the Great Lakes area which, in boreal regions, becomes mostly a Jack Pine Forest.

REMARKS: Sometimes called the "Pine Barrens," Pine-Oak Forests are far more interesting than they first appear. Among the birds present, Pine Warblers and Chipping Sparrows, sounding somewhat similar, sing from the upper branches, while Prairie Warblers sing their upscale *zee, zee, zee* from small oaks and other lower vegetation. The Rufous-sided Towhee, a large and colorful member of the sparrow tribe, scratches with both feet simultaneously in the dried lit-

Coastal Pitch Pine Forest. This Cape Cod (Massachusetts) forest exists on low-nutrient, highly sandy soil, accounting for the scraggly appearance of the trees. The understory is blueberry and huckleberry.

ter, searching for insect food. A walk through a summer pineland forest can flush up a family of Northern Bobwhite or a sleepy Whip-poor-will. At night the repeated calls of the Whip-poor-wills and the deep hooting of a Great Horned Owl provide one of the finest choruses in nature.

Both the Northern Pine-Oak Forest and the Southern Mixed Pine-Oak Forest (see below) are maintained largely by the effects of frequent fires. Were it not for fire, and the fact that pines, by virtue of their fire-resistant bark, withstand fire better than oaks, these forests would slowly change to become increasingly mesic, dominated by oaks, hickories, and Red Maple. The Northern Pine-Oak Forest is actually a unique part of the Oak-Hickory Forest that never quite becomes dominated by oaks and hickories because of the combination of dry sandy soil and frequent natural fires. The pine species that thrive have *serotinous* cones that open only when exposed to heat from fire. Fire also kills oak seedlings and releases minerals from the burned leaf litter to the soil, making a natural fertilizer for the seedling pines. Pines are not shade tolerant. With total protection from fire, slow-growing oaks gradually overtop, shade, and outcompete the pines. Research in New Jersey has shown that, on the average, any given tract of the Pine-Oak Forest now burns about every 65 years, compared with intervals of 20 years earlier in the century.

Several "scrubby" oak species are abundant in this forest type. Bear Oak rarely grows higher than 20 ft. and can develop into a very thick "shrub" layer. Blackjack Oaks have extremely thick leathery leaves, the undersurface of which is brown and hairy. These leathery leaves aid in reducing water loss during the heat of the day. Post Oak leaves, which have deep-rounded lobes, are also leathery. Post Oak and Blackjack Oak grow to heights of 50 and 30 ft., respectively. The tallest of the scrubby sand-plain oaks is Chinkapin Oak, which can reach 80 ft. in height. This species has elongate leaves with large notches.

Southern Mixed Pine-Oak Forest (Pl. 12)

Indicator Plants

TREES: Longleaf Pine, Loblolly Pine, Shortleaf Pine, Slash Pine, Virginia Live Oak, Turkey Oak, Post Oak, Myrtle Oak, Laurel Oak, Southern Red Oak, Common Persimmon, Southern Catalpa, hickories, hawthorns, Southern Bayberry, Carolina Holly.

SHRUBS: Saw-palmetto, Southern Bayberry, Odorless Bayberry, Winged Sumac.

HERBACEOUS AND GRASS SPECIES: Little Bluestem grass, Yellow Stargrass, Wiregrass, Spanish Moss (an epiphyte), Colicroot.

Indicator Animals

BIRDS: Brown-headed Nuthatch, Pine Warbler, Yellow-throated Warbler, Northern Parula, Red-cockaded Woodpecker (local), Red-headed Woodpecker, Bachman's Sparrow (uncommon), Wild Turkey, Northern Bobwhite, Loggerhead Shrike, Painted Bunting, Eastern Bluebird, Chuck-will's-widow, Red-shouldered Hawk, Black Vulture, Turkey Vulture, Common Ground-dove, Mourning Dove.

MAMMALS: Nine-banded Armadillo, Virginia Opossum, White-tailed Deer, Gray Fox.

Description

The Southern Mixed Pine-Oak Forest is the characteristic forest of the southern Coastal Plain. It occurs on sandy and/or dry soils or on sites that are frequently exposed to naturally occurring fires. As in the Northern Pine-Oak Forest, the species richness is low; many tracts are characterized by only one or two tree species. Unlike the Northern Pine-Oak Forest, however, this forest often has poor shrub growth and much open grass. The understory is often poorly defined or non-existent but may be dense in areas recently exposed to fire. Grasses usually dominate the herbaceous layer but palmettos are often thickly developed as well. Many hardwood species tend to invade, especially on moist soils protected from fire. Pine species are raised for timber, often in single-species plantations.

SIMILAR FOREST COMMUNITIES: This forest is closely related to both the Oak-Hickory Forest and the Southern Hardwood Forest. Periodic fire

Fig. 16. Longleaf Pine seedling.

plus very dry sandy soils are the factors which give the pines an advantage over hardwoods. Some overlap in species also occurs with the Northern Pine-Oak Forest (southern New Jersey through Maryland, Virginia, and North Carolina).

RANGE: Coastal areas and upland sandy soils of the Southeast, from extreme southern New Jersey and Virginia south through the Gulf states to eastern Texas but excluding extreme southern Florida.

REMARKS: As in the Northern Pine-Oak Forest, fire is the force responsible for the existence of this forest community. Longleaf Pine demonstrates unique adaptations to survive and reproduce in a fire-prone environment. Seeds lie dormant in the soil until fire burns the grass and litter. Germination quickly follows but the Longleaf Pine seedling grows very slowly above ground. Instead, it devotes most of its energy to rapidly growing a thick, deep root, which extends 2½ in. deep after only a week's growth. Above ground, the seedling is covered by a dense protective umbrella of thick needles, giving it an almost "hairy" appearance. This growth form is called the grass stage, and it persists for about 3–7 years while the root continues to grow. The thick, hairlike needles that superficially resemble a grass clump protect the seedling from fire, though Longleaf Pine seedlings do fall prey to rabbits and Cotton Rats. During the grass-stage years, prodigious growth continues underground as a large, energy-rich taproot develops out of the reach of fire. After the grass stage, the tree adds height very rapidly, outgrowing the danger of fire along the ground.

Loblolly Pine, often mixed with Shortleaf Pine and various hardwoods, is another abundant sandy soil species. This species, as well as Longleaf, Shortleaf, Slash, and Virginia pines, is highly intolerant of shade and is easily outcompeted by oaks, hickories, Sweetgum, and Tuliptree. However, the frequency of fire in the region makes it possible for Loblolly and the other species to remain abundant.

Perhaps the most picturesque tree of the South is the Virginia Live Oak. Draped with clumps of Spanish Moss, low-spreading Live Oaks provide a sense of serenity as well as beauty. Though the tree is seldom taller than 50 ft., its crown may spread 150 ft., and its trunk may be 6–7 ft. wide. Live Oaks produce long tapered acorns, up to 390 per pound, and Wild Turkeys feed heavily on them. Should the top of a Live Oak be killed, the tree will develop many root sprouts and regenerate. Unlike the pines, this species is intolerant of fire, though otherwise quite well adapted to the sandy coastal soils. Because of its susceptibility to fire, however, Live Oak tends to occur only on the moister soils. It is resistant to salt spray and is therefore abundant along the coast and on coastal islands, from Louisiana to the Carolina Outer Banks.

Forests exclusively of Turkey Oak occur on higher elevations of the Coastal Plain, along the Georgia–South Carolina border and around Aiken, South Carolina. Unlike many oaks, this hardy oak is well adapted to periodic fire. It spreads rapidly by growth of underground runners following fire. Turkey Oak can grow so densely that it effectively shades out virtually all other species.

Among the mammals frequently encountered is the Nine-banded Armadillo, related to the tropical sloths and anteaters. Like the abundant Virginia Opossum, Armadillos originated in the tropics and have expanded their range northward. The Opossum has established itself in New England and the Armadillo is becoming abundant in the Southeast. The Armadillo's hide is thick and bone-like, and the animal protects itself by curling up, tucking in its soft underparts. Unfortunately, both the Opossum and the Armadillo are frequently struck by automobiles as they try to cross roads at night.

Southern Mixed Pine-Oak Forests support several birds in addition to the more widespread northern pineland species. Brown-headed Nuthatches, moving about with warblers, chickadees, and titmice in noisy bands of up to 20 individuals, nest in fire-charred stumps and feed heavily on pine seeds. The more widely distributed Yellow-throated Warbler sings a clear *tew, tew, tew* from the higher pine boughs. Northern Parula warblers buzz from the Spanish Moss, and Eastern Bluebirds can be found nesting in decaying pine stumps. Open pinelands also host the striking Red-headed Woodpecker, which nests in small, loosely formed colonies, often along roadsides, where they are fond of perching on telephone poles. Another woodpecker, much reduced in population and localized in range, is the Red-cockaded Woodpecker, a semicolonial, zebra-striped bird with large white cheeks. It covers the outside of its nest hole with pine pitch. The rather drab Bachman's Sparrow can be found perched atop a pine singing its trilling song, which bears a resemblance to the Field Sparrow's song. Both the Bachman's Sparrow and Red-cockaded Woodpecker are found only in Southern Mixed Pine-Oak Forests (the former name of the Bachman's Sparrow was Pinewoods Sparrow). Unfortunately, the sparrow, like the woodpecker, has become much less common.

White-cedar Swamp Forest (Pl. 10)

Indicator Plants

TREES: Atlantic White-cedar (abundant, Southeast and Coastal Plain), Northern White-cedar (abundant, interior states, boreal region),

Red Maple, Tamarack (American Larch), Black Spruce (boreal), Balsam Fir (boreal), Balsam Poplar (boreal).

SHRUBS: *Atlantic White-cedar Swamps:* Sweet Pepperbush, Winterberry, Inkberry, Sheep Laurel, Leatherleaf, Swamp White Azalea, various blueberry species. *Northern White-cedar Swamps:* Swamp Fly Honeysuckle, Red-osier Dogwood, Northern Holly, Dwarf Raspberry.

VINES AND HERBACEOUS SPECIES: Greenbrier, Sphagnum Moss (abundant), Starflower, Goldthread, various orchid species (especially in Northern White-cedar Swamps).

Indicator Animals

BIRDS: Cedar Waxwing, Brown Creeper, Northern Saw-whet Owl, Alder Flycatcher, Hermit Thrush, Northern Waterthrush, Canada Warbler, Black-throated Green Warbler.

Description

Atlantic White-cedar forms dense, boglike stands in wet areas along the coast and is replaced by Northern White-cedar inland in the Northeast. Northern White-cedar usually occurs on limestone soils and is not always a swamp species. The presence of a water-soaked mat of Sphagnum Moss plus shrub species characteristic of bogs and acid soils makes the White-cedar Swamp easy to identify. In more northern areas, Tamarack (American Larch) and occasionally Black Spruce, both bog species, will occur. These forests are well shaded and dense, and travel through them is not easy.

Atlantic White-cedar Swamp. In some acidic, hydric areas, Atlantic White-cedar (or Northern White-cedar, in some locations) may be virtually the only tree species to occur.

SIMILAR FOREST COMMUNITIES: See Northern Swamp Forest (p. 88), where Atlantic White-cedar sometimes occurs, and Boreal Bog and Boreal Forest (p. 62). Atlantic White-cedar mixes among the Northern Pine-Oak coastal plain forest, occupying sites with a high water table.

RANGE: Atlantic White-cedar Swamps occur mostly along the coast, from southern Maine through the Southeast to the Mississippi River. Northern White-cedar Swamps range through the area of the Northern Hardwoods and southern Boreal Forest, and are common around the Great Lakes region.

REMARKS: White-cedar Swamps share many characteristics with bogs. They typically form in kettles or depressions where standing water accumulates. An abundance of Sphagnum Moss makes the water quite acidic, and a layer of thick, undecomposed peat builds up. White-cedar Swamps are excellent places to search for some of the rarer northern orchid species. Walking through such a swamp is a wet and muddy endeavor—it is often best to hop from one cedar root to the next. As cedar swamps fill, trees such as Balsam Fir and Black Ash invade, crowding the cedars. These forests share many of the same undergrowth species as the Boreal Forest and Bogs.

Northern White-cedar also grows on uplands with calcareous (lime-rich) soil, but this population is thought by some to be a different race, genetically distinct from the swamp-inhabiting population.

MIXED APPALACHIAN FOREST COMMUNITIES

The Appalachian Mountains formed approximately 200 million years ago. Today, though smoothed and rounded by their years of erosion, the Appalachians still display a complex topography that supports a high richness both of total species and of forest communities within a relatively small area. Within Great Smoky Mountains National Park in eastern Tennessee it is possible to recognize 130 tree species and 7 forest communities separated on the basis of elevation and amount of soil moisture. For instance, along upper elevations of the Appalachians, such as Clingman's Dome at an elevation of 6643 ft., is found the Appalachian Extension of the Boreal Forest. Red-breasted Nuthatches, Black-throated Green Warblers, Golden-crowned Kinglets, and Northern Juncos forage in Red Spruce and Fraser Fir trees. At Sugarlands (elevation 1460 ft.) grows a diverse forest of Tuliptree, American Holly, various oaks, Yellow Buckeye, White Basswood, American Hornbeam, and various rhododendrons. Louisiana Waterthrushes patrol the streamsides and Worm-eating Warblers sing from the rhododendron clumps.

Forest Diversity in the Appalachians

The Mixed Appalachian Forests represent a great complex of forest communities, often called the **MIXED MESOPHYTIC FOREST.** In addition to tree species typical of other regional forests, this area hosts several species that give the area its own identity.

The following forest communities, described elsewhere in this chapter, are all found in the area of the Mixed Appalachians:

Boreal Forest: High elevations.

Northern Hardwood Forest: Cool, well-drained slopes.

Oak-Hickory Forest: Warm exposed areas; very common throughout the region.

Beech-Maple Forest: Cool, moist shaded areas; includes an abundance of Eastern Hemlock.

Pine-Oak Forest: Sandy exposed areas.

Northern Riverine Forest: Streambanks.

Appalachian Cove Forest: Unique to the region. Tremendously lush and diverse. See next section.

It is often difficult to know exactly what forest type you are in, since there is extensive overlap in this region. You may be in a forest that seems like a Cove Forest but contains enough Beech and Sugar Maple that you wonder if it shouldn't be called Beech-Maple rather than Cove. You may be in what seems to be a Pine-Oak Forest, but there are enough hickories to make you unsure whether you're actually in an Oak-Hickory Forest. Such overlap is common because soil type, soil moisture, and climate vary gradually throughout the region and species ranges overlap extensively. There is rarely an absolute boundary between two forest types.

Another feature unique to the Mixed Appalachian Forests is the diversity of salamanders: the southern Appalachian region is inhabited by 27 species (plus many subspecies) of salamanders—more than occur in any other area of North America. Many species are restricted to a particular mountain range.

Appalachian Cove Forest (Pl. 13)

Indicator Plants

CANOPY TREES: White Basswood, Carolina Silverbell, Tuliptree, Yellow Buckeye, Sugar Maple, Red Maple, Yellow Birch, Beech, White Ash, Bigleaf Magnolia, Allegheny Chinkapin, Bitternut Hickory, Eastern Hemlock. Up to 20 other species may be found in the canopy. The *high species richness* within a single tract of forest is an indicator of this forest community.

UNDERSTORY TREES: Eastern Redbud, Box-elder, Sourwood, Fraser Magnolia, Mountain Maple, Flowering Dogwood, Sassafras, Witch-hazel, Yellowwood.

SHRUBS: Great Rhododendron, Mountain Laurel, Flame Azalea, Fetter-bush.

HERBACEOUS SPECIES: More than 1500 herbaceous species occur in the southern Appalachians. Among the most common in the Cove Forests are the Wood Anemone, Mayapple, Red Trillium (Wake-robin), White Trillium, Wild Ginger, Canada Violet, Large-leaved White Violet, Hepatica, Wild Geranium, Wild Leek, Cut-leaved Toothwort, Foamflower, Dutchman's-pipe, Squirrel-corn, Solomon's-seal, False Solomon's-seal, American Ginseng.

Indicator Animals

BIRDS: Hooded Warbler, Kentucky Warbler, Worm-eating Warbler, Louisiana Waterthrush, Acadian Flycatcher, Carolina Chickadee, Tufted Titmouse, Carolina Wren, Yellow-throated Vireo, Pileated Woodpecker, Wild Turkey. The Cove Forest supports a high diversity of nesting birds, including the Wood Thrush, Ovenbird, Summer Tanager, and Rose-breasted Grosbeak, in addition to those named above.

MAMMALS: The Black Bear, widely distributed in other parts of North America, occurs quite commonly in Cove Forests and surrounding areas. White-tailed Deer are also very common.

SALAMANDERS: *Cove Forest streams and banks:* Seal Salamander (brownish), Jordan's Salamander (orange cheeks). *Mountain slopes (3200–5000 ft.):* Yonahlossee Salamander (orange-red back), Mountain Dusky Salamander. *Cove and mountain streams (under rocks and logs):* Black-bellied Salamander (7½ in. long), Spring Salamander (pinkish red, speckled), Two-lined Salamander (yellowish brown). Many other species occur but are confined to specific mountain ranges.

Description

The Cove Forest conveys a sense of primeval wilderness. Two trees, White Basswood and Carolina Silverbell, are the key indicator species, sharing the forest with many other, more widely distributed species. The forest is very lush, with Tuliptrees over 100 ft. tall and a colorful understory of Redbud, Fraser Magnolia ("Umbrella Tree"), and Flowering Dogwood. Most trees are deciduous, though Eastern Hemlock may be occasionally present. The shrub layer is thickly developed and colorful in spring with dense clumps of Great Rhododendron and Mountain Laurel. All

Appalachian Cove Forest. A mesic (rich, moist) forest of Appalachian bottomlands, noted for its many different shrub, wildflower, and tree species, including White Basswood and Carolina Silverbell. Most are broad-leaved species, but some needle-leaved species, such as Eastern Hemlock, also occur.

things seem abundant: trees, shrubs, birds, and even salamanders. The obvious high species richness of trees and shrubs is the best field mark of this forest.

SIMILAR FOREST COMMUNITIES: The Cove Forest shares many tree species with the Northern Hardwood and Oak-Hickory forests but is easy to distinguish from those two forests by the presence of many additional species, especially White Basswood, Silverbell, and Tuliptree.

RANGE: The Great Smoky Mountains area of eastern Tennessee (especially within Great Smoky Mountains National Park) northward through the eastern third of Kentucky, southeastern Ohio, most of the mesic areas of West Virginia, and southwest Virginia.

REMARKS: The highest species richness of trees in North America occurs in the Cumberland and Allegheny mountains, especially within the confines of Great Smoky Mountain National Park. Species richness decreases to the west and north. The Appalachian Cove Forest has 25–30 tree species, though you will not find all of these in any single stand. Normally, any Cove Forest will support 6–8 dominant species. To put the diversity in perspective, however, note that in the tropics, there can be more than 100 tree species in a single forest tract equivalent in size to a Cove Forest. Tropical plant species richness far exceeds that in the temperate zone. Cove Forests are well developed in the Great Smokies, but also occur throughout the Appalachians, especially along ravines.

Yellow Buckeye (also called Sweet Buckeye) is a common canopy tree of Cove Forests. This species is much more limited in distribution than its close relative Ohio Buckeye, because Yellow Buckeye requires rich bottomland loams along streambanks. Yellow Buckeye leafs out very early in spring, producing yellowish white flowers that develop into fruits in a rounded, leathery pod 2–3 in. in length. Seeds are dispersed in September, some by animals and others by falling from the tree and floating along streams. The largest Yellow Buckeye in Great Smoky Mountains National Park was 85 ft. tall, with a crown spread of 54 ft. and a circumference of 15 ft. 11 in. An interior forest species, Yellow Buckeye is not to be found in open fields among the pioneer species.

White Basswood is considered an important indicator tree of Cove Forests. A smaller-leaved species than American Basswood, it is otherwise similar and tends to replace American Basswood in the South, especially on rich bottomland soils of mixed mesophytic forests.

Carolina Silverbell and Eastern Redbud are two small but colorful trees that characterize the Cove Forest. Silverbell can grow to 80 ft. and thus enter the canopy, but Redbud is confined to the subcanopy, as it rarely exceeds 50 ft. in height. Silverbell produces delicate white, bell-like flowers in the spring, and Redbud, as the name implies, makes bright pink-red blossoms, which festoon the branches before the leaves develop.

Tuliptree has a wide range and is part of many eastern forests but is perhaps most stately in the Cove Forest. The straight grayish trunk of the tree can ascend up to 150 ft. The tallest on record reached 190 ft., but heights of around 120 ft. are much more normal. Though the bark is initially smooth, it becomes quite furrowed as the tree ages. Its distinctive, tuliplike flowers are greenish, with orange at the base. It can grow in a wide variety of soil types and is a common successional species in old fields in the region (see p. 198). Tuliptree is not very shade tolerant but grows quickly, "outrunning" the competition to reach the canopy. It produces many seeds annually, but most are infertile, with empty seed coats. It produces bumper crops of fertile seeds only at irregular intervals.

Bird watching is very rewarding in Cove Forests. The crow-sized Pileated Woodpecker carves large oval nest holes in trees and swoops through the forest, challenging the birder to get a good view of it. Along with the Pileated, you can find Downy, Hairy, and Red-bellied woodpeckers, as well as the Common Flicker. Along with American Redstarts and several other wood warblers, flycatchers of varying sizes dart out from the canopy, "hawking" for insects. The largest flycatcher is the Great Crested, which shares the canopy with the Eastern Wood-pewee and the

Acadian Flycatcher. The understory, especially in areas with rhododendrons and azaleas, hosts the buzzing song of the Worm-eating Warbler, and, in lush undergrowth, you will find the brilliant Hooded Warbler. The soft gobble of a Wild Turkey flock can be heard among the more raucous dawn chorus of bird song.

Appalachian Heath Balds

These unique areas consisting of mountain slopes covered by heaths are best seen in Great Smoky Mountains National Park, usually in areas of complex topography, on south-facing slopes, at elevations between 1220 and 1525 meters. A total of 478 Heath Balds occur within the boundaries of Great Smoky Mountains National Park. Dominated by the early-blooming, purple-flowered Catawba Rhododendron and the later-blooming Rosebay Rhododendron, with its red, pink, or white blossoms, plus a scattering of deep orange Flame Azalea, these Heath Balds, when in flower, are a glorious sight in spring.

The reason why Heath Balds exist is an ecological mystery, although most who have studied the area presume that periodic fires have been an important force in favoring the continued presence of the thick shrubby heaths. These plants are now so dense that tree seedlings do not get sufficient light to grow above the heaths and shade them out. Many sites that seem climatically identical to Heath Balds support forest and not heaths. Therefore, many believe that occasional fires, occurring by chance (perhaps caused by lightning), produced the balds. Still, no one knows for certain exactly what causes these unique balds.

Fig. 17. Catawba Rhododendron flower.

Southern Hardwood Forest (Pl. 14)

Indicator Plants

CANOPY TREES: Southern Magnolia (plus other magnolia species, including Sweetbay and Cucumbertree), Virginia Live Oak, Common Persimmon, Pecan, White Oak, Laurel Oak, Redbay, Pawpaw, American Beech, Black Tupelo (Black Gum), Sweetgum, Hackberry, Sourwood, various hickories.

UNDERSTORY TREES: Sweetbay, Hophornbeam, American Hornbeam, Eastern Redbud, American Holly, Red Mulberry, various hawthorn species.

SHRUBS: Sparkleberry, Common Sweetleaf.

HERBACEOUS SPECIES AND AN EPIPHYTE: Carolina Jessamine, Trumpet-creeper, Southern Harebell, Wild Cucumber, Spanish Moss (a herbaceous epiphyte or air plant—see p. 104).

Indicator Animals

BIRDS: Summer Tanager, Painted Bunting, Northern Mockingbird, Loggerhead Shrike, Carolina Wren, Red-bellied Woodpecker, Orchard Oriole, Yellow-throated Warbler, Blue-gray Gnatcatcher, Northern Cardinal, Tufted Titmouse, Carolina Chickadee, Ruby-throated Hummingbird, Chuck-will's-widow.

MAMMALS: Virginia Opossum and Fox Squirrel.

Description

The Southern Hardwood Forest includes several broad-leaved evergreen species as well as several oak species plus the Pecan. The flora includes many genera that are basically tropical in origin. The finest development of this forest community is seen on moist sites; drier sites support a mixture of oak and pine (see Southern Mixed Pine-Oak Forest). The most obvious field mark of this forest is the abundant presence of Spanish Moss draped from the trees.

SIMILAR FOREST COMMUNITIES: Many species are shared with the Oak-Hickory Forest, which is found on drier soils. In very moist sites this forest intermingles with Baldcypress Swamp and Southern Mixed Hardwood Swamp forests. On dry sandy soils it intermingles with Southern Mixed Pine-Oak Forest.

REMARKS: The Southern Hardwood Forest, with its richness of evergreen plants, many of which produce brilliant large flowers, its abundance of picturesque Spanish Moss, and its high diversity of bird

Southern Hardwood Forest. Tall oaks, Persimmons, and Pecans and other hickories characterize this broad-leaf forest, which occurs on mesic, loamy soils throughout the Deep South. Spanish Moss, an epiphytic bromeliad, drapes the branches. Photo shows a forest gap.

species, is one of the most exciting of eastern forests. Ruby-throated Hummingbirds probe the orange-red flowers of Trumpet-creeper vines, and Northern Mockingbirds, Loggerhead Shrikes, and Painted Buntings sing from telephone wires nearby. Mockingbirds sing throughout the night, combining their voices with a chorus of Chuck-will's-widows, and perhaps a Barred or Great Horned Owl. Two bright red birds, the Cardinal and the Summer Tanager, are both closely associated with this forest, though the Cardinal, along with the Tufted Titmouse and Northern Mockingbird, is rapidly expanding its range northward.

The Virginia Opossum, also expanding its range to the north, is the only marsupial found in North America. It gives birth to babies so tiny that a dozen will fit on a tablespoon. These babies migrate from the birth canal to the teats, located in a pouch on the mother's lower abdomen, where they complete their development. The Opossum has a prehensile tail that acts as a fifth limb, enabling the animal to hang upside down from a branch. When threatened, the Opossum will often feign death until the danger passes.

Throughout the Deep South, gracefully draped Spanish Moss provides a distinctive, aesthetically pleasing look to the forest. Not a true moss, Spanish Moss is more closely related to the pineapple; both are members of a group of plants called bromeliads. Spanish Moss is sometimes confused with parasitic plants, but it is not a parasite and causes no harm to the tree upon which it hangs. Most bromeliads are *epiphytes,* or "air plants," species that grow on the surface of tree branches, absorb minerals from rainwater, and photosynthesize as normal plants do. Spanish Moss is covered with minute scales with which it takes up moisture. It is frequently seen entangling itself on telephone wires. It reproduces by making tiny flowers.

The Southern Magnolia, like Spanish Moss, is of tropical origin. Magnolia has a simple branching pattern and large thick, waxy, unlobed leaves, very similar to those of many tropical rainforest trees. Magnolia flowers are large and showy, white or pale yellowish, and quite fragrant. The overall structure of the flower is considered to be primitive, and the genus *Magnolia* is believed to be one of the most ancient groups of flowering plants. Flowers develop into upright fruits that resemble cones but are composed of numerous small pods, each containing 1–2 red seeds, which hang in the pod by threads. Southern Magnolia is at its finest on moist, well-drained loamy soils, and is rarely found in pure stands. The average tree reaches a height of 60–80 ft., though magnolias as tall as 125 ft. have been recorded. Life expectancy is anywhere from 80 to 120 years. Southern Magnolia is evergreen, with thick leathery leaves characteristic of many typically tropical trees, but some magnolias, like Sweetbay and Cucumbertree, are deciduous.

Southern Magnolia is one of about 70 species in the genus *Magnolia.* Other species are found from the southeastern United States through Central America and the West Indies. Interestingly, magnolias are also found in eastern Asia from the Himalayas to Japan. This distribution, shared by some other trees of the eastern forest such as the Eastern Sycamore, strongly supports the idea that the continents of North America and Asia were united at the time these plants evolved. The continents have since drifted apart, through the ongoing dynamic process of plate tectonics.

The Common Persimmon is another species that belongs to a group more typical of tropical areas, especially Africa and Indo-Malaya. The tree has dark brown to blackish bark in scaly plates or blocks, a characteristic shared with other trees in this family, the ebony family. Leaves are oblong or oval and very leathery. Male and female flowers occur on separate trees, and the round whitish fruit is soft, juicy, and sweet. Good seed crops are produced every 2 years or so and seeds are disseminated by birds and animals, as well as by flooding. Persimmon is a very good competitor, persisting as a sapling in the shaded understory for many years until it can respond to light gaps and grow to canopy stature. It is also a good stump-sprouter and can resprout after fire.

Indicator Plants

TREES: Green Ash, Carolina Ash, White Ash, Hackberry, Eastern Sycamore, Eastern Cottonwood, Swamp Cottonwood, Sweetgum, Black Tupelo (Black Gum), Pecan, American Elm, Winged Elm, Water-elm, American Hornbeam, Swamp White Oak, Red Maple, Box-elder, River Birch, Pin Oak, various willows.

SHRUBS: Roughleaf Dogwood, Buttonbush, Southern Mock-orange, Southern Arrowwood, Swamp Dogwood, American Holly, Witch-hazel.

HERBACEOUS SPECIES AND VINES: Trumpet-creeper, wild grapes, Poison-ivy, Wild Hydrangea, Trout-lily, Dutchman's-breeches, Jack-in-the-pulpit, Spring-beauty, American Wisteria.

Indicator Animals

BIRDS: Prothonotary Warbler, White-eyed Vireo, Wood Duck, Yellow-billed Cuckoo, Louisiana Waterthrush, plus the same species listed on p. 103 for the Southern Hardwood Forest.

Description

This forest is identified by an abundance of Green and Carolina ashes, elms, Cottonwoods, Sugarberry, Sweetgum, and Black Tupelo. Pecan is also present, but only in the Mississippi River valley. Species composition of the forest varies with locality. Many species also occur in the Northern Riverine Forest, where the southern ashes and Sweetgum are far less numerous and Pecan is absent. Vines usually are prolific along the water courses. The forest occupies floodplain habitats and shares many species with the Southern Mixed Hardwood Swamp Forest and Baldcypress Swamp (see below).

SIMILAR FOREST COMMUNITIES: See Northern Riverine Forest, Southern Mixed Hardwood Swamp Forest, and Baldcypress Swamp Forest.

RANGE: This riverine floodplain forest occurs widely throughout the southeastern part of North America. It overlaps in range with the Northern Riverine Forest from Pennsylvania and New Jersey southward through Maryland, Virginia, and West Virginia.

REMARKS: Riverine forests not only provide a richness of plants and animals but also perform important ecological functions. Due to both the long growing season and the deposition of rich silts during the annual flooding cycle, Southern Riverine Forests have very high rates of photosynthesis, which means that they make a great deal of organic material in the form of leaves, stems, and so on. This material is eventually recycled, some of it going into the river itself and downstream, and the remainder being used by plants along the banks. Riverine floodplain forests reduce nutrient loss by efficiently recapturing nutrients from silt and debris deposited by floods and from decomposing litter. The ability of riverine forests in general to absorb water as well as nutrients is an important contributor to natural flood control. Many of the most abundant species, such as Green Ash, are well adapted to tolerate prolonged immersion during flooding. Studies have shown that Green Ash, which usually reaches heights of 50–60 ft., remains healthy even when flooded for up to 40% of the growing season. Ash seeds are distributed by wind but may also fall into rivers, where they are transported by water and deposited during flooding, an important method of long-distance dispersal.

The Pecan is one of the finest of southern trees. It grows on young alluvial soils, especially well-drained loams. It is not common in the eastern region but is characteristic of the bottomland forests of the Mississippi River valley, where it associates with Eastern Sycamores, American Elms, and Roughleaf Dogwoods. Pecan is intolerant of competition, requiring high levels of light to get established. It is also highly susceptible to damage by fire. A very stately tree, Pecan can reach heights of 180 ft., though 100 ft. is much more common. A large Pecan tree can have a diameter of 6–7 ft. Because Pecans grow along banks of rivers and streams, they are often laden with vines, and their crowns, like those of many tropical trees, can be so heavily covered with vines that they become deformed.

Pecan is a member of the hickory family and produces seed crops annually. A healthy mature tree can yield 2–3 bushels of Pecan nuts each year. Seeds are disseminated by animals and flood waters.

Baldcypress Swamp Forest (Pl. 16)

Indicator Plants

TREES: Baldcypress, Pondcypress, Redbay, Swamp Tupelo, Water Tupelo, Black Willow, Swamp Cottonwood, Atlantic White-cedar, American Elm, Water Hickory, Common Persimmon, Red Maple, Carolina Ash, Green Ash, Box-elder, Eastern Sycamore.

SHRUBS: Buttonbush, Fox Grape, Poison-sumac, Carolina Rose, Wild-raisin.

HERBACEOUS SPECIES: Spanish Moss, various orchid species, Yellow Loosestrife, Sphagnum Moss.

Indicator Animals

BIRDS: Prothonotary Warbler, Northern Parula warbler, various heron and egret species, Barred Owl, Wood Duck, Wood Stork, White

Ibis, Limpkin (Deep South only), Pileated Woodpecker, Yellow-billed Cuckoo, plus those birds listed for the Southern Hardwood Forest (p. 103).

MAMMALS: Marsh Rabbit, Nutria.

REPTILES: American Alligator, Cottonmouth.

DESCRIPTION

This is a very wet forest principally composed of one tree species, Baldcypress, although many other species may be less abundantly represented. Pondcypress occurs only in ponds and lakes. The cone-shaped cypress "knees" protruding above the water and the branches thickly laden with Spanish Moss give this forest community a very distinct appearance. Prothonotary Warblers are conspicuous with their brilliant orange plumage and loud whistled song. Baldcypress forests are normally too wet to allow for easy travel on foot. A boat is necessary to see the forest well.

SIMILAR FOREST COMMUNITIES: The Baldcypress Swamp Forest occurs throughout the range of the Southern Hardwood and Southern Mixed Pine-Oak forests. On higher ground, the species commonly found in those two forests intermingle with the swamp species. In addition, see Southern Riverine Forest.

RANGE: This forest is found adjacent to lakes, ponds, slow streams, and lowland areas of standing water or hydric soils throughout the southern Coastal Plain from Maryland to Texas, including Florida.

REMARKS: Large tracts of Baldcypress Swamp, such as are found in the Okefenokee region in Georgia, remain as refuges for such spectacular species as the Bobcat and the Cougar (Florida Panther), as well as many rare orchid species that grow on the cypress branches among the Spanish Moss. Large colonial wading birds such as herons, egrets, White Ibis, and Wood Stork locate their breeding colonies in Baldcypress Swamps. The mysterious Limpkin, known for its penetrating call, stalks fish in cypress swamps in the Deep South. The Ivory-billed Woodpecker, now probably extinct in North America but recently found to persist in Cuba, was once common in Baldcypress Swamps throughout the South. This, the largest of North American woodpeckers, requires very large territories plus an abundance of dead trees, within which it searches out certain species of bark beetles. Forest cutting, swamp drainage, and removal of dead trees have led the Ivory-bill to extinction. The Pileated Woodpecker is similar but proved far more adaptable than the Ivory-bill and is an abundant resident of southern forests.

Several large reptiles still maintain population strongholds in southern swamp forests, especially in Baldcypress Swamps. The

American Alligator, which can reach a length of 15 ft., is one of the few vocal reptiles. Males emit a deep bellow during mating season. The female builds and protects a raised nest where from 10 to 50 eggs are laid. After guarding the eggs for 9–10 weeks, the female remains protective of the young after they hatch. Alligators are carnivores with wide-ranging appetites: they eat fish, turtles, various riverine birds and mammals, and even other alligators! The Cottonmouth, or Water Moccasin, is a member of the poisonous pit-viper family (see p. 377). This thick-bodied snake can grow to 6 ft. and is often seen sunning itself on a log. Close approach is not recommended.

Baldcypress is unique to very wet southern soils, where dense mud normally occurs. It is intolerant of dry soils but grows well on loams, though it is not common on such soils, probably due to competition from other plants. Trees grow at a rate of about 1 ft. annually, reaching a mature height of 100 ft. The picturesque knees are thought to help the tree absorb oxygen, which is in very low supply in the water-logged soil.

Southern Mixed Hardwood Swamp Forest (Pl. 17)

Indicator Plants

TREES: Black Tupelo (Black Gum), Water Tupelo, Sweetgum, Red Maple, Swamp (Nutmeg) Hickory, Water Hickory, Eastern Sycamore, Swamp Chestnut Oak, Overcup Oak, Cherrybark Oak, Water Oak, Willow Oak, Pawpaw, Sweetbay, Sourwood, Deciduous Holly.

SHRUBS: Swamp-privet, Spicebush, Buttonbush, Hobblebush, Swamp Dogwood, Wild-raisin.

HERBACEOUS SPECIES AND VINES: Spanish Moss, Virgin's-bower, Heartleaf Ampelopsis, Trumpet-creeper, Japanese Honeysuckle, various wild grape species, Solomon's-seal, Spring-beauty, Pipsissewa.

Indicator Animals

BIRDS: Prothonotary Warbler, Northern Parula, various heron and egret species, Barred Owl, Wood Duck, Wood Stork, White Ibis, Limpkin (Deep South only), Pileated Woodpecker, Yellow-billed Cuckoo, plus those birds listed for the Southern Hardwood Forest (p. 103).

MAMMALS: Marsh Rabbit, Nutria.

REPTILES: American Alligator, Cottonmouth.

Description

This is a high-diversity forest where any combination of indicator

Sweetgum. The star-shaped leaves make Sweetgum, or Liquidambar, easy to identify.

species may predominate, though tupelos should be found in most places. The presence of several hydric (water-adapted) oak species is a good indicator. Understory and shrub growth is usually dense and many vines are present. The bird community is similar to that of Baldcypress Swamps and Southern Hardwood Forests. Some Southern Swamp Forests contain large colonies of herons and egrets.

SIMILAR FOREST COMMUNITIES: Both the Southern Riverine Forest and the Southern Swamp Forest share many similarities. Sweetgum is an abundant species in both, as are many of the understory, shrub, and vine species. Swamp forests abound with tupelos and have more oak species and fewer floodplain species, such as Silver Maple, American Hornbeam, and willows.

RANGE: This community occupies wet sites from Virginia and Maryland through the Carolinas, Georgia, Louisiana, Arkansas, and eastern Texas.

REMARKS: Two tupelo species, Black Tupelo (also known as Black Gum, or Sourgum) and Water Tupelo, abound in this forest type. Tupelos are deciduous and, like magnolias and sycamores, are found in China and other parts of Asia. Fruits are bluish and highly favored by birds. Water Tupelo is more water tolerant than Black Tupelo and, like many tropical trees, often develops buttressed roots. Black Tupelo reaches heights of 100 ft., while Water Tupelo rarely tops 70 ft. Black Tupelo's range takes it all the way into southern Maine and Michigan, but Water Tupelo is truly a tree of the Deep South, occurring in Coastal Plain swamps north to North Carolina, and following the Mississippi River valley to southern Illinois.

Many water-adapted oak species are present in southern

swamps, including Water Oak, Swamp Chestnut Oak, Cherry bark Oak (a variety of the Southern Red Oak), Willow Oak, and Overcup Oak. These oaks not only require moist soils but are also generally intolerant of both fires and competition from other species. They survive by finding refuge in the bottomland swamps. Two hickories, Swamp (Nutmeg) Hickory and Water Hickory, also are found among the oaks, making the Southern Mixed Hardwood Swamp Forest in some places a wetter version of the upland Oak-Hickory Forest.

The Pawpaw grows as a small tree or as a dense thicket of shrubs. It grows very tasty, large fruits, which are consumed by large birds such as Wild Turkeys, as well as by many mammals. The Pawpaw is a temperate member of the otherwise tropical custard-apple family.

Subtropical Forest (Pl. 18)

Indicator Plants

TREES: Coconut Palm (introduced), Florida Royalpalm, Florida Cherrypalm, Jamaica Thatchpalm, Cabbage Palm, Paurotis, Gumbo-limbo, West Indies Mahogany, Manchineel, Florida Strangler Fig (begins as a vine), Tallowwood, Black-calabash, Butterbough, Wild Tamarind, Papaya, Paradise-tree, Bahama Lysiloma, Jamaica Caper, Everglades Velvetseed, Sweetbay, Redbay, Laurel Oak, Virginia Live Oak, Myrtle Oak, Southern Magnolia, Myrtle Dahoon, Dahoon, Yaupon Holly.

UNDERSTORY TREES AND SHRUBS: Saw-palmetto, Longleaf Blolly, Darling-plum, Southeastern Coralbean, Eastern Baccharis.

HERBACEOUS SPECIES: Sawgrass, various orchids, Spanish Moss, Wild Pineapple *(Tillandsia)* and other bromeliads, various ferns, vines such as Passionflower.

Indicator Animals

BIRDS: Swallow-tailed Kite, Gray Kingbird, Smooth-billed Ani (uncommon), Scrub Jay (local), Painted Bunting, Black-whiskered Vireo, Red-bellied Woodpecker, Summer Tanager.

REPTILES AND AMPHIBIANS: Green Anole, other anole lizards, American Alligator, Cuban Treefrog.

BUTTERFLY: Zebra Longwing.

SNAIL: Liguus Wood Snail.

Description

The Subtropical Forest barely reaches North America. It occurs on raised areas of land called *hammocks,* which occur as forest islands among the Everglades sawgrass. Palms are the most distinctive trees present, but broad-leaved evergreens such as Gumbo-limbo and Mahogany are also apparent. Leaves on broad-leaved species are usually oval in shape and thick and waxy. Epiphytes and vines, including Strangler Fig (see below), are prolific. Many orchid species occur, some of which are quite rare. The ground is extensively covered by various fern species (Strap Fern, Leather Fern).

SIMILAR FOREST COMMUNITIES: See Mangrove Swamp Forest, Everglades, and Southern Mixed Pine-Oak Forest.

RANGE: Southern Florida, especially raised hammocks within Everglades National Park.

REMARKS: The Subtropical Forest is an intermediate type between the deciduous forests of the temperate zone and tropical evergreen forests. The growing season lasts virtually all year and the forest exhibits many of the characteristics of tropical forests, such as abundant epiphyte and vine growth, trees with oval waxy leaves, and an abundance of colorful butterflies and reptiles. Leaves terminate in sharp, pointed "drip-tips" that allow rain to run off rapidly and reduce mineral loss from the leaves. Some trees have buttressed root systems.

The most obvious reptiles of the Subtropical Forest are the anole lizards. The most abundant species, the Green Anole, can be seen scurrying about over the dry leaf litter or climbing tree trunks. Green Anoles are usually, but not always, green. They have the ability to change color, becoming very buffy-brown. Males have a prominent flap of pinkish skin, the dewlap, hanging below their throats. This flap is used in various head-bobbing ceremonies associated with territorial encounters among males. Green Anoles range throughout southern forests but other anole species are found only in southern Florida.

The brilliant, almost neon-bright Zebra Longwing butterfly is another specialty of the Subtropical Forest. This insect flies very slowly and is very obvious when perched, seemingly a tempting target for a Gray Kingbird or other insectivorous bird. However, the butterfly's gaudiness serves to warn would-be predators that the insect is poisonous. As a caterpillar, it feeds on Passionflower vine and absorbs toxic compounds present in the Passionflower. For more about warning coloration, see p. 345.

The Florida Strangler Fig begins its life as a seed dropped by a bird that has eaten the fig fruit. The seed must be dropped on a branch of a tree in order to germinate and become a vine. The vine grows toward the tree trunk and thickens, becoming increasingly woody. Vine tendrils wrap around the tree trunk and eventually fuse together, preventing the tree from continued expansion. Because it can no longer grow, the original host tree often dies, leaving the fused woody Strangler Fig alone as one of the Subtropical Forest's most picturesque trees.

Many of the Subtropical Forest species are West Indian in origin and there are many introduced species from the far corners of the world's tropics.

Mangrove Swamp Forest (Pl. 19)

Indicator Plants

TREES: Red Mangrove, Black Mangrove, White Mangrove, Buttonwood, Sea-grape, Coconut Palm.

Indicator Animals

BIRDS: White-crowned Pigeon, Black-whiskered Vireo, Roseate Spoonbill, Wood Stork, Brown Pelican, Magnificent Frigatebird, many heron, egret, and ibis species.

MAMMALS: Key Deer, Manatee.

Description

Mangroves comprise a dense, often impenetrable, thicket of small spreading trees that grow where the tropical sea meets the land. Mangrove islands dot tropical estuaries, and interlacing channels, often populated by Manatees, weave throughout the vast coastal forest. All mangrove species have thick oval leaves. Red Mangrove has numerous aerial or prop roots, and Black Mangrove produces an abundance of short, pencil-like pneumatophore (breather) roots that protrude upward from the thick muddy sediment. Many saltwater invertebrate species such as fiddler and land hermit crabs occur abundantly. Many colonial wading birds such as egrets nest among the mangroves.

Red Mangrove and Magnificent Frigatebird. Red Mangrove forests provide ideal nesting sites for colonies of tropical frigatebirds, whose wingspreads reach 7 ft.

Sprouting Coconut. Coconuts float and can remain at sea for many months without losing viability. Thus coconuts are common throughout the global tropics.

SIMILAR FOREST COMMUNITIES: See Subtropical Forest.

RANGE: Mangrove Swamp Forests occur in extreme southern coastal Florida (especially Everglades National Park) and the Florida Keys.

REMARKS: Ecologically, Mangrove Swamp Forests are woody versions of northern salt marshes. They protect the coast from storm damage (especially from hurricanes), and serve as a valuable nursery for fish and invertebrates. Mangrove leaves are an important energy source for the marine food chain. When a mangrove leaf drops from the tree it is colonized and eaten by numerous bacteria, fungi, and protozoans. These microbial plants and animals in turn form an important food source for zooplankton present in tropical estuaries, and the zooplankton are a critical food for juvenile fish and larval invertebrates. Much of the energy captured by the mangroves is thus passed into the marine food chain and helps support such habitats as the coral reef.

The Red Mangrove is found throughout the tropical world. It drops long, podlike seedlings, called "sea pencils," into the sea, where they float horizontally for many miles. Eventually the seedling takes on a vertical orientation and will root on a tropical beach or sandbar. Red Mangroves are pioneer species that actually initiate the process of bar stabilization and island formation as they colonize. Once Red Mangroves have become established, other species invade.

The Manatee, though it looks like a sluggish seal, belongs to an entirely different group of aquatic mammals, the sirenians. Unlike seals, which are carnivorous, Manatees are herbivorous and feed on such plants as Water Hyacinth, a major pest species (imported weed) that clogs channels. Unfortunately, Manatees have become uncommon, and, though they are fully protected by conservation law, some fall prey to accidents with boat propellers. Manatees normally exist in small social groups.

THE EVERGLADES

The Everglades are part of southern Florida, and, though not a forest, the region is worthy of inclusion for its uniqueness.

The Everglades are a vast wet grassland comprised mostly of sawgrasses as well as such familiar marsh grasses as tules and cattails. Drier areas have cordgrass and various palmetto species, particularly Saw-palmetto, Paurotis, and Florida Silverpalm. High, relatively dry areas called hammocks support the Subtropical Forest species described above.

The Everglades contain a wealth of mammals, birds, reptiles, amphibians, and fish. Birds such as the Swallow-tailed Kite, Bald Eagle, Wood Stork, White Ibis, Anhinga, Purple Gallinule, Limpkin, Pileated Woodpecker, and many heron and egret species are all easily seen in Everglades National Park. The rare Everglade Kite (now officially named the Snail Kite), though far from abundant, still survives in the Everglades, where it feeds exclusively on the large apple snails. The Bobcat and Cougar (Florida Panther) still roam the Everglades, and the American Alligator is a common sight. The Eastern Diamondback Rattlesnake is one of several poisonous snakes that add an element of danger to those who carelessly make their way through the hammocks.

The Everglades. Sometimes called "river of grass." Sawgrass predominates, interrupted by elevated hammocks of pines and other trees.

INDICATOR PLANTS

TREES: Mesquite, Virginia Live Oak, Ashe Juniper, Catclaw Acacia, Sweet Acacia, Blackjack Oak, Bluejack Oak, Post Oak.

SHRUBS: Creosotebush (Rio Grande).

GRASSES: Seacoast Bluestem, Wintergrass, Three-awns, Plains Bristlegrass, Buffalo Grass, Little Bluestem, various grama grasses.

CACTUS: Prickly-pear.

INDICATOR ANIMALS

BIRDS: Scissor-tailed Flycatcher, Eastern Kingbird, Western Kingbird, Eastern Meadowlark, Roadrunner, Scaled Quail, Wild Turkey, White-winged Dove, Boat-tailed Grackle, Blue Grosbeak, Lark Sparrow, Turkey Vulture, Black Vulture, Red-tailed Hawk, Golden-cheeked Warbler (very local on the Edwards Plateau), Verdin (in mesquite).

MAMMALS: White-tailed Deer (abundant), Collared Peccary (see Fig. 18), Black-tailed Jackrabbit, Nine-banded Armadillo.

DESCRIPTION

A savanna is an open, parklike grassland with scattered trees. In eastern Texas there is extensive dry savanna dominated along the coast by Live Oak and Mesquite, along with Seacoast Bluestem and Wintergrass, and inland by Live Oak and Ashe Juniper mixed among Little Bluestem grass and Buffalo Grass. To the south, in the area of the lower Rio Grande valley, the area becomes more desertlike, with Creosotebush and acacias as well as Mesquite

Savanna. Semi-arid grasslands with scattered trees and shrubs.

and Live Oak. Throughout the area various species of prickly-pear cactus are common and many herbs, especially composites, abound. Trees are short and widely scattered among the grasses. Roadside birding is very rewarding, with Scissor-tailed Flycatchers and both Eastern and Western kingbirds on telephone wires, and Eastern Meadowlarks and Lark Sparrows on fence posts. Scaled Quail and Roadrunners often dart across the highway.

SIMILAR FOREST COMMUNITIES: None, though southern coastal Pine-Oak Forests bear a superficial resemblance.

RANGE: Gulf Coastal Plain from Arkansas to the Rio Grande, including the Edwards Plateau of south-central Texas.

REMARKS: This is an area where forest meets grassland. Due to aridity (only 20–30 in. of precipitation fall annually) and probably also to fires and grazing from cattle (and formerly Bison), this area is dominated by various short and medium to tall grasses, along with a few hardy tree species. Trees are typically evergreen and are widely spaced and short of stature, rarely more than 25 ft. tall.

Birds are rather easy to observe since the area is so open. The Roadrunner, a unique member of the cuckoo family, chases lizards and snakes, reminding one of a tiny feathered dinosaur as it races along. Wild Turkey flocks are easier to see here than in any other area and the small gray Scaled Quail becomes common as you near the Rio Grande valley. The graceful Scissor-tailed Flycatcher flies from a telephone wire to chase insects. The rarest bird of the area is the well-named Golden-cheeked Warbler, which nests only in the Ashe Junipers of the Edwards Plateau.

White-tailed Deer are abundant, and toward the lower Rio Grande valley you can see America's version of the wild pig, the Collared Peccary (Fig. 18). Bands of peccaries forage through the savanna, snorting and rooting among the trees and grasses.

Fig. 18. Collared Peccary.

INDICATOR PLANTS

GRASSES: Side-oats, Grama Grass, Blue Grama, Big Bluestem, Little Bluestem, and Indiangrass, plus many species of composites and legumes. Among the most common are Pasqueflower, Leadplant, Bergamot, Pussytoes, Prairie Buttercup, Prairie Coneflower, Birdsfoot Violet, Shooting-star, Compass-plant, Wild Indigo, Common Lespedeza, many clover species, various gayfeathers, Rattlesnake-master, Bedstraw, and Prairie Oxalis. No trees or shrubs, except along streams and rivers.

INDICATOR ANIMALS

BIRDS: Horned Lark, Eastern Meadowlark, Mourning Dove (abundant), Upland Sandpiper, Ring-necked Pheasant (introduced), Greater Prairie Chicken, Barn Swallow, Red-tailed Hawk.

MAMMALS: Thirteen-lined Ground Squirrel, Black-tailed Prairie Dog, American Bison (formerly abundant—now restricted to several herds located farther west).

DESCRIPTION

Though this book focuses on forests, the range of the book includes the region of dynamic interface between eastern forests and midwestern prairie grassland. It is therefore important to take a brief look at the easternmost components of the great midwestern prairie. A prairie is a grassland-herb ecosystem, where woody

Fig. 19. Thirteen-lined Ground Squirrel.

Prairie. Primarily grasslands, prairies also have numerous species of wildflowers, especially legumes.

vegetation is confined to streamsides and riverbank floodplains. Tall and medium-sized grasses give the ecosystem its characteristic look, but mixed among the grasses is a high diversity of wildflowers, mostly from the composite family (sunflowers, asters, goldenrods) and the legumes (Wild Indigo, lespedezas, Birdsfoot Violet, various clovers). There is a dramatic flowering of wildflowers in the spring. Ground squirrels (see Fig. 19) and prairie dogs are common sights, Eastern Meadowlarks and Horned Larks sing from fence posts, and Barn Swallows skim the skies for insects.

SIMILAR PLANT COMMUNITIES: See Northern Savanna (p. 120).

RANGE: Central Kansas, eastern Nebraska, western and central Iowa, western and southern Minnesota, parts of southern Wisconsin, tiny parts of southern Michigan and northwestern Illinois. Prairie remnants extend well north into Canada and formerly were much more evident, but most prairie has been converted to agriculture or forest.

REMARKS: The American Prairie Grassland has largely been transformed into the nation's breadbasket. With the opening of the West in the last century, tall and medium-sized prairie grasses gave way to the plow. Today, it is not easy to find examples of the prairie that used to be. Still, some areas do persist, and they are worth a visit, especially to enjoy the plethora of wildflowers that add a diversity of color to the waving sea of yellow grass stems.

There is an ecotone (see Fig. 5) between the eastern forests and the midwestern prairie. Why does forest give way to grassland as you travel from east to west across the continent? One factor is rainfall. Prairies receive 20–40 in. of precipitation annually, whereas forests receive 40 in. or more. Also, the hot summers in the Midwest can cause plants to become water-stressed as evaporation exceeds precipitation. Clearly, one reason the prairie exists is the drier climate. Grasses are better adapted physiologically than trees to thrive in the dryness. However, this cannot be the only reason, because in the area where forest meets prairie annual precipitation is about 40 in., so the area could theoretically support either forest or prairie. Periodic, relatively frequent burning is thought to be the major factor in sustaining the prairie against encroachment by the forest. The earliest European explorers of the prairie region described spectacular lightning-set prairie fires. Fire kills trees but only burns the grass to ground level. The thick, matted root systems of prairie grasses quickly regenerate the shoots and the prairie again turns green. In areas where fire has been minimized, trees and shrubs readily invade the prairie. Grazing by Bison also may have contributed to the maintenance of prairie over forest. Grasses can tolerate grazing pressure far better than trees and grow back rapidly after being chewed. The eastern section of the prairie owes its existence to a combination of hot temperatures in the summer (where evaporation exceeds precipitation), grazing pressure, and, mostly, to periodic burning.

Northern Savanna

Indicator Plants

TREES: Bur Oak, Black Oak, White Oak, Northern Pin Oak, Blackjack Oak, Jack Pine (northern region), Bigtooth Aspen, Norway Pine (introduced).

SHRUBS: Redroot, Black Huckleberry, Late Lowbush Blueberry, Sweetfern.

GRASSES AND HERBS: Many grasses, composites, and legumes occur, some of which are also common to the Prairie (see above). Most characteristic of the wildflowers are Flowering Spurge, Leadplant, Frostweed, Common Strawberry, and Starry False Solomon's-seal.

Indicator Animals

BIRDS: Savannah Sparrow, Clay-colored Sparrow, Connecticut Warbler (uncommon).

Description

This habitat reminds one of a combination forest and prairie, with mixtures of woody plants, grasses, and many herbaceous species, some of which are characteristic of the prairie. Trees are usually widely spaced and scrubby in appearance. Oaks are most common, but, to the north, Jack Pines invade. Oak savannas are often local, described as island-like "oak openings" in a sea of prairie. To the north, near the range of the Boreal Forest, Jack Pine intermingles with and eventually replaces oaks.

SIMILAR FOREST COMMUNITIES: Though the range is quite different, this savanna looks somewhat like the Texas Savanna. Also see Jack Pine Forest and Prairie.

RANGE: Local in Minnesota and Wisconsin.

REMARKS: The Northern Savanna represents the intermingling of elements from three major communities, the Prairie, the Boreal Forest, and the Eastern Deciduous Forest. Poor soils and periodic fire appear responsible for the existence of this scrubby grassy community. To the south are oak barrens, savannas dominated by Bur Oak, Black Oak, Blackjack Oak, and some White Oak. To the north, the boreal element is represented by Jack Pine. Throughout, prairie composites, legumes, and roses grow among the grasses and short trees. Some researchers believe that forest was once much more extensive throughout the areas that now support Scrub Oaks, but that frequent fires since settlement have permitted the Scrub Oaks to attain a competitive edge over the more mesic (moisture-loving) tree species.

Among the birds most sought after by birders is the elusive Connecticut Warbler. The northern Jack Pine–Scrub Oak savannas provide one of the best breeding habitats for this species.

Fig. 20. Connecticut Warbler.

BOREAL FOREST (P. 62)

The "Spruce-Moose" forest of conifers, but with some hardwoods. Spruce needles are prickly, fir needles softer. Spruce cones dangle below the branch; fir cones are upright.

WHITE SPRUCE

To 75 ft., with square-shaped, dark, 1-in.-long needles curving toward the branch tip.

WHITE-WINGED CROSSBILL

A 6½-in. finch. Both sexes have 2 prominent wing bars. Male rosy red with black wings; female olive, streaked. Crossed bill visible only at close range. Male sings atop conifers.

BALSAM FIR

To 60 ft. Aromatic, with flattened, short, dark green needles that tend to curve upward.

GRAY JAY

Large (11–13 in.). Dark gray on the back, wings, and tail, with a black patch on the back of its head.

TAMARACK or AMERICAN LARCH

To 80 ft. A deciduous conifer; soft, blue-green needles turn bright yellow in fall, then drop off. Needles grow in small bunches along branch, giving tree a feathery look. Abundant around bogs.

BLACKBURNIAN WARBLER

One of several common wood warblers in Boreal Forest. Male is streaked with black and has a brilliant orange throat and facial stripes. Female's throat yellow, not orange. Both sexes have a prominent white wing patch.

PAPER BIRCH

To 80 ft. Bright white, peeling bark. Leaves are light green (yellow in fall), heart-shaped, toothed, and sharply pointed.

RED SQUIRREL

Bright reddish body and tail, white underneath. White around eyes. Very noisy, often chattering and thrashing tail.

MOOSE

Largest hoofed mammal in North America. Blackish brown with long legs and long, drooping face. Males have a broad rack of flat antlers and a black "beard." Often seen around lakes and bogs.

CANADA MAYFLOWER

Also called Wild Lily-of-the-valley. Ovate, green, unlobed leaves clasp stem and have parallel veins. Flowers bloom in spring, forming a small white cluster. Berries white at first, then red.

WHITE-WINGED
CROSSBILL
male
WHITE
SPRUCE
GRAY
JAY
BLACKBURNIAN
WARBLER
male
BALSAM
FIR
TAMARACK
(AMERICAN LARCH)
RED
SQUIRREL
MOOSE
PAPER
BIRCH
CANADA
MAYFLOWER

PLATE 5

BOREAL BOG (P. 67)

Bogs are areas of standing water, very acidic, with an abundance of Sphagnum Moss.

BLACK SPRUCE

To 60 ft. tall (often shorter); abundant along boggy borders and high elevations. Dark green, 4-sided needles do not feel prickly.

OLIVE-SIDED FLYCATCHER

Large-headed, 7–8-in. flycatcher, often seen sitting atop a conifer or dead tree. Grayish patches on either side of breast, large bill, whitish tufts behind wing. Call is a strident *hip, three-cheers.*

LABRADOR-TEA

Evergreen shrub with alternate, elongate, oval leaves that feel very leathery. Leaves with rolled edges; green above but orange and fuzzy below, aromatic when crushed.

LEATHERLEAF

Evergreen shrub, similar to Labrador-tea, but leaves have tiny teeth along the margins and are yellowish, not orange below. Not aromatic.

NASHVILLE WARBLER

Small (4¾ in.), yellowish bird with a white eye-ring and a gray head. Song a whistled *seebit, seebit, seebit, chitititititi.*

RUSTY BLACKBIRD

A robin-sized blackbird with yellow eyes. Plumage turns rusty in fall. Song suggests a squeaky hinge.

NORTHERN GREEN ORCHIS

One of several northern orchids. Long, slender leaves. Flowers small, yellowish green, clustered on a long spike.

ROUND-LEAVED SUNDEW

Small, rounded leaves with sticky hairs that trap insects.

PITCHER-PLANT

Large, funnel-shaped leaves with wide lips. Leaves heavily veined, with hairs that point inward and trap insects.

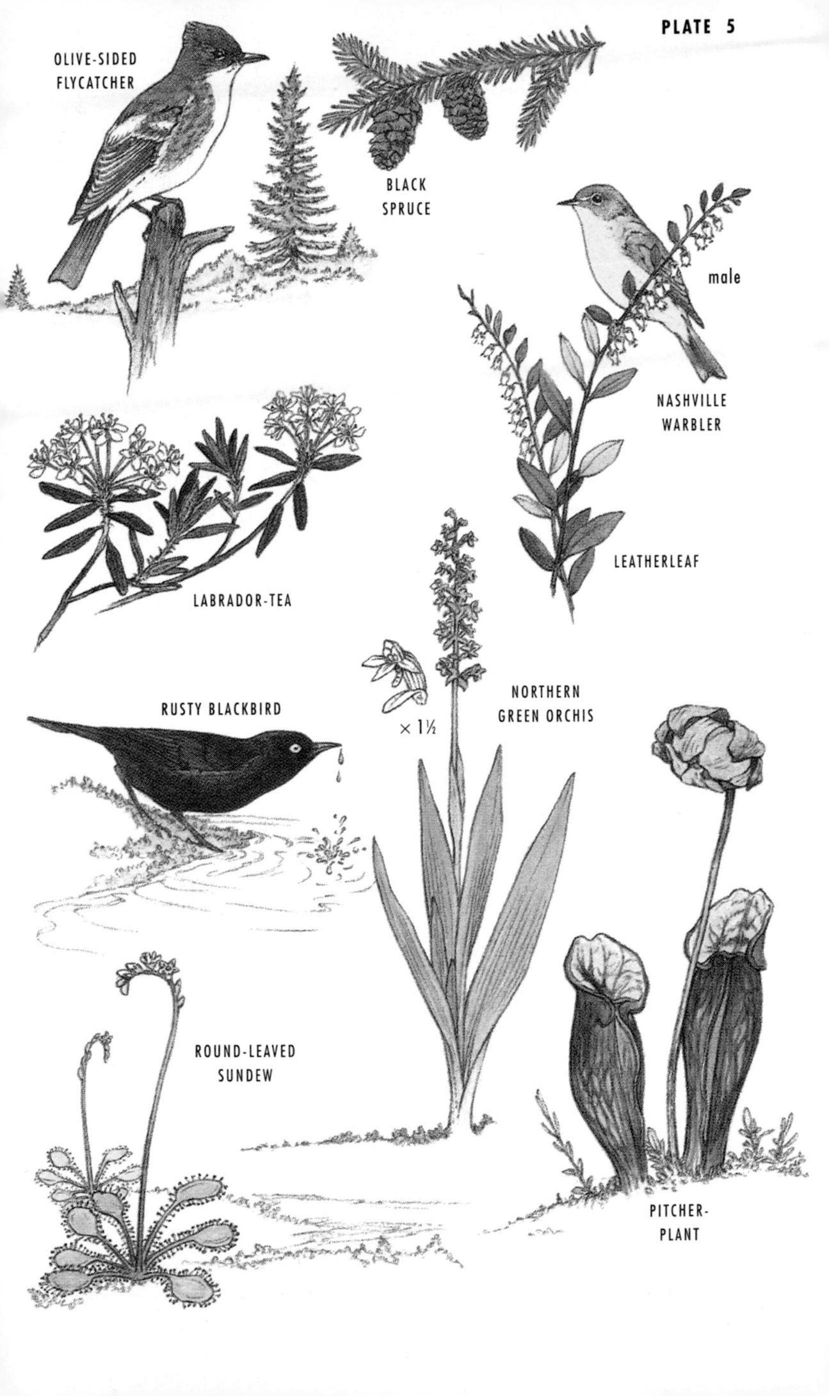
OLIVE-SIDED
FLYCATCHER
BLACK
SPRUCE
male
NASHVILLE
WARBLER
LABRADOR-TEA
LEATHERLEAF
RUSTY BLACKBIRD
× 1½
NORTHERN
GREEN ORCHIS
ROUND-LEAVED
SUNDEW
PITCHER-
PLANT

NORTHERN HARDWOOD FOREST (P. 72)

This forest is mostly composed of Sugar Maple, Yellow Birch, and American Beech, but with abundant White Pine and Eastern Hemlock.

YELLOW BIRCH

A canopy species up to 100 ft. tall. Shiny, peeling, paperlike, yellowish bark. Leaves oval; dark green above, with tiny teeth.

YELLOW-BELLIED SAPSUCKER

An 8–9-in. woodpecker with a large white wing stripe, visible in flight and when perched. Sapsuckers drill trees for sap, making horizontal rows of holes.

EASTERN HEMLOCK

A stately conifer with small cones that hang from branch tips. Dark green needles, very short and flattened.

NORTHERN (SLATE-COLORED) JUNCO

A sparrow-sized bird of the understory. White outer tail feathers.

WHITE PINE

To 150 ft., though individuals taller than 100 ft. are rare. Soft, bluish green, 3–5-in. needles in bundles of 5. Also see Pl. 35.

BLACK-THROATED BLUE WARBLER

Male dark blue above, with black on the throat and sides of breast. Breast white. Female brownish. Common in understory. Song a husky, buzzy trill.

STRIPED MAPLE

Bark striped white and brownish green. Leaves large and 3-lobed, with tiny teeth. Leaves have long tips. An understory shrub or tree, rarely exceeding 40 ft.

SOLITARY (BLUE-HEADED) VIREO

A methodical insect-catcher of the subcanopy and canopy. Greenish olive with a blue-gray head and white "spectacles" around the eyes. Two wing bars.

PAINTED TRILLIUM

One of several trilliums common in this forest. Blooms in spring, early summer. White petals with red at base.

WHITE-THROATED SPARROW

A largish sparrow (6½–7 in.) with bright white throat, white eyestripe and crown, a spot of yellow in front of the eyes. Frequents understory with conifers. Song a plaintive, whistled *Old Sam Peabody, Peabody, Peabody.*

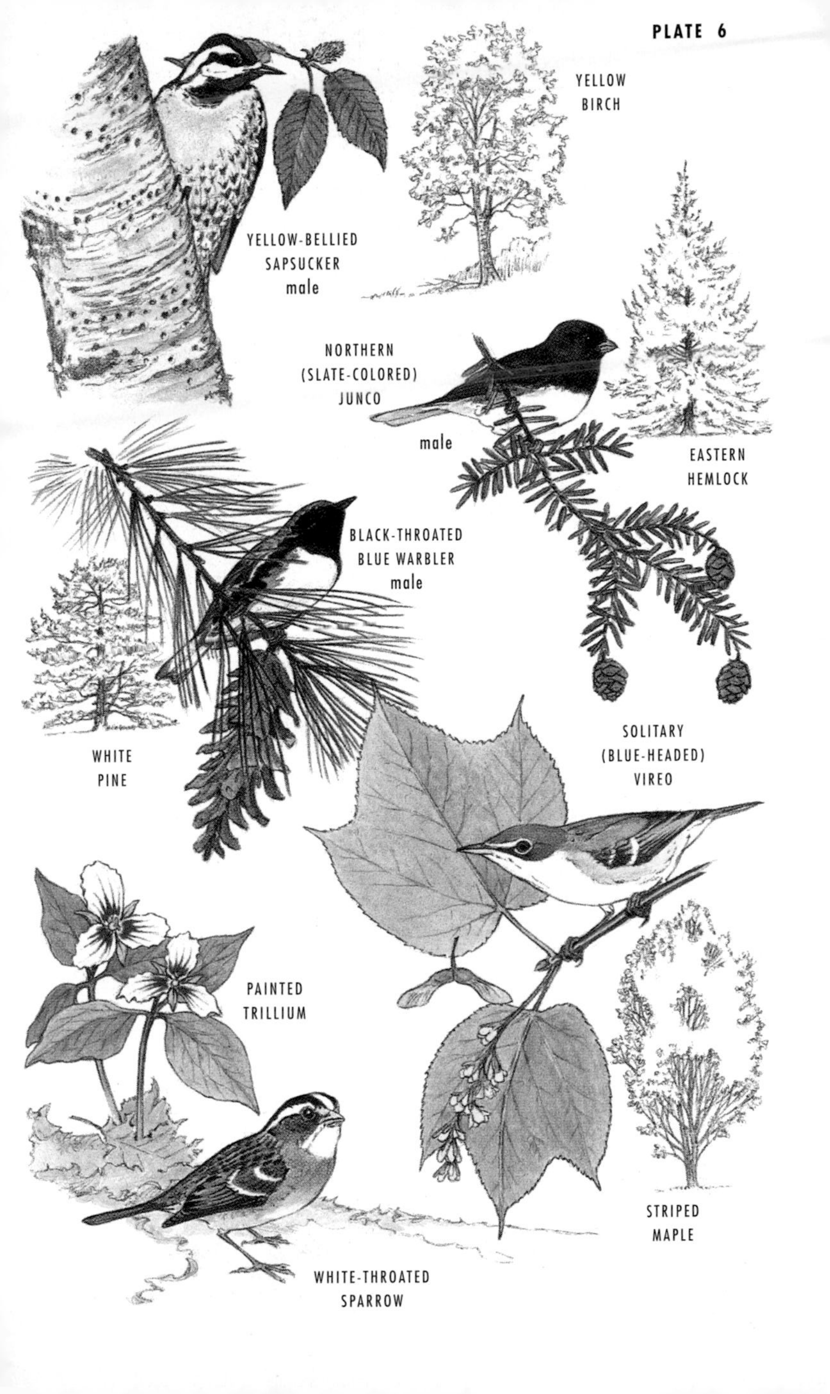
YELLOW BIRCH
YELLOW-BELLIED SAPSUCKER male
NORTHERN (SLATE-COLORED) JUNCO male
EASTERN HEMLOCK
BLACK-THROATED BLUE WARBLER male
WHITE PINE
SOLITARY (BLUE-HEADED) VIREO
PAINTED TRILLIUM
STRIPED MAPLE
WHITE-THROATED SPARROW

BEECH-MAPLE AND MAPLE-BASSWOOD FORESTS (PP. 77, 80)

These forests extend the dominance of hardwoods southward from New York through the Great Lakes states to the Appalachians. Many species characteristic of the Northern Hardwood Forest are also present in these forests.

AMERICAN BEECH

To 80 ft., often with a wide trunk and spreading horizontal branches. Large individuals frequently surrounded by root saplings. Smooth, light gray bark. Alternate leaves, elongate and oval, with large teeth along margins. Leaves feel papery and turn yellow, then brown in fall. Many leaves can remain on the tree during winter. Beech nuts in prickly, burred fruits.

CERULEAN WARBLER

A treetop warbler, hard to observe in dense foliage. Male is pale blue above, with a white throat, white breast, and black "collar" across neck, along with black stripes on sides of breast and belly. Two white wing bars. Female pale blue-green, without collar or streaking. Song a buzzy whistle.

SUGAR MAPLE

To 100 ft.; abundant throughout our area. Leaves opposite, with 5 deep lobes and few teeth; leaves turn brilliant yellow, orange, and/or orange-red in fall. Bark gray, furrowed. Seeds in long, winged fruits.

ALTERNATE-LEAF DOGWOOD

The only dogwood without opposite leaves (compare with Flowering Dogwood, Pl. 1). A small understory tree (usually not over 25 ft. tall) with white, flat-topped flower clusters in spring. Fruits bright red.

BLACK-CAPPED CHICKADEE

Abundant throughout northern states. A small (5 in.), grayish bird identified by its black cap and bib, and its *chicka-dee-dee-dee* call.

AMERICAN BASSWOOD

To 100 ft.; trunk often a yard or more in diameter. Alternate, heart-shaped leaves, up to 5 in. long. Small, yellowish white flowers hang in clusters from the middle of each slender, long bract. Fruits are clusters of small yellowish berries, also hanging from bracts. Bark gray and smooth, becoming furrowed in old trees. Also called American Linden.

PLATE 7

CERULEAN WARBLER
male

AMERICAN BEECH

SUGAR MAPLE

BLACK-CAPPED CHICKADEE

ALTERNATE-LEAF DOGWOOD

fruit

AMERICAN BASSWOOD

OAK-HICKORY FOREST (P. 81)

A widespread, nut-producing forest of mixed oaks and hickories, typically on drier soils or south-facing slopes, with many additional tree species in more mesic (moist) areas. Rich wildflower diversity in spring.

NORTHERN RED OAK

To 90 ft. Leaves alternate, with 7–11 lobes, each sharply pointed; dark green above, light green below. Acorn egg-shaped, with a small cap. Similar **Black Oak** has more deeply lobed leaves with 7–9 lobes.

BITTERNUT HICKORY

To 80 ft. Leaves compound and alternate, each consisting of 9 oval leaflets with tiny teeth. Nut in a thick, brown, rounded husk.

SCARLET OAK

To 80 ft. Leaves alternate, usually with 7 (occasionally 9) deep lobes with sharp points; shiny green in summer. Acorn small, with a cap covering about 50% of the nut.

BLUE JAY

Large (11–12 in.); mostly blue above and white below, with a prominent crest and black necklace. Call a scolding *jay! jay!* Local abundance varies with the size of the nut crop.

SHAGBARK HICKORY

To 100 ft. Gray bark peels in long strips. Favors moist areas. Leaves compound and alternate, with 5 finely toothed leaflets.

WHITE OAK

To 100 ft. Alternate leaves with 5–9 smooth, rounded lobes. Acorn rounded, about 25% enclosed by cup. Bark whitish gray and scaly, often easily peeled.

EASTERN GRAY SQUIRREL

Gray above, white below. Bushy tail. Melanistic individuals occur.

PIGNUT HICKORY

To 80 ft. Alternate compound leaves with 5 leaflets, each narrow and finely toothed. Nut in a dark brown, oval husk.

WILD TURKEY

More slender than a domestic turkey, with rusty tail tips (white in barnyard fowl). Feathers shiny bronze.

MAYAPPLE or MANDRAKE

Widespread; often carpets the forest floor in spring. Blossom single, on a nodding stem; large, white, with a yellow center. Leaves large (12 in.), umbrella-like.

NORTHERN
RED
OAK
BITTERNUT
HICKORY
SCARLET
OAK
BLUE JAY
SHAGBARK
HICKORY
WHITE OAK
EASTERN
GRAY
SQUIRREL
PIGNUT
HICKORY
WILD
TURKEYS
MAYAPPLE
(MANDRAKE)

NORTHERN RIVERINE (FLOODPLAIN) FOREST (P. 85)

A forest bordering rivers, often on floodplains.

EASTERN SYCAMORE

Stately tree, to 100 ft. tall, with a wide trunk and large patches of peeling, white and brown bark. Leaves maple-like but alternate, with 3–5 shallow lobes. Fruits are round balls, hanging from long stalks. Also abundant in South.

BELTED KINGFISHER

A large (13 in.), big-headed bird with a ragged crest and a large bill. Female has rufous breast band. Dives head-first for fish. Strident rattling call.

SILVER MAPLE

To 80 ft. Leaves opposite, deeply lobed with large teeth; silvery below. Bark silvery.

EASTERN COTTONWOOD

To 100 ft. tall, with wide (3–4 ft.) trunk and spreading branches. Leaves alternate, on long stalks; triangular, with prominent teeth. Female (pistillate) and male (staminate) flowers are on separate trees. Seeds blow from catkins; each seed is attached to a tuft of cottony fibers.

RIVER BIRCH

To 80 ft. Leaves alternate, arrowhead-shaped, with double-toothed margins. Seeds in cone-like strobiles, near branch tips. See also Pl. 15.

MINK

A slender, uniformly brown, weasel-like animal of river banks. Thick, hairy tail, about one-third the length of the body. Minks are smaller and less social than River Otters.

BLACK WILLOW

A small (20–40 ft.) tree, usually with several spindly trunks. Leaves alternate, very long, with fine teeth. **Sandbar Willow** is similar, but its leaves have larger, more widely spaced teeth.

SPOTTED SANDPIPER

A slender, 7½-in., brown bird with a pointed posterior, dove-like head, and densely spotted breast. Walks, bobbing its tail or "teetering" as it moves.

GREEN-BACKED HERON

A crow-sized heron, reddish brown on the neck, dark green on the back and wings. When taking flight, it emits a loud *skyow.*

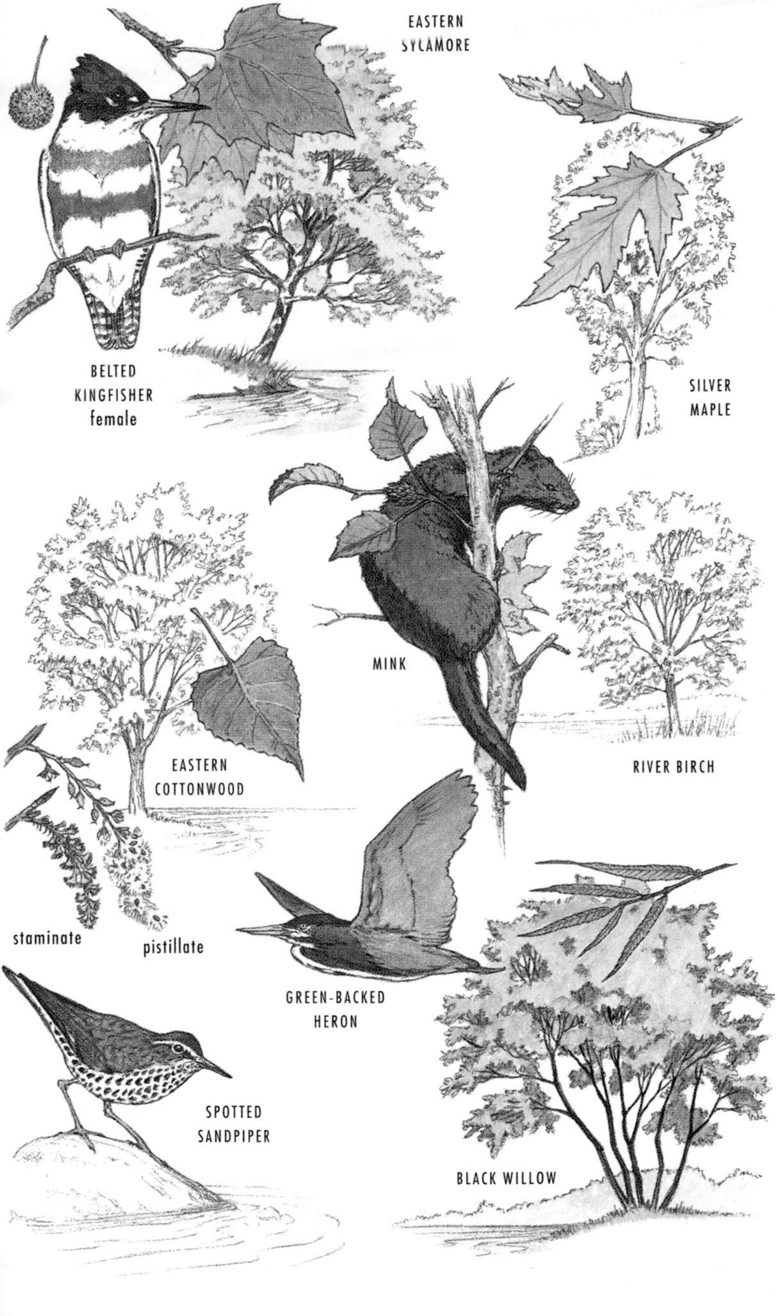

PLATE 9

EASTERN SYCAMORE

BELTED KINGFISHER female

SILVER MAPLE

MINK

EASTERN COTTONWOOD

RIVER BIRCH

staminate

pistillate

GREEN-BACKED HERON

SPOTTED SANDPIPER

BLACK WILLOW

NORTHERN SWAMP FOREST (P. 88)

This forest occurs on wet soils and is quite variable in species composition. Some species, such as Atlantic White-cedar and Red Maple, may be present in single-species stands.

WILLOW OAK

To 80 ft. Leaves unlobed, elongate; margins untoothed, often turned slightly upward. Not present in New England.

CEDAR WAXWING

Soft brown, crested, with black on face and a yellow tail band. Very fond of fruit trees. Voice a high-pitched, thin *seeee*.

ATLANTIC WHITE-CEDAR

To 80 ft. Evergreen conifer with scaly, blue-green foliage. Small bluish cones. Coastal. **Northern White-cedar** occurs in northern and inland states and is similar but has brown cones.

BLACK TUPELO or BLACK GUM

To 100 ft. Leaves alternate, oval; margins smooth. Shiny green in summer. Common in southern swamps as well as in North. See also Pl. 49.

RED MAPLE

Often occurs in single-species stands in northern swamps. Opposite leaves with 3 prominent lobes. Leaves brilliant crimson in fall. Seeds with reddish wings. See also Pl. 1.

SWEET PEPPERBUSH

A dense shrub with alternate, oval, toothed leaves. Flowers are small and white, arranged in upright clusters.

SPICEBUSH

A shrub with highly aromatic leaves and twigs. Broken leaves emit a pleasant pungent odor. Leaves alternate, oval with smooth margins. Tiny yellow flowers open before foliage. See also Pl. 49.

SKUNK CABBAGE

One of the earliest spring wildflowers. Flower opens before leaves and consists of a cloaklike spathe shielding a cone-like spadix of tiny flowers. Leaves very large, cabbage-like. Broken leaves emit a very strong, foul, pungent odor.

WILLOW
OAK
ATLANTIC
WHITE-CEDAR
CEDAR
WAXWING
BLACK
TUPELO
(BLACK GUM)
RED MAPLE
SWEET
PEPPERBUSH
× 2
SPICEBUSH
SKUNK
CABBAGE

NORTHERN PINE-OAK FOREST (P. 90)

A scrubby coastal forest of dry, sandy soils, with mixed pine and oak and an understory of heaths.

PITCH PINE

A medium-sized pine (to 60 ft.) with stiff needles (3–5 in. long) in bunches of 3. Needle tufts often grow directly from trunk. Cones rounded, remain on tree when open.

VIRGINIA PINE

To 60 ft. Needles stiff, 1½–2 in., in bundles of 2. Cones egg-shaped, remain attached when open.

PINE WARBLER

5–5½ in. Yellow below, olive above, with 2 white wing bars. Song a continuous, deliberate, whistled trill.

PRAIRIE WARBLER

A 5-in. yellowish warbler with black streaking on face and sides, most pronounced on males. Upperparts olive, back faintly streaked with red. Song a buzzy, upscale *zee zee zee*. Frequents the understory.

BLACKJACK OAK

Often short, but can grow to 50 ft. Leaves widest near tip, shiny, with sharply pointed lobes. Acorns terminate in sharp point.

BEAR OAK

A shrubby oak, rarely more than 20 ft. tall. Leaves thick, shiny, underside silvery. Lobes sharply pointed.

DWARF HUCKLEBERRY

A dense shrub with shiny, leathery leaves with tiny yellow resin dots (use hand lens). White or pinkish, bell-like flowers, in clusters. Fruits are hairy black berries.

RUFOUS-SIDED TOWHEE

A cardinal-sized, colorful sparrow of the understory with white outer tail feathers. Female is brown above, where male is black. Call a sharp *chewink!* Song a musical *drink your tea-a*.

BEARBERRY

A ground-hugging, evergreen heath with small, light green, oval, smooth, waxy leaves. Spreads widely over ground. White or pale pink flowers, in clusters. Berries bright red.

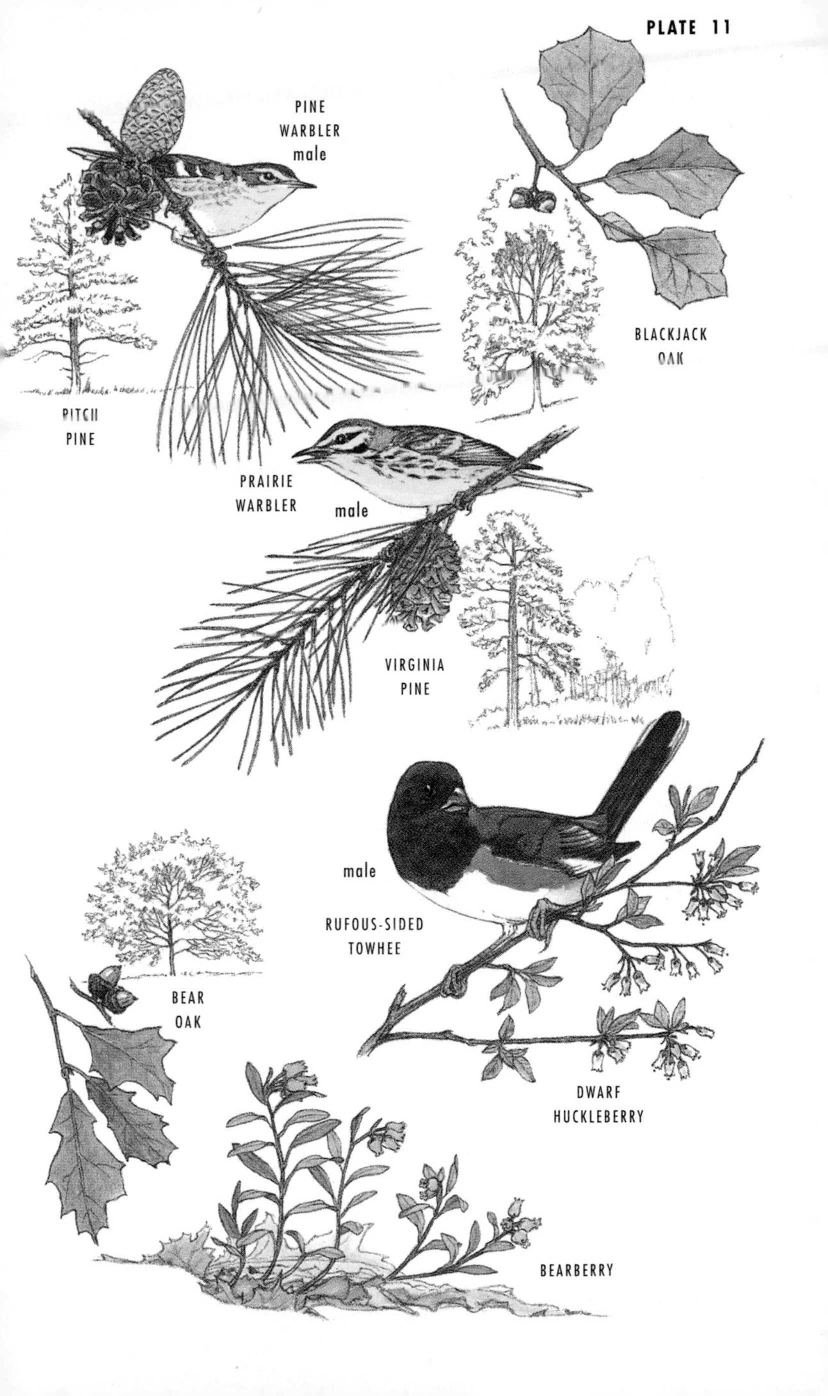
PINE
WARBLER
male
BLACKJACK
OAK
PITCH
PINE
PRAIRIE
WARBLER
male
VIRGINIA
PINE
male
RUFOUS-SIDED
TOWHEE
BEAR
OAK
DWARF
HUCKLEBERRY
BEARBERRY

SOUTHERN MIXED PINE-OAK FOREST (P. 92)

This forest covers much of the southern Coastal Plain. Found on dry, sandy soils, it is subject to frequent fires and supports a high diversity of pines, oaks, and other hardwood species.

LONGLEAF PINE

To 100 ft. Needles bluish green, to 18 in. long, in bundles of 3. Cones rusty, prickly, to 10 in. long.

LOBLOLLY PINE

To 100 ft. Yellowish green needles, up to 9 in. long, in bundles of 3. Prickly, reddish brown cones.

RED-COCKADED WOODPECKER

Lives in noisy family groups of up to 6. About 8½ in. White cheeks, black cap, and black and white striped back. Male has a tiny red spot—the cockade—on the upper part of the cheek.

BROWN-HEADED NUTHATCH

Small, about 4½ in. Probes for insects on pine bark and tips of needle clusters. Brown head, white throat and breast, white neck spot, pale blue back. Often in noisy flocks. Sexes similar.

SHORTLEAF PINE

To 100 ft. Pyramid-shaped crown. Soft, flexible needles, 3–5 in. long; usually in bundles of 2, sometimes 3. Cones oblong. Reddish brown bark in scaly plates.

YELLOW-THROATED WARBLER

5–5½ in. Bright yellow throat, black face patch, black side streaks. Sings *tew, tew, tew* loudly from canopy. Sexes similar.

TURKEY OAK

A shrubby oak, to 30 ft. Leaves with 7 deep, sharply pointed lobes. A few teeth on each lobe. Acorns oval, with cup covering about one-third.

SOUTHERN BAYBERRY

A bushy evergreen shrub (but may grow to 40 ft.) with elongate, toothed leaves, shiny yellow-green above, orange below. Aromatic when crushed. Berries blue-gray, waxy. Also called Waxmyrtle.

SAW-PALMETTO

Dense shrub, occasionally to 25 ft. Fanlike leaves, 25 in. in diameter, with sharp points. Leaf stalks lined with spines.

NINE-BANDED ARMADILLO

Covered with thick, bony plate. Body with little hair. Face pointed, with prominent ears. Long, stiff tail. Curls into ball when threatened. Active at night. Burrower.

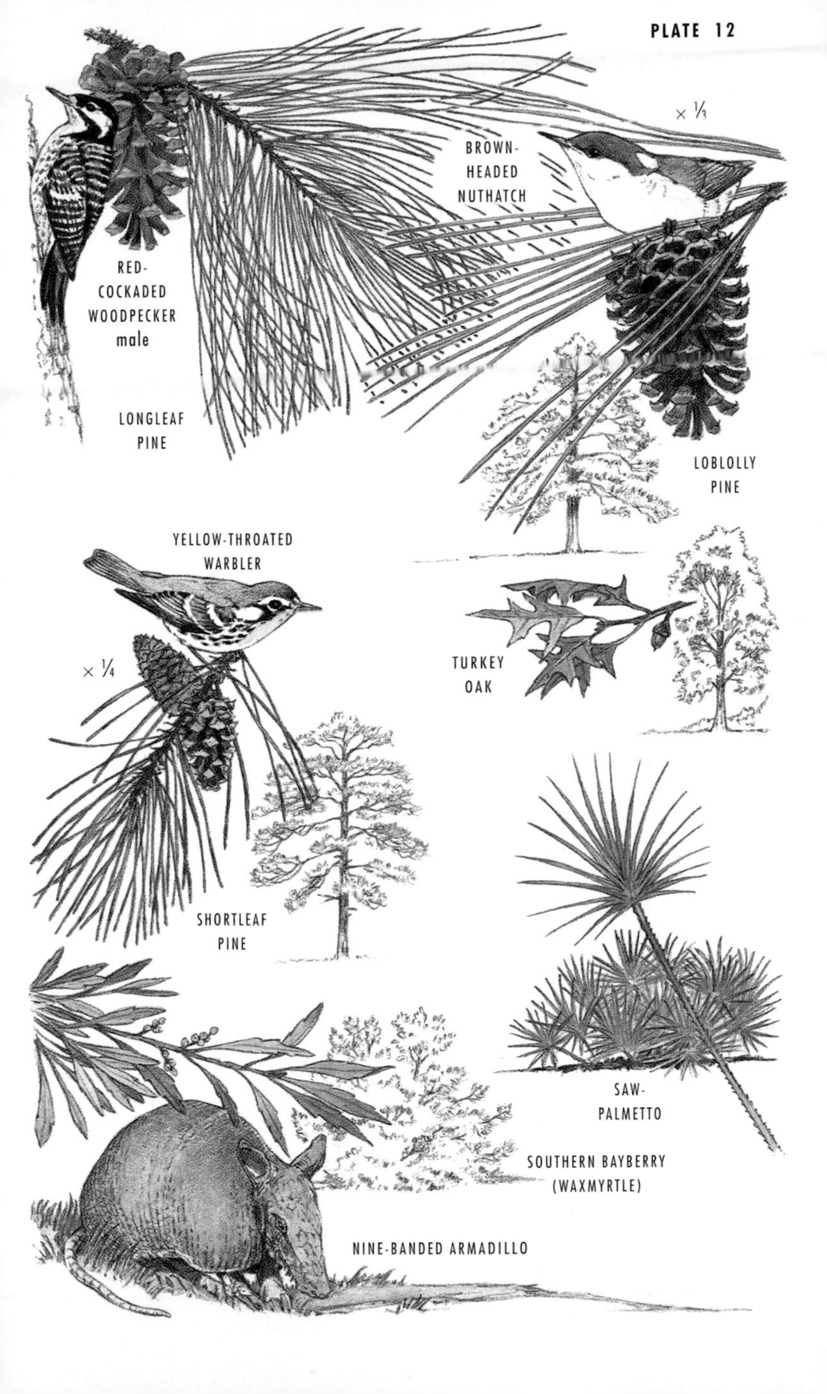
× 1/3
BROWN-
HEADED
NUTHATCH
RED-
COCKADED
WOODPECKER
male
LONGLEAF
PINE
LOBLOLLY
PINE
YELLOW-THROATED
WARBLER
× 1/4
TURKEY
OAK
SHORTLEAF
PINE
SAW-
PALMETTO
SOUTHERN BAYBERRY
(WAXMYRTLE)
NINE-BANDED ARMADILLO

APPALACHIAN COVE FOREST (P. 98)

A diverse, lush forest of moist sites in the Appalachians.

YELLOW BUCKEYE

To 90 ft. A wide-spreading tree with opposite, palmately compound leaves, each with 5 toothed leaflets. Yellow flowers, in elongate clusters near branch tips.

YELLOWWOOD

A smallish (to 60 ft.) tree with alternate, compound leaves. Leaflets oval, untoothed. Flowers white; seeds in brown pods.

TULIPTREE

Also called Yellow Poplar. Very tall (occasionally 150 ft.), straight-trunked tree. Leaves bright shiny green, up to 10 in. wide, alternate, with 4 shallow lobes. Flowers tuliplike.

WHITE BASSWOOD

To 80 ft. Similar to American Basswood (Pl. 7), but leaves more elongate and more finely toothed. Bark grayish, not brownish. Flowers yellowish white, hanging in clusters from bract.

CAROLINA SILVERBELL

Normally a small tree (30 ft.) or shrub, but can reach 80 ft. Leaves alternate, elongate, with tiny teeth. Flowers white, bell-shaped, in clusters drooping below branches.

FLAME AZALEA

Not really a Cove Forest species but a shrub of grassy balds throughout Appalachians. Brilliant orange-red, tubular flowers in spring. Leaves ovate, leathery.

HOODED WARBLER

Male is bright yellow below, olive above, with black "hood" and yellow face. Female duller yellow, without black throat. White tail spots in both sexes. Song a loud, whistled *wheeta,* repeated, ending in a sharp *tee-o.* Understory species.

AMERICAN GINSENG

One of scores of Cove Forest spring wildflowers. Three compound leaves, each with 5 toothed leaflets. Tiny white flowers in a ball-shaped cluster. Berries red. Root thick.

WORM-EATING WARBLER

A brown understory warbler with black facial streaks. Prefers rhododendron slopes. Song a repeated dry buzzing.

EASTERN REDBUD

A small tree (to 40 ft.) with large, alternate, heart-shaped leaves. Bright pink flowers open before leaves. Seeds in brown pods.

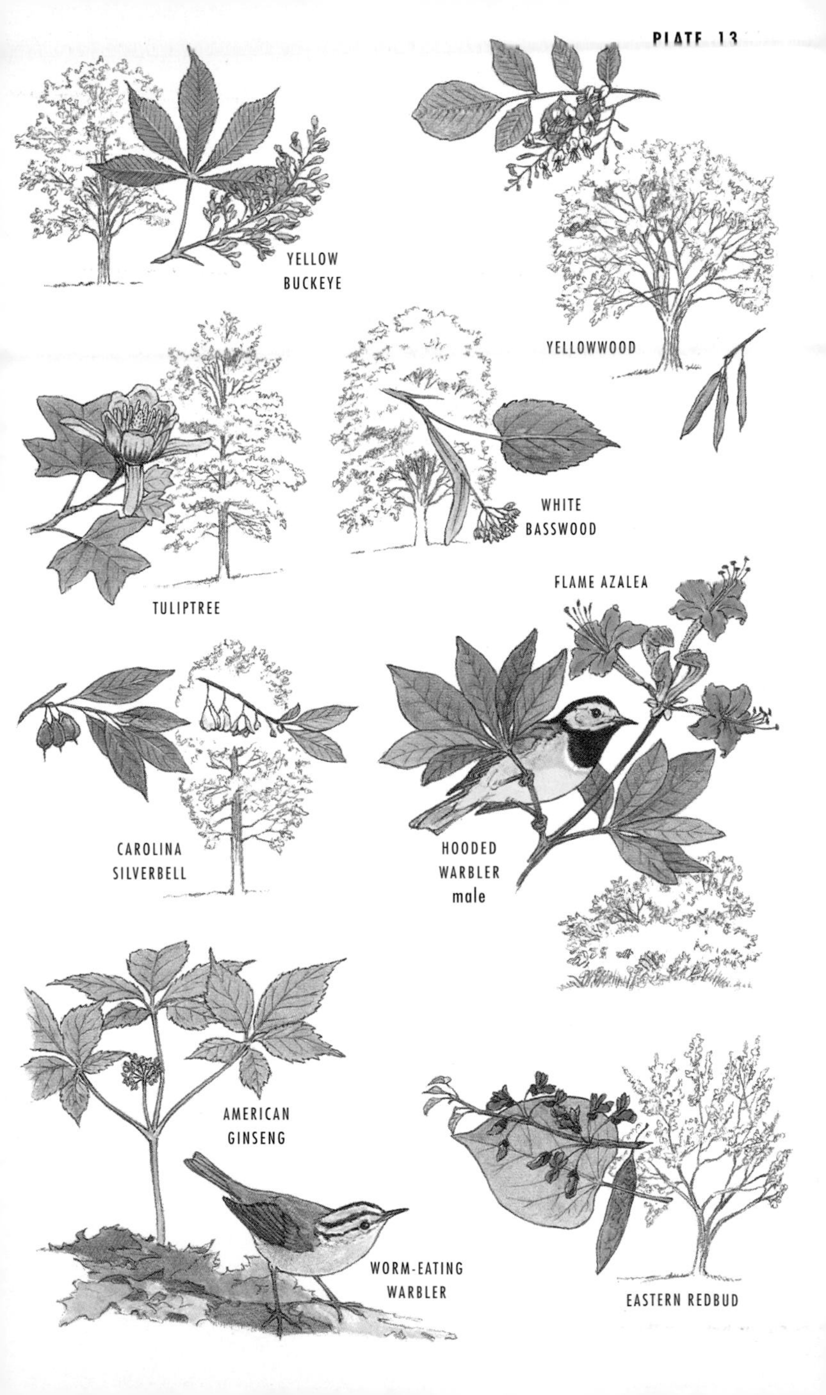
YELLOW
BUCKEYE
YELLOWWOOD
TULIPTREE
WHITE
BASSWOOD
FLAME AZALEA
CAROLINA
SILVERBELL
HOODED
WARBLER
male
AMERICAN
GINSENG
WORM-EATING
WARBLER
EASTERN REDBUD

SOUTHERN HARDWOOD FOREST (P. 103)

A forest of mixed deciduous and evergreen species, characteristic of upland and coastal areas.

TRUMPET-CREEPER
Bright orange-red, trumpetlike flowers. See also Pl. 31.

VIRGINIA LIVE OAK
A beautiful spreading crown. To 50 ft. Leaves evergreen, oval, shiny green, untoothed, hairy underneath.

SPANISH MOSS
A wispy air plant, suspended in long, gray-green strands from trees and telephone wires. Tiny, pale green flowers. Single strands may be up to 25 ft. long.

TUFTED TITMOUSE
A 6-in. gray bird with a crest. Wash of rufous on flank.

SOUTHERN MAGNOLIA
To 80 ft. Evergreen, alternate leaves, very shiny, oval, untoothed but with edges curled under. Large, white, fragrant blossoms.

NORTHERN MOCKINGBIRD
Slender gray bird (9–11 in.) with large white wing patches. Sings from wires and often sings at night. Mimics other bird species.

LAUREL OAK
To 80 ft. Shiny green, alternate, slender, oval leaves. Basically evergreen, though it sheds many leaves in early spring.

RED-BELLIED WOODPECKER
Ladder-backed. Male has red on neck and top of head; female has red neck only. Shows small white wing patches in flight.

SWEETBAY
To 60 ft. Deciduous in northern areas, evergreen in Deep South. Leaves alternate; light, shiny green above, very pale below; oval, untoothed. Flower like that of Southern Magnolia.

COMMON PERSIMMON
To 70 ft. Deciduous. Leaves alternate, oval, untoothed, with a slight point at tip. Fruits orange-red.

PECAN
To 100 ft. Alternate, compound leaves with 11–17 sharply pointed, oval, yellowish green leaflets. Nut has dark brown husk.

VIRGINIA OPOSSUM
Gray fur, pointed snout, naked tail. Cat-sized. Nocturnal. Good tree-climber, using prehensile tail. Plays dead when threatened.

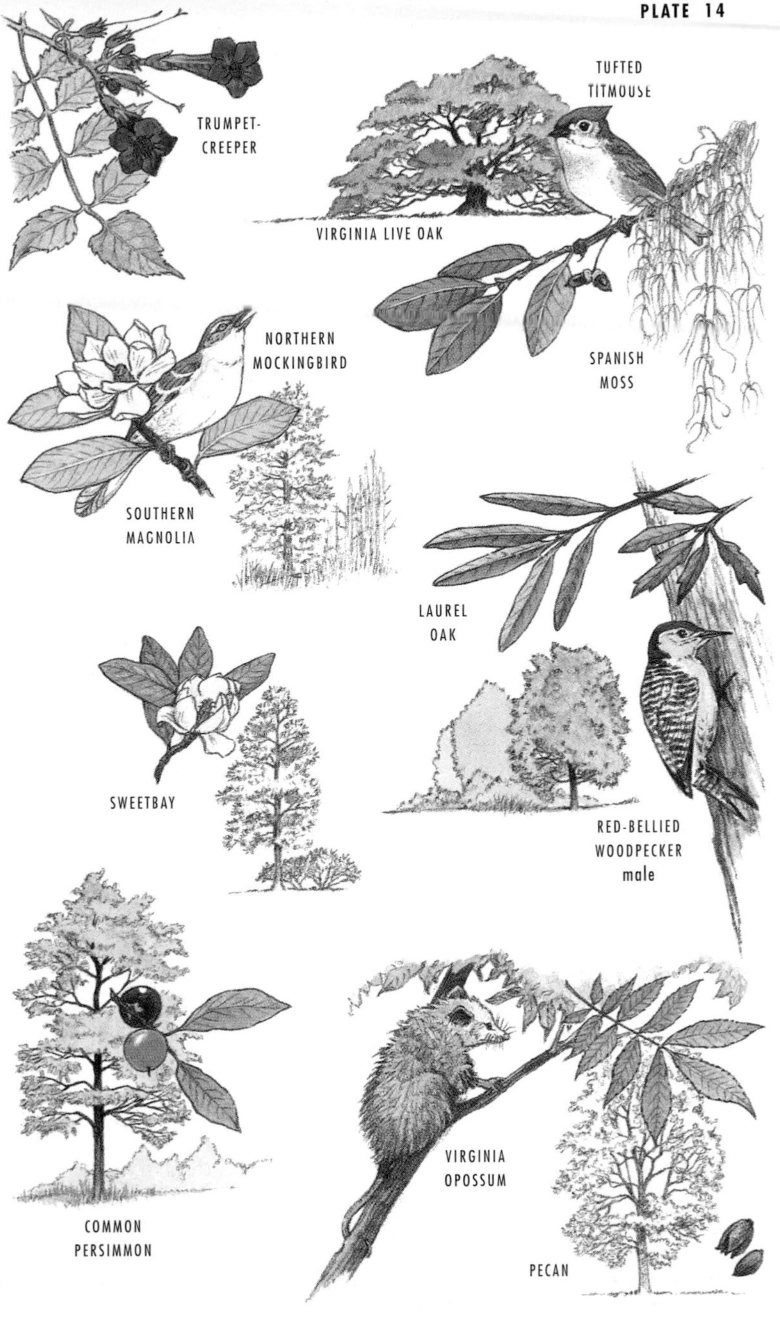

TRUMPET-CREEPER

TUFTED TITMOUSE

VIRGINIA LIVE OAK

SPANISH MOSS

NORTHERN MOCKINGBIRD

SOUTHERN MAGNOLIA

LAUREL OAK

SWEETBAY

RED-BELLIED WOODPECKER male

COMMON PERSIMMON

VIRGINIA OPOSSUM

PECAN

SOUTHERN RIVERINE FOREST (P. 106)

These species line river banks and floodplains in southeastern and central states.

BOX-ELDER or ASHLEAF MAPLE
To 60 ft. The only maple with compound leaves. Leaflets oval, pointed, with large teeth.

SOUTHERN ARROWWOOD
A dense shrub of river banks, thickets, swamps. Opposite leaves, oval, with large teeth. Large rounded clusters of white, 5-petaled flowers. Fruits deep violet, blue, or blackish.

NORTHERN CARDINAL
Male red with a crest. Female reddish brown. Song a repeated, loud, melodious whistle.

SWAMP WHITE OAK
To 70 ft. Wide spreading crown, often with branches drooping. Leaves oval, with shallow, rounded lobes. Leaves green above, pale whitish below.

BARRED OWL
A large owl, often seen in daytime. Brown eyes, horizontal barring on breast. Calls, *who cooks for you?*

SMOOTH (TAG) ALDER
A treelike shrub with multiple trunks. Leaves alternate, widely oval, toothed. Male flowers in drooping catkins, female flowers in short cones. Similar to **Speckled Alder** of far northern areas.

RIVER BIRCH
To 80 ft. Leaves are alternate, arrowhead-shaped, with double-toothed margins. Seeds in cone-like strobiles, near branch tips.

WOOD DUCK
Male multicolored, female grayish. Voice an emphatic *whee-a,* given in flight. Nests in tree cavities, bird boxes.

SWEETGUM
To 130 ft. Leaves starlike, with deep lobes and pointed tips. Fruits hang on stalks in ball-like clusters.

BUTTONBUSH
Thick shrub with leaves in whorls of 3, or opposite. Leaves oval, untoothed, pointed. Flowers in ball-like clusters.

LOUISIANA WATERTHRUSH
Like a small Spotted Sandpiper (see Pl. 9) but with a white eye-stripe, throat, and streaked (not spotted) breast.

BOX-ELDER
(ASHLEAF
MAPLE)
NORTHERN
CARDINAL
male
SOUTHERN
ARROWWOOD
BARRED OWL
SWAMP
WHITE OAK
SMOOTH
(TAG)
ALDER
RIVER BIRCH
WOOD DUCK
male
BUTTONBUSH
SWEETGUM
LOUISIANA
WATERTHRUSH

BALDCYPRESS SWAMP FOREST (P. 107)

Tall Baldcypresses laden with Spanish Moss give this swampy forest an eerie, primeval look. Cypress "knees" are protruding root systems.

REDBAY

To 60 ft., with a widely spreading, often irregular crown. Leaves evergreen, alternate, elliptical, aromatic.

NORTHERN PARULA

A small, treetop warbler that nests in Spanish Moss. Males have a reddish band across yellow breast. Song a buzzy trill. Also in northern forests.

BALDCYPRESS

A majestic tree, up to 120 ft. tall, that often dominates southern swamps. Very soft needles, in 2 rows on twig, turn brown and drop in fall. Cones very rounded. Bark is shed in strips.

PONDCYPRESS

A distinct variety of Baldcypress, with needles that are much more scale-like. Common to stagnant swamps.

FOX GRAPE

One of many vines that grow abundantly in southern swamps. See Pl. 31.

PROTHONOTARY WARBLER

A brilliant orange-yellow warbler of southern swamps. Males bright glowing orange-yellow, females greenish yellow. Nests in cavities. Song a one-pitched *seet-seet-seet-seet.*

PILEATED WOODPECKER

Crow-sized woodpecker with red crest and white under wings. Voice is a loud, strident *keek-keek-keek.* Widely distributed in forests throughout the East, but abundant in southern swamps.

SWAMP TUPELO

A variety of tupelo (see also Pl. 10) common in southern coastal swamps. Leaves oblong, with sharp point.

AMERICAN ALLIGATOR

Hard to misidentify. Long, rounded snout; bulky body; long tail. Often basks on bank. Sometimes swims with only its eyes and nostrils above water.

COTTONMOUTH or WATER MOCCASIN

A large (to 6 ft.), bulky pit viper with dark diamond patterning. Mouth white. ***Poisonous.***

male
NORTHERN
PARULA
WARBLER
REDBAY
SPANISH
MOSS
BALDCYPRESS
FOX GRAPE
PROTHONOTARY
WARBLER
male
AMERICAN
ALLIGATOR
PILEATED
WOODPECKER
male
SWAMP
TUPELO
PONDCYPRESS
COTTONMOUTH

SOUTHERN MIXED HARDWOOD SWAMP FOREST (P. 109)

This is a high-diversity forest with many oak species. Species composition varies widely from site to site.

SWAMP OAK

To 80 ft., with a clearly defined oval crown. Leaves have many shallow lobes with pointed tips, turn reddish brown in fall.

OVERCUP OAK

To 80 ft., with a broadly spreading crown. Leaves have deep, irregular lobes, widest in middle and not sharply pointed at tips; turn yellowish brown to red in fall.

SUMMER TANAGER

Males completely red with a white bill and no crest. Females greenish yellow. Common in oaks throughout southeastern forests.

WATER OAK

To 100 ft. Leaves oblong, thickest in middle, unlobed, wavy on edges.

SOUTHERN RED OAK

To 80 ft. Leaves widest in middle, with deep lobes and sharp points. Not strictly a swamp-dweller, also common on uplands and in mesic forests.

AMERICAN HOLLY

To 70 ft. Evergreen, dark green leaves with sharp points. Fruit bright red. Also common in riverine forests.

PAWPAW

To 30 ft., but often grows as a shrub. Leaves alternate, large (to 12 in.) and oval, with point at tip; turn yellow in fall. Flower pinkish red. Fruit large (to 5 in.) and fleshy, yellowish green.

CAROLINA WREN

Small, active, chestnut-colored bird with a white eye-stripe. Tail often cocked. Song a melodious *cheedala, cheedala!* Very common in woodland understory throughout the East, except in northern states.

male
SUMMER
TANAGER
SWAMP
OAK
OVERCUP
OAK
WATER
OAK
SOUTHERN
RED OAK
AMERICAN
HOLLY
CAROLINA
WREN
fruit
PAWPAW

SUBTROPICAL FOREST (P. 111)

Confined to Florida, this forest is a mixture of West Indian, Central American, and North American species.

COCONUT PALM

To 100 ft.; fronds 18 ft. long. Coconuts are green at first, then brown.

FLORIDA ROYALPALM

Straight, up to 100 ft., trunk ending in a wide green stem from which 10–12-ft.-long fronds radiate. Clusters of small purple fruits grow from stem. Native.

FLORIDA STRANGLER FIG

A woody vine that grows around host trees such as mahogany. The many grayish tendrils surround the host tree and fuse. Vine hardens and thickens, preventing growth of host tree. Dark, shiny, green leaves are oval with sharply pointed tips. Native.

GREEN ANOLE

Arboreal lizard that can change body color from green to brown. Sharply pointed snout and loose flap of reddish skin (dewlap) hanging from throat.

PAPAYA

Large (24 ft.), palmate, evergreen leaves, deeply lobed. Large pinkish white, tubular flowers. Oval, greenish orange fruit.

GUMBO-LIMBO

To 60 ft. Bright orange-red bark, very smooth. Leaves compound. Leaflets very shiny; oval, with sharply pointed tips.

BROMELIAD

Bromeliads are tropical air plants (epiphytes) that grow on the branches of trees. Tiny white flowers enclosed in red bracts. Leaves stiff and sharp, very similar to those of pineapple.

ZEBRA LONGWING

Heliconius butterflies are abundant in the New World tropics; this is one of the few to reach North America. Slender wings, black with metallic yellow stripes. Flies slowly.

WEST INDIES MAHOGANY

To 60 ft. Compound leaves with smooth, untoothed leaflets with pointed tips. Bark rough, brown.

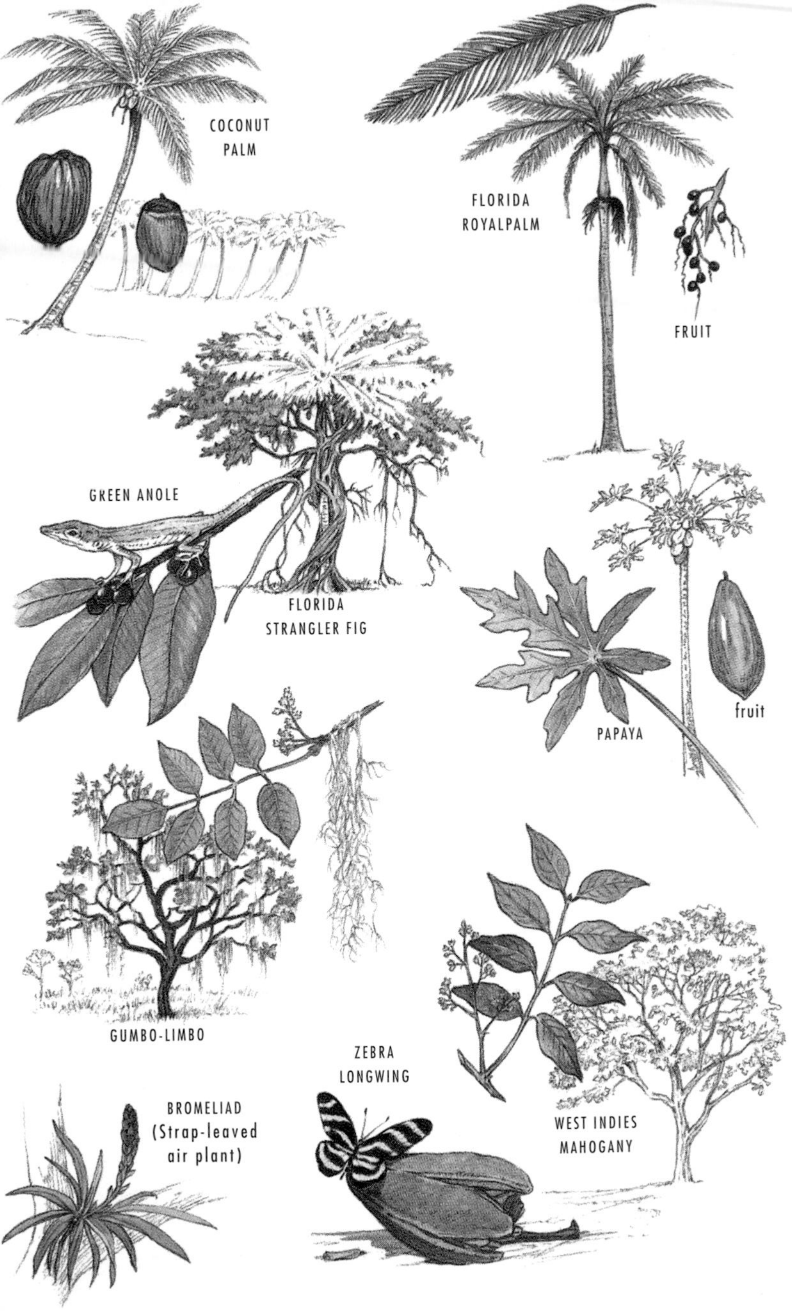

COCONUT PALM

FLORIDA ROYALPALM

FRUIT

GREEN ANOLE

FLORIDA STRANGLER FIG

PAPAYA

fruit

GUMBO-LIMBO

ZEBRA LONGWING

BROMELIAD (Strap-leaved air plant)

WEST INDIES MAHOGANY

MANGROVE SWAMP FOREST (P. 113)

This coastal forest lines the Florida Keys and the southern Florida mainland, especially Everglades National Park. Mangroves are tropical trees, tolerant of salt water, mostly small in stature.

BUTTONWOOD or BUTTON-MANGROVE
To 60 ft. Elliptical, untoothed, leathery leaves. Tiny greenish flower clusters. Brown, buttonlike fruits. Grows on raised areas, not directly in salt water.

BLACK MANGROVE
To 30 ft. Grows in dense thickets on mudflats beyond the line of Red Mangroves. Roots send up scores of fingerlike pneumatophores. Leaves opposite, elliptical, leathery. Small, white flower clusters at branch tips.

WHITE MANGROVE
To 60 ft. Opposite, oval, leathery leaves. Flowers small, white, at branch tips. Fruits brown. Does not grow directly in salt water.

RED MANGROVE
Most aquatic of the mangroves. Almost shrublike (20 ft. tall) but can grow up to 80 ft. in true tropics. Reddish prop roots grow directly in sediment covered by salt water. Leaves elliptical, thick, leathery. Delicate, pinkish white flowers. Long green seedpods. Plate shows seedling anchored in sediment.

SEA-GRAPE
Very rounded, thick, leathery leaves. Widely spreading tree, grows to 25 ft. Fruits clustered like grapes at branch tips.

ROSEATE SPOONBILL
Pink-white, heronlike bird with a long, flat, spoonlike bill. Adults have very rich red shoulders, pink wings and tail. Legs red.

WOOD STORK
Very large white bird with a decurved bill and dark naked head. Wings lined with black, visible in flight.

MANATEE
A large gray mammal of the mangroves and weed-infested waterways. Doglike head; paddle-like forefeet; large, flattened tail. Feeds on underwater vegetation, especially Water Hyacinth.

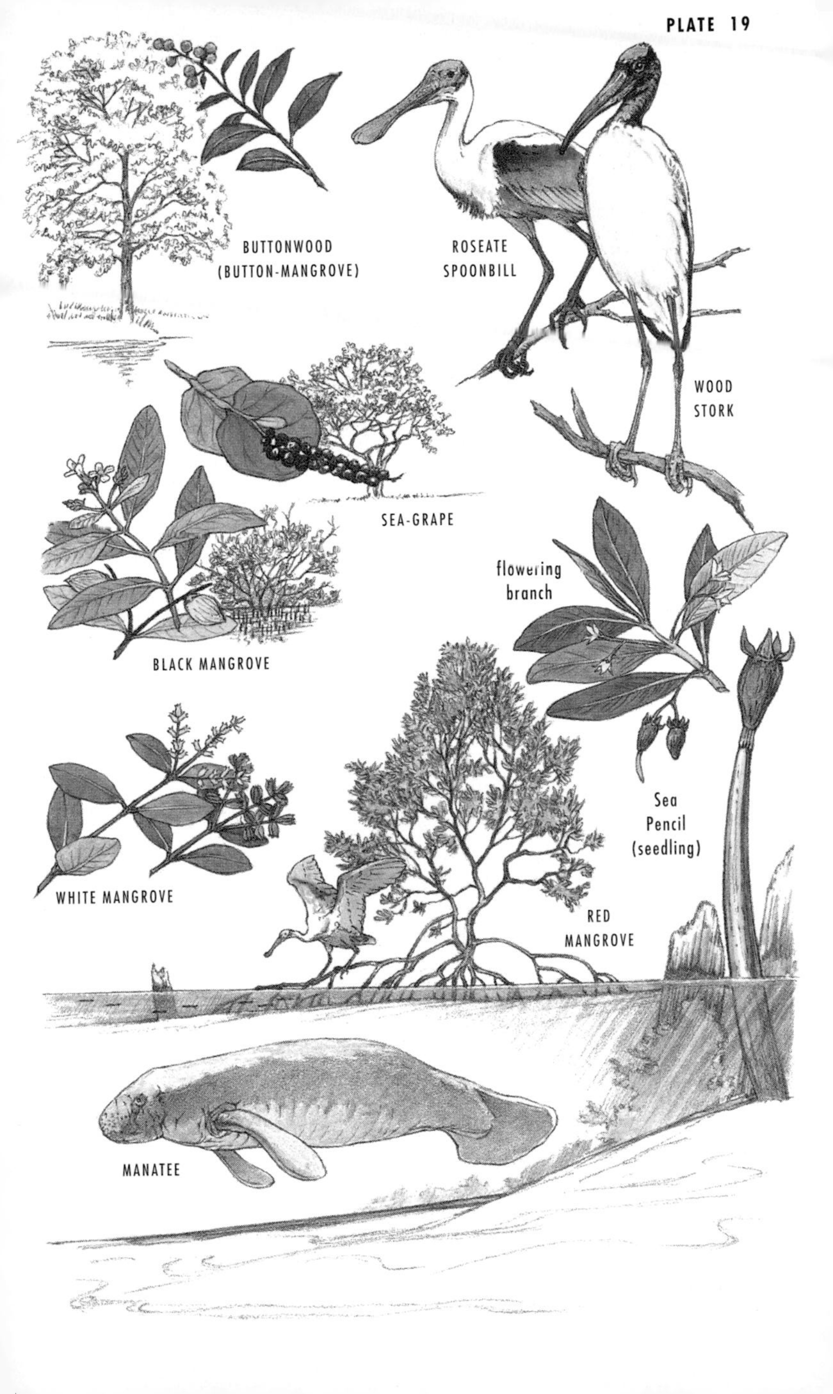
BUTTONWOOD
(BUTTON-MANGROVE)
ROSEATE
SPOONBILL
WOOD
STORK
SEA-GRAPE
flowering
branch
BLACK MANGROVE
Sea
Pencil
(seedling)
WHITE MANGROVE
RED
MANGROVE
MANATEE

Disturbance and Pioneer Plants

Eastern North America is a patchwork of landscapes. Rather than an unbroken tract of mature forest, there are woodlots of various ages, fields, orchards, and, of course, human habitations. Most habitat alteration has been the work of humans. *Homo sapiens* undoubtedly ranks as the single most important influence on nature today. However, long before the first Asian peoples traveled eastward across the Bering land bridge to settle in the New World, and long before the first Europeans set out westward on their voyages of discovery and settlement, natural disturbances were routine events, producing a diversity of landscapes.

What Is Disturbance?

Disturbance results from any factor that alters a landscape. Fire, wind, disease epidemics, and many other causes of disturbance are common in nature. Landscapes may be temporarily or even permanently changed as a result.

Ecological disturbance has three basic components—frequency, predictability, and magnitude. Frequent low-level fires keep certain pines dominant in areas that, in the absence of fire, would change to become oak forests (see Chapter 3). Disturbances that occur fairly regularly, such as seasonal fires, result in plants and animals gradually becoming adapted to the disturbance. Jack Pine (see p. 71) requires fire to release its seeds. Kirtland's Warbler nests only in young Jack Pine Forests, making it also dependent on periodic fire. A highly unpredictable disturbance, such as the sudden incursion of a species new to the community (like the Gypsy Moth), can create short-term havoc. The magnitude of the disturbance influences the rate and extent of recovery. Small-magnitude disturbances permit the community to return to its previous state. Large-magnitude disturbances may permanently change things. Evidence increasingly indicates that

Landscape Diversity. Aerial view of landscape patches typical of much of eastern North America. Varying-sized forest islands are connected by corridors, with agricultural and old fields interspersed.

a comet or asteroid struck the earth 65 million years ago, disturbing the planet's climate and hastening extinction of dinosaurs. The plant and animal communities on earth were permanently altered and another fauna, the mammals, diversified.

The single largest disturbance to occur in North America within the past several thousand years has been the Ice Age, four periods of glaciation. This major climatic event drastically affected all habitats and reshuffled ecological communities. The last major glacier to invade was the Wisconsin, which began its retreat approximately 20,000 years ago. During that glaciation, so much of the earth's water was ice that sea level was about 330 ft. lower than it is today. Following glacial retreat disturbed habitats reorganized. Initially the landscape of New England resembled arctic tundra, a flattened landscape of lichens and mosses with small isolated clumps of stunted willow and birch. As the climate warmed and glacial retreat continued, tundra was gradually replaced by spruces mixed with sedge, with the land eventually taking on more of the look of today's Boreal Forest (see below). Finally, to areas that had once had two miles of ice atop them, the oaks, maples, and pines returned and the present-day deciduous and mixed conifer forests became recognizable.

During the Ice Age, when Woolly Mammoths, American Mastodons, and Sabre-toothed Cats roamed eastern North America, what became of the forest types we now recognize? The answer, in brief, is that these forests did not exist as such in their present locations, and the ranges of the various species were widely scattered in the southern latitudes. Approximately 18,000 years ago, northern pines, oaks, and hickories were confined to the Southeast. Boreal Forests of spruce and fir existed as far south as northern Mississippi, Alabama, and across the Carolinas. Spruce was concentrated to the north, fir to the south. Jack Pine, now found primarily on poor glacial soils in boreal regions, was widespread in the East south to Georgia, and White Pine and Tamarack (American Larch) were present in Louisiana. Mesic forest species were confined to refuges with favorable microclimatic conditions, such as cool moist ravines. Following glacial retreat, plant and animal species did not move north together, as a community, but individually. American Chestnut moved north quite slowly, probably because its heavy seeds are dispersed by mammals; as a result, Chestnut did not return to New England until 2000 years ago. In contrast, White Pines reinvaded Massachusetts about 9000 years ago. American Beech reached upper New York state 7000 years ago and its western limit in Michigan and Wisconsin 3000 years ago. It has been calculated that maples and elms, both with wind-dispersed seeds, moved north following glacial retreat at rates of about 650 and 800 ft. annually, reaching their current status in the Great Lakes region 4000 years ago. Modern forest communities represent rearrangements of many species that were not together until relatively recently.

Disturbance is a permanent feature among the processes of nature. This chapter tells how to recognize and evaluate the forces that alter landscapes and describes the process of *ecological succession,* or *vegetation development,* whereby landscapes recover from disturbance.

What Causes Disturbance?

WEATHER FACTORS: All organisms, whether plant, animal, or microbe, are subject to the constant effects of their external environments.

Windthrow. The forces of isolated winds or major hurricanes can bring down canopy-sized trees, creating gaps in the forest.

Prolonged heat, cold, or drought can affect some species more than others. Small insectivorous birds such as the Golden-crowned Kinglet may be significantly reduced in population by a winter cold snap or severe ice storm. Such an event is a small-scale disturbance since it affects only one or a few species. A heavy wet snowfall can break tree branches or cause whole trees to topple, creating forest gaps (see p. 34). Clumps of fallen trees scattered throughout a forest represent a medium-scale weather disturbance. Finally, large-scale disturbances are caused by major weather events. On September 21, 1938, a severe hurricane flattened forests throughout southern and central New England. Other major hurricanes affected this region in 1815 and 1635. These events cause major regional disturbance and most parts of the landscape take years to recover.

In forests, *windthrow* is an important disturbance factor, creating gaps of varying sizes that permit light to enter, churning up the soil and providing new sites where seedlings and saplings can grow.

FIRE: Many habitats experience fire as a recurring event and have adapted to it. Michigan Jack Pine Forests, Cape Cod Pitch Pine–Oak Forests, and Georgia Longleaf Pine Forests all represent habitats where fire has a major influence. Learn to recognize the evidence that fire has occurred (see p. 30). Severe fires causing immense damage are almost always accidents of human carelessness. Fires caused by lightning usually do only moderate damage unless there are strong winds. In habitats where periodic fire is normal, recovery tends to be relatively quick. Plant species such as Pitch Pine and Longleaf Pine focus their life cycles around fire (see pp. 92 and 94).

Microhabitat. Following root decay, mosses invade and colonize a raised area of forest created by windthrow.

Purple Loosestrife Among Cattails. The alien Purple Loosestrife has spread rapidly in freshwater marshes, often outcompeting native cattails.

FLOODING: Forests border rivers and streams. During spring, when winter snow melts and rains soak the landscape, flood waters rise and cover the forest adjacent to flooding rivers. Flooding may undercut soil and remove areas of forest. But flooding is a dynamic process that establishes a rough balance between destruction and creation. Though they can cause much damage, floods also deposit rich soils, which are quickly colonized by plants and animals. Sediment bars created during flooding provide a foundation for the eventual reestablishment of bottomland forest. Many species of plants are well adapted to withstand periods of immersion during the flooding season.

ALIEN SPECIES: Species interact, constantly exerting influences upon one another. When a species enters an area to which it is not native, it often cannot cope with its new environment and quickly becomes extinct. However, it may succeed and exert a very dramatic effect upon the ecology of that area. Without natural predators, parasites, and competitors to restrain it, the invading species may become disproportionately abundant. The European Starling has in all likelihood been a significant force in reducing populations of cavity-nesting birds such as woodpeckers, Eastern Bluebirds, Tree Swallows, and others. Bird communities have changed as a result of the addition of the Starling. Gypsy Moths were accidentally released in eastern Massachusetts shortly after the beginning of the 20th century. Within 30–40 years, this species caused major cyclic destruction of oaks, pines, and other tree species throughout the north and central United States. The blight that destroyed the American Chestnut in effect eliminated a dominant tree species from major tracts of eastern forest (see p. 82). Kudzu-vine, a native of Japan, was brought to the United States in 1911.

Since then it has spread rapidly through the Southeast, destroying whole woodlots by covering and shading the trees. In northeastern freshwater marshes, Purple Loosestrife is rapidly replacing native cattails and, in thickets and woodlots, Asiatic Bittersweet is outcompeting our native climbing Bittersweet.

HUMAN ACTIVITIES: Perhaps the most dramatic ecological event of the past 300 years, since the time of settlement by Europeans, has been the massive cutting of virtually the entire eastern forest. Forests were felled to supply lumber for fuel and building material and to create pasture and agricultural land. Extremely few are the areas of forest that have never experienced the ax and saw. Prior to Europeans, Native Americans set fires regularly to underbrush throughout the East. These fires effectively opened the forest to make hunting, travel, and farming easier. As the Midwest became our primary agricultural base, much of the cut-over eastern forests have grown back. This process of forest reestablishment continues today. Many eastern forests are now managed for lumber production and tree plantations are common in many areas. Managed forests may be clear-cut or certain trees may be selectively removed. Whatever management technique is used, management represents yet another type of disturbance.

POLLUTION EFFECTS: In the strictest sense, pollution belongs with the discussion of human activity above. However, pollution is very distinct from human disturbance caused by cutting for farming or lumber. Air pollution may affect forest habitats in ways that are as yet poorly understood. Components of air pollution react in the atmosphere to create *acid rain*, the effect of which is to remove chemical nutrients from the soil. The cumulative effects of the soil becoming increasingly acidic and poorer in nutrients could significantly affect the species composition of forests experiencing this stress.

OTHER FACTORS: Disturbance may be caused by geologic events such as earthquakes and volcano eruptions. The eruption of Mt. St.Helens on May 18, 1980, flattened forests and created floods and mudflows. Although recovery has begun, the area will show the effects of the eruption and its aftermath for many years to come.

ECOLOGICAL SUCCESSION: THE PROCESS OF VEGETATION DEVELOPMENT OVER TIME

All disturbances eventually end. Some types of disturbance such as farming may occur continuously on a single site for years. Eventually, however, the farm may be abandoned. Other disturbances, like a fire or hurricane, could take only a few hours to alter the landscape. Once a disturbance has ceased, a process of ecological change begins. The disturbed area represents a new habitat available for plants and animals to colonize. Often, the physical conditions of disturbed habitats are severe. A burned field may bake in the sun without benefit of shade from trees. Winds, uninterrupted by tree branches and trunks, cause high rates of evaporation and water stress for plants. Nonetheless, many species are well adapted to exploit disturbed environments.

Since newly disturbed areas are often not very hospitable, the first species to invade the area following the disturbance are physiologically tough. They grow quickly and reproduce before other species enter and shade them out. These colonizing species, called *pioneer plants*, are replaced by others that take longer to complete their life cycles. Finally, species occur that continue to grow in the area indefinitely until the next disturbance. This process of vegetation development, the replacement of species by other species over time, is called *ecological succession*.

Succession is roughly predictable in any given area. An abandoned field will be invaded by certain species before others appear. A pattern for the region will emerge. The succession will usually result in the regrowth of species identified as most characteristic of the area. Remove an Oak-Hickory Forest down to the bare soil and, in time, another Oak-Hickory Forest is likely to grow back. Prior to this, however, a series of species will probably grow on the site, ranging from Common Ragweed and Poison-ivy to Staghorn Sumac, Gray Birch, and Eastern Red Cedar.

Old Field Succession

The most common form of ecological succession in eastern North America is called *old field succession*. You can observe old field succession anywhere from Florida to Canada, from coastal North Carolina to the midwestern boundary between prairie and forest. Abandoned farmland, burned forest, or even roadsides are populated by species adapted to live and reproduce in disturbed areas. Succession is a process involving many very different species. By observing old field succession you can learn much about the ways plant species adapt to differing environments.

Field Marks of Succession

Succession is most easily understood as a series of developing plant communities in which various species populations are usually replaced by others over time. The most characteristic successional communities in eastern North America are described below.

Selected Indicator Species

It is impossible in a field guide of this size to illustrate and describe all of the plant species found in old fields throughout eastern North America. The mixture of species occurring on a given site is strongly influenced by chance. There is very much a "lottery" element to vegetation change. The species included here are only some of the most common that you are likely to encounter almost anywhere in the region. Rather than focus entirely on identification, I stress the characteristics, or adaptations, these species have developed to survive and reproduce in their open field environments. To identify the many species not included here, refer to *A Field Guide to the Wildflowers* and *A Field Guide to Eastern Trees.*

Native and Alien Species

Fields undergoing vegetation change usually contain many alien (introduced) plant species as well as those native to North America. The growing conditions in open fields require colonizing species with seed longevity, wide dispersal powers, and the ability to grow quickly and reproduce (see below). These conditions are met by many herbaceous and grass species, mostly from Europe, that have been brought accidentally (and occasionally on purpose) since the time of European colonization. Alien species are often very good competitors against native species. When they are both successful and undesirable, we call them weeds. It is not uncommon for aliens to outnumber native species in newly opened fields. However, as vegetation change proceeds, native species gradually replace the aliens. This is

Butter-and-eggs. One of many European species that successfully colonized much of eastern North America. Colorful flowers attract pollinating insects.

because native trees and shrubs shade out the early herbaceous invaders. Native species, of course, have the advantage of already being present in large numbers. In total, approximately 18% of the species occurring in the Northeast are aliens.

Why do some introduced species fare so well in competition with native species? Many aliens experience what can be called an "ecological release" from the competitors, predators, and parasites of their native land. In open sites, they have advantages over native species that are restrained by other species that can compete with them or eat them.

Early Old Fields

The term "old field" refers to any abandoned field or disturbed terrestrial habitat that has a well-developed soil base. Old field succession often begins with bare soil. Not all levels of disturbance are sufficiently severe to remove all vegetation down to the bare soil; succession also occurs when a pasture or agricultural field is abandoned. But, when bare ground is available, it is quickly colonized by herbaceous plants whose seeds were present in the soil. These species are often thought of as obnoxious weeds but are more properly termed *pioneer species,* and they develop quickly in abundant sunlight.

The Pioneer Communities

Early vegetation changes in an old field become evident when the field is covered entirely by plants such as Common Ragweed, Horseweed, Fireweed, or Crab Grass (see below). Such fields are

Rapid Old Field Succession. Field on left is three years post-abandonment, and is dominated by Daisy Fleabane, a common composite. Field on right is two years post-abandonment and dominated by stunted Common Ragweed.

usually only one year old (following disturbance or abandonment). Fields two and three years or older include many species of asters and goldenrods plus numerous alien species. Goldenrods have dense conspicuous clusters of bright yellow flowers. Individual species may have plume-like, elm-branched, clublike, wandlike, or flat-topped shapes (see *A Field Guide to Wildflowers* and this volume, p. 176). Asters are daisy-like flowers, often densely clustered. Species vary in flower color: most asters are violet or blue but some are white and a few are yellow. Grasses are also common among the goldenrods and asters, and other herbaceous species occur as well. The pioneer plants that were dominant during the first year tend to disappear. To see all the species you need to search carefully. Ragweed or Horseweed, both annuals often dominating first-year fields, may be present only as stunted plants beneath the goldenrod and aster stems. Woody species become apparent by the third or fourth year, but only as seedlings and saplings.

Newly exposed soil is anything but devoid of life. It is estimated that arable ground contains as many as 100,000 dormant seeds per square meter. Among them are the seeds of many colonizing herbaceous and grass species. Many of these seeds have extreme longevity and are able to germinate and grow when the soil is exposed. For example, archeologists in Denmark have successfully germinated a few relic seeds from two annuals that were found in a dig estimated to be 1700 years old (one was Lamb's-quarters, common in our area). Seeds from most species found later in succession are much shorter lived than early successional species. As vegetation change continues, seeds are often brought by wind and birds (see p. 272 and Chapters 7 and 8).

Occasionally, bare soil will be quickly invaded by wind-dispersed seeds of maple or birch, and a dense stand of these species may develop, rather than the goldenrods and asters described above. Vegetation change is always subject to chance events.

The Perennial Herbaceous Plant Community

This community has an abundance of perennial herbaceous and grass species. Goldenrods and asters no longer uniformly cover the field. Grasses become very prominent and usually include bunch grasses such as Broom-sedge. Seedlings of shrub and tree species begin to germinate but are very tiny and easily overlooked. This stage occurs when the field is from 3 to 10 years old.

The Perennial Herbaceous-Woody-Plant Community

These fields range in age from 10 to 60 years or more post-abandonment. Herbs and grasses become much less obvious. Clumps

Butterflyweed. Within 4–5 years an old field becomes a mixed community of herbs, legumes, and grasses. Butterflyweed, of the milkweed family, is conspicuous in this photograph and typifies the kinds of wildflowers that populate successional fields.

Woody Invasion. Within 10–12 years the old field is populated by many invasive woody species, such as Eastern Red Cedar, shown here. The field soon becomes a patchwork of woody thickets and open, grassy areas.

Canopy Development. As succession continues, canopy tree species overtop species of lesser stature. This mature Eastern Red Cedar will soon be shaded out by the developing broadleaf canopy.

Woody Closure. Eventually woody species replace all grasses and open-area herbaceous species and the field becomes a young, densely populated woodlot. Many of these species will eventually be replaced by slower growing, more shade-tolerant species.

Young Forest. Though the canopy is developed and the forest is closed, the trees remain small in stature, their boles closely spaced. Much light still penetrates to the herb layer.

Closed Forest. A mature forest has more widely spaced trees of larger size. Photo, taken in spring (prior to full canopy leafout), shows a mature White Oak (right) among an understory of Flowering Dogwood with Mayapple as the dominant herb species.

of crowded trees and shrubs shade the ground. Patches of shrubs such as sumacs, Multiflora Rose, Shadbush, and Meadowsweet may be present, as well as Northern Bayberry, huckleberries, and Sweet-fern. Vines such as Poison-ivy and Virginia Creeper may grow among the woody species. Large patches of trees and shrubs are interrupted by areas of grass. The habitat is very patchy in appearance. Indicator tree species (see following sections) include various pines, Eastern Red Cedar, aspens, birches, cherries, Sweetgum, or Tuliptree, depending on location and site characteristics. Very old fields begin to look like woodlands, with dense clumps of thin-boled (slender) trees.

Other Field Marks of Vegetation Disturbance and Change

Uniform-aged Stands

Uniform-aged stands are not uncommon. Species such as White Pine, Quaking Aspen, Black Walnut, or various birches, maples, and southern pine species begin to sprout from seeds simultaneously following disturbance, especially fire. When grown, these trees are all relatively neatly spaced and equal in height, resembling a plantation.

Stump- and Root-sprout Growth

Should the above-ground parts of trees in a forest be removed, as by clear-cutting or a violent windstorm, the trees may survive below ground as root systems. Many species, such as the oaks and maples, regrow their above-ground parts from surviving root systems. Often termed "coppice shoots," these regenerated trees are recognized by their characteristic cluster of thin trunks emanating from a common base. Some species, such as Black Locust and Quaking Aspen, grow quickly from root sprouts. The underground network of roots periodically sends up new shoots or ramets (see p. 32).

Old Field Succession — A Detailed Look

Below are descriptions of some of the most abundant annuals in our region.

NATIVE ANNUALS PL. 25

COMMON RAGWEED, the bane of hay-fever sufferers, is one of the finest examples of an annual. The wind-blown pollen is a principal cause of late-summer hay fever. The tiny green flowerheads consist of small male flower clusters on spikes and larger, less con-

Common Ragweed. The tiny green flowers are inconspicuous, as they are wind pollinated and thus do not act to attract insect pollinators.

spicuous female flowers at the leaf bases and branch forks. Flowers, though they occur on long spikes (male flowers only), are not showy and insects are not attracted to them. Pollen is dispersed by wind (see p. 272). Leaves are deeply and ornately lobed, an adaptation that may have evolved to permit maximum exposure to sunlight with minimal cost of tissue production. Common Ragweed can grow nearly 5 ft. tall but is often much shorter. Ragweed growing in a field one year following abandonment or disturbance will typically be very tall. However, the following summer, only stunted Ragweed plants will be seen. Decomposing Ragweed poisons the soil, stunting the growth of newly germinated Ragweed seeds, as well as those of other plants. Ragweed seeds are heavily fed upon by birds, especially sparrows.

HORSEWEED is extremely common in early old fields, roadsides, and other disturbed areas. The flowers are inconspicuous and greenish white or yellowish. Seeds, with slender pale bristles, are dispersed by wind. Horseweed grows occasionally as a winter annual. Its seeds germinate in late summer or early autumn and the plant overwinters as a rosette. Flowers bloom the following spring. Like Ragweed roots, Horseweed's decomposing roots inhibit its own offspring as well as other plants.

DAISY FLEABANE has flowers resembling miniature daisies, and this species is often confused with asters (see composites, p. 175). A winter annual, Daisy Fleabane blooms throughout spring and summer.

TUMBLEWEED is a spreading plant with small, oval leaves and small greenish flowerheads abundantly located at branch tips and leaf axils. This species, which probably was originally native to the

Great Plains, colonized the East as forests were cut and open areas became abundant. Its seeds are scattered by wind-blown tumbling plants, a characteristic common to other species of the western Great Plains.

BEGGAR-TICKS, or **STICKTIGHTS** (Pl. 20), are very common. About 25 species occur in our area, but *Bidens frondosa* is the one most frequently encountered. The conspicuous yellow flowerheads are densely packed with elongate hairy seeds called achenes, easily recognized by their barbed spines that help the seed cling to the fur of passing mammals. Most of the 25 Beggar-ticks are found in moist soils, but *B. frondosa* occurs in dry open fields and along roadsides as well. Not all Beggar-ticks are annuals; some are perennials.

SPURGE, or **EYEBANE,** is recognized by its opposite leaves, usually with red blotches. Milky sap oozes from the stem when broken. Flowers are very small and easily overlooked because they lack both petals and sepals. Petals and sepals are most useful to plants as insect attractors for pollination. Their absence in Spurge attests to the plant's dependence on wind pollination. Both male and female flowers occur, ranging in color from pale whitish to green. Each flower produces 3 seeds. This species occurs in disturbed areas and dry soils throughout the area. Another common species is Milk-purslane, which is a prostrate herb.

PENNSYLVANIA SMARTWEED, or **KNOTWEED,** like all members of its genus (*Polygonum*), is recognized by its conspicuous jointed stems and flower shoots. About 11 species are common in our region and, though a few are perennials, most are annuals. Aquatic species occur as well as terrestrial ones. The tiny pink flowers are in dense spikes at the branch tips. Other species have rose-colored or white flowers. Seeds are black and shiny. Approximately 30 species of birds, including many blackbirds and sparrows, feed on upland smartweeds. Bird-dispersal is a very important factor in the distribution of the smartweeds.

POOR-MAN'S-PEPPER is also called **PEPPERGRASS.** It is a member of the mustard family and can grow as either an annual or winter annual. The flowers are inconspicuous, green or white in color. Seeds are in tiny heart-shaped pods, each containing 2 seeds. This species prefers dry soils. About 25 species of Peppergrass occur, many of which are alien, and some of which are biennials.

ALIEN ANNUALS **PLS. 25, 26**

GREEN AMARANTH (Pl. 25) is also sometimes called **PIGWEED** (but see Common Lamb's-quarters below). Many amaranth species have clusters of flowerheads growing on a central spike, a growth form that you will see commonly among old field herbs. Flower spikes are on the plant from August through October. As many as 100,000 to 200,000 seeds can be produced on a single plant. **COMMON LAMB'S-QUARTERS** (Pl. 26) is also frequently referred to as **PIGWEED** (see Amaranth, above). It is very common throughout our region. Lamb's-quarters is most easily recognized by its red-streaked stems and arrow-shaped, broadly toothed leaves. Its flowers, inconspicuous green clusters on the upper part of the main stem, are pollinated by wind. Many birds, including doves, finches, sparrows, and Horned Larks, eat the seeds, and as many as 75,000 seeds may occur on a single plant.

COMMON CHICKWEED (Pl. 26) is a cosmopolitan species found in open fields with rich soil. It has opposite, oval-shaped leaves that are not toothed. The star-shaped flowers are surrounded by hairy sepals. The growth form is short and reclining and the plant spreads by roots sent out from the stem joints. Blackbirds, quail, doves, and sparrows feed heavily on the seeds.

Several species in the mustard family are among the most commonly encountered alien annuals in old fields. **WILD MUSTARD,** or **CHARLOCK** (Pl. 26), is an early-blooming species recognized by its yellow, four petaled flowers in clusters at the branch tips, and smooth seedpods. The leaves are alternate and lobed, with short teeth. Similar species, also alien, include **WHITE MUSTARD,** which has hairy pods and more deeply lobed leaves, and **BLACK MUSTARD,** a winter annual that has lance-shaped upper leaves.

Spice mustard is obtained from tiny, black, oily seeds. Mustard seeds are long-lived in the seed bed. Under favorable conditions, the seeds germinate and Wild Mustard may cover an entire early old field. Seeds are eaten by many birds as well as voles and cottontail rabbits.

COW-CRESS, or **FIELD PEPPERWEED,** and **SHEPHERD'S PURSE** (Pls. 20, 26) are also members of the mustard family. Both are very similar to Poorman's-pepper (see above) but Cow-cress has very flattened pods and unlobed large basal leaves, and Shepherd's Purse has basal leaves arranged in a dandelion-like rosette.

Two plants in the tomato family are widely distributed alien annuals. **COMMON NIGHTSHADE** is identified by its triangular-shaped, opposite leaves and flowers that dangle in clusters at the branch tips. The flowers have 5 white petals and prominent yellow stamens. Black berries form in late summer. Nightshades, many of which are poisonous to humans, number approximately 1 000 species, mostly tropical. Not all are annuals and native species occur as well as alien. The succulent berries, miniature "tomatoes," are eaten by many birds, including doves, quail, thrashers, thrushes, meadowlarks, and finches. Among the mammals, raccoons and skunks feed on the berries.

The other common member of the tomato family is **JIMSONWEED** (Pl. 26). This unmistakable plant features very large, trumpetlike flowers, white to violet in color. Pollination is by insects, not wind. Leaves are alternate, large, and toothed, and the plant may grow to 5 ft. The root is thick but shallow. Seeds are pitted, blackish, and kidney-shaped, contained in conspicuous green spiked pods. Like many plants with tropical origins, Jimsonweed is poisonous, an adaptation that significantly reduces grazing pressure (see p. 259).

BLACK BINDWEED, or **WILD BUCKWHEAT,** is an alien member of the highly successful buckwheat family (Polygonaceae). The plant grows prostrate, with wide, alternate, arrowheadlike leaves and elongate clusters of greenish white flowers. This species is very similar to **CLIMBING FALSE BUCKWHEAT,** a native species.

CRAB GRASS is one of several abundant alien annual grasses notorious for invading lawns as well as old fields. The central spike is thin, with several thin flowering spikes radiating horizontally, like "fingers."

The Life Cycle of Annuals

Bare soil is usually colonized first by herbs and grasses called annuals, whose seeds were present in the soil (see above). There are relatively few species of annuals, so many first-year fields are totally covered by only one or two species. Among the most common annuals are Common Ragweed, Horseweed, and Crab Grass.

Annuals complete their entire life cycles in a single growing season. They sprout from seed in spring, grow to maturity and flower by late summer, drop seed in the fall, and survive the winter as seeds. Adult plants perish entirely after the first frost. Some species are *winter annuals.* These plants set seed in late summer and quickly germinate in the waning heat of fall days. They overwinter as compact, prostrate rosettes, tightly hugging the ground. In spring they flower and set seed, repeating the process. Populations of annuals grow quickly following a disturbance that exposes the soil. Seeds of annuals are long-lived and remain dormant in the soil until they encounter suitable conditions for germination.

Annuals, because of their brief life cycles, are prodigious reproducers. Pollen is usually windblown, enhancing the probability of cross-pollination, though some annuals occasionally self-pollinate. Many annuals have inconspicuous small green flowers that bloom from mid- to late summer (see p. 166). An individual plant may set thousands of seeds. The probability of any given seed germinating, growing, and reproducing is very small, but because there are thousands of seeds per plant, some are successful. Seeds of annuals are like lottery tickets. Many are made and distributed but few are winners.

Annuals are adapted to exploit disturbed open environments. They are physiologically hardy and able to withstand high temperatures and water stress better than most other plants. Although hardy, annuals are poor competitors against other, more long-lived plant species. They require much light and are easily outcompeted by other herbs and grasses that soon invade their open field environments (see following sections). Some of them produce chemicals that inhibit the growth of other species, a characteristic called *allelopathy*, but these chemicals also poison the soil sufficiently to inhibit the growth of the very annuals that produced the toxins, a characteristic called *autotoxicity*.

Best described as *opportunistic species*, annuals are characterized by wide dispersal powers, exceptional seed longevity, and the ability to complete their life cycles rapidly in temporary, high light-intensity habitats. They are, in all ways, ephemeral.

BIENNIALS **PL. 21**

QUEEN ANNE'S LACE, or **WILD CARROT,** is an alien biennial originally from Europe, now abundant in our region. It is easily identified by its flattened wide "plate" of minute white flowers, which together give the lacy look responsible for the plant's name. The flower cluster is called an umbel, and the plant is a member of the parsley family (Umbelliferae). Its leaves are deeply and elaborately lobed, like those of Ragweed. The plant grows to a height of 3 ft. and is very conspicuous when in flower, attracting numerous insect pollinators. The thick, carrotlike root, which even smells like a cultivated carrot, is developed during the plant's first year.

COMMON MULLEIN (Pl. 20) and **MOTH MULLEIN** are both alien members of the snapdragon family. Common Mullein is recognized by the large flattened rosettes of elongate, gray-green leaves, covered with a dense woolly layer of tiny hairs. The leaves feel velvety. The densely clustered, yellow, five-petaled flowers are borne on very tall (up to 6 ft.) spikes. Seed capsules each contain many seeds. Moth Mullein is a shorter plant (up to 3 ft. tall) and lacks the dense flower clusters. Flowers, which are usually yellow but may be white, are more separated and have conspicuous orange anthers. Leaves are less woolly. Both mullein species thrive on dry sandy-gravelly soils.

COMMON BURDOCK is a biennial alien member of the composite family (see p. 174). The plant superficially resembles thistles. Purple flowerheads grow from rounded green burs. Leaves are alternate, oval, and unlobed, and form a rosette in the first year.

The taproot is very large. Burdock is most common in rich soils.

SPOTTED KNAPWEED, like Burdock, is an alien species superficially resembling a thistle. About 12 species of knapweeds commonly occur in our region, and some are called star-thistles. Spotted Knapweed is among the most widely distributed. The flowers are pink-purple, located at the tips of the multibranched stem. Leaves are alternate, untoothed, and deeply lobed. This species is common in dry sandy soils.

BULL THISTLE and **NODDING THISTLE** (Pl. 28) Are aliens. Both have large, conspicuous, purple-red flowerheads atop very spiny bracts. Nodding Thistle heads are at the tips of long drooping stems. Leaves of both species are spiny, with numerous sharp points. Thistles are avoided by grazing animals like goats, sheep, and cattle and can therefore dominate a field. Goldfinches are particularly attracted to thistle seed.

COMMON EVENING-PRIMROSE is a native biennial. This tall herb is characterized by a cluster of yellow, four-petaled flowers that have X-shaped stigmas. The elongate greenish seedpods are found on the stem beneath the flowers. The stem is reddish and hairy. The elongate, lance-shaped leaves form winter rosettes.

Life Cycle of Biennials

Biennials have a unique life cycle that allows them to exploit short-lived old field environments. They grow and reproduce over a two-year period, spending the first year putting down mostly root tissue, surviving on the surface as prostrate rosettes. During the plant's second growing season, it flowers, then dies. Unlike many of the annuals, biennials are rarely pollinated by wind. Most have conspicuous, brightly colored flowers that attract a diversity of insect pollinators. Of the three growth forms exhibited by plants in early successional fields—annual, biennial, and perennial—the biennials have the fewest species. However, some biennial herbs are among the most abundant and best-known weeds.

Perennial Herbs of Old Fields (Pl. 20)

Most flowering plants, including virtually all trees and shrubs, most grasses, and most herbs, are perennials. These plants usually are long-lived and capable of reproducing annually and repeatedly, given favorable conditions. Though opportunistic annuals and biennials often dominate early old fields, they are ephemeral: the perennials soon move in and take over.

Wind-pollinated flowers are usually small and inconspicuous, often green or pale white. Insect-pollinated flowers are colorful and conspicuous and are positioned in clusters or dense flowerheads, often atop spikes emerging from the central stalk. Old field herbs usually produce an abundance of flowers, with seeds that number in the thousands per plant.

SEEDS: Seeds are usually minute, although some species have rather large seeds. Many are equipped for dispersal by having devices that can attach to animals or catch the wind.

LEAVES AND STEMS: Many old field herbs have elongate, lanceolate, or very narrow leaves. Leaves are often toothed and often covered by fine hairs called trichomes. Leaves are often lobed, some very elaborately. Basal leaves are usually much larger than leaves on the stalk. Many species form rosettes. Stems tend to grow tall and are frequently covered with fine hairs. Some plants are multibranched and bushlike, but the majority have a single prominent stem.

ROOTS AND RHIZOMES: In general, annuals and biennials have smaller root systems than perennials. While most annuals and biennials have little or no taproot, many perennials have a prominent taproot. Some perennials lack a taproot but have dense fibrous roots. Many perennials spread by rootstocks or by underground stems called rhizomes.

Adaptations of Perennial Herbs of Old Fields

Herbs that grow in old fields are adapted for sunny, open habitats. Reproduction occurs relatively soon after germination. These species grow quickly and produce abundant seeds, many of which are able to remain viable in the soil for years to come. Wind-pollinated flowers form on outer branches, thus maximizing the effectiveness of wind dispersal of the dustlike pollen. Insect-pollinated flowers sit atop tall stems or form dense clusters easily visible to searching insects. Seeds are occasionally dispersed by wind but the wind usually does not carry the seeds very far from the parent plant and thus puts the offspring in direct competition with its parent. Wind dispersal is most effective when the seed has a parachute-like attachment to help it remain airborne. Most old field herbs, however, tend to be dispersed by passing through an animal's gut or by hooking onto an animal.

Many old field perennials reproduce asexually by sending out rootstocks or underground horizontal stems called rhizomes. Once the plant gets established it is able to spread and becomes a genet (see p. 32). Each ramet (above-ground individual) can reproduce sexually. From one initial plant results a clone of sexually reproducing plants, a characteristic that greatly enhances the number of seeds produced.

Leaves are angled, an adaptation that maximizes their exposure to the sun's rays for photosynthesis while minimizing absorption of heat (see p. 324). At the same time, many leaves are lobed or dissected. These shapes may represent a compromise between the plant's need to maximize its photosynthesis but to minimize the cost of growing leaves, so that most energy can be put into reproduction. Many leaves are covered with fine hairs that can act to break up wind flow over the leaf surface and thus reduce evaporation stress. Toothed and hairy leaves may also aid in reducing herbivore pressure from both insects and mammals. Many plants, especially perennials, load their leaves and stems with toxins, which help deter grazing herbivores.

Rosettes, prostrate on the ground, enable winter annuals, biennials, and perennials to survive winter's severe weather. Rosette leaves absorb maximum amounts of sunlight and are usually very hairy, which helps reduce evaporation.

Taproots are energy-storage organs, so that reserves produced in summer can be used to survive winter underground. Some taproots are poisonous, a protection against herbivorous insects and worms in the soil.

Composites (Pls. 23, 27, 28)

Goldenrods and asters (Pl. 23) are among the many members of the composite family. The composite family stands out conspicuously as the most diverse family of the flowering plants. In North America alone, 292 genera of composites occur. Globally, composites comprise approximately one-tenth of all plant species. The composite family is to flowering plants what the passerines

New England Aster. This flowerhead shows clearly the disk and ray flowers which, together, make up the complex inflorescence, a unique characteristic of composites.

are to birds; both are unrivaled evolutionary success stories.

Composites occur abundantly in most habitats. They are particularly prevalent in open fields, meadows, and disturbed areas. In addition to the goldenrods and asters, many other common successional plants, such as fleabanes, daisies, dandelions, thistles, hawkweeds, Boneset, ironweeds, and blazing-stars, are composites. Many common cultivated plants, such as zinnias, sunflowers, dahlias, artichokes, and lettuce, are also composites.

Most composites are herbs but a few are shrubs, vines, or trees. All important species in our area are herbaceous (green, not woody).

The flower is the key to recognizing a composite. After locating some of the common species illustrated on the plates examine the flowers closely, if possible with a hand lens. The anatomic feature responsible for their evolutionary success is the unique flowerhead. In composites, what appears at first glance to be but a single flower, is in reality a dense cluster of many individual flowers, hence the name "composite." Several hundred tiny flowers are grouped tightly together to present a single display.

The typical composite flowerhead is well illustrated by an aster or fleabane. Two types of minute flowers can be seen. In the center are tiny tubular "disk" flowers, while around the edge are rings of strap-shaped "ray" flowers, commonly called petals. Each minuscule disk flower contains both stamens and pistils. Some composites, like ironweeds, Boneset, and the thistles, have only tube-type disk flowers. Others, like the hawkweeds and Chicory, have heads of only ray-type flowers. Most composites, however, have both basic types. In some, like the daisies, tube-type and ray-type flowers are colored differently, so the flower appears to have

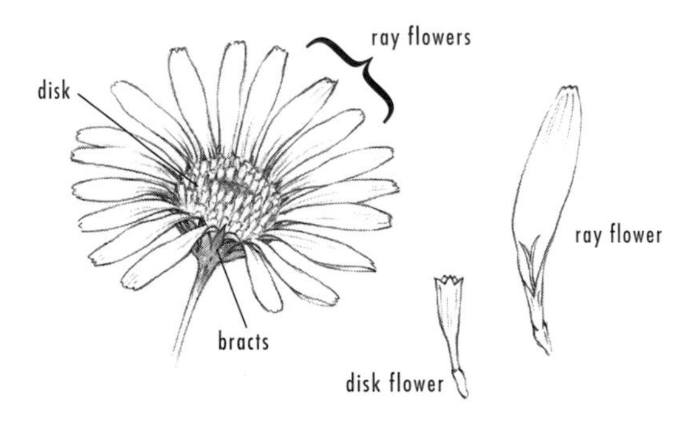

Fig. 21. Parts of a composite flower.

a center part of one color surrounded by "petals" of a different color. In others, like the goldenrods, disk and ray-type flowers are the same color.

Although each flowerhead is composed of literally hundreds of individual flowers, most composites contain numerous flowerheads. Thus the number of actual flowers present on a single plant commonly numbers in the thousands.

Adaptations of Composites

Composite flowers have been compared to the pile on a rug. Such an analogy goes far in explaining the evolutionary advantage composites have. The flowerheads are like rugs for insects, the pollinators. With a few exceptions such as Common Ragweed, a wind-pollinated annual (see above), composites are pollinated by insects, and a single insect can make contact with hundreds of flowers simultaneously by visiting a single flowerhead. The efficiency and ease of cross-pollination is greatly facilitated by the composite flower structure.

Cross-pollination in turn facilitates genetic variability. Genes, the hereditary units, are rearranged in new combinations as a bee pollinates several different Daisy Fleabane plants on a single food-gathering trip. Seeds produced by these plants contain this genetic variability and, as a result, may be better able to exploit variable environments. Some seeds may contain genes best suited for dry soil and some for wet soil. Some may be more resistant to disease than others. The high efficiency of fertilization and cross-pollination provides composites with ample genetic variability for adapting to changing habitats.

GOLDENRODS OF OLD FIELDS **PL. 23**

Approximately 90 species of goldenrods are native to North America. All are perennials. They inhabit old fields, meadows, roadsides, and woodlands (see p. 329 for a comparison of woodland, wetland, and old field goldenrods). Blooming takes place from midsummer through autumn, depending upon the species. In most species the flowerheads are bright yellow and arranged in clusters atop the stem. The shape of the cluster is helpful in identification: the common shapes are plume-like, elm-branched, wandlike, clubbed, or flat-topped (see *A Field Guide to Wildflowers*). The seeds, with hairy projections, are a valuable winter food source for finches, sparrows, and rabbits. All goldenrods have alternate leaves, but leaf sizes and shapes, presence or absence of teeth, and degree of hairiness vary with species.

TALL GOLDENROD grows up to 6 ft. tall. The flower cluster is plume-like and the leaves are narrow and toothed. The stem is hairy.

Similar species include **SWEET GOLDENROD, LATE GOLDENROD,** and **CANADA GOLDENROD.**

LANCE-LEAVED GOLDENROD is flat-topped and has very narrow, untoothed leaves with conspicuous parallel veins. The stem is smooth. The plant grows to a height of 4 ft. This species is found mostly in moist soils. Similar species include **SLENDER FRAGRANT GOLDENROD** and **OHIO GOLDENROD.**

Rough-stemmed Goldenrod is usually plume-like but may also have elm-branched flower clusters. The leaves feel quite rough and have deeply toothed edges. The stem is very hairy. This widely distributed species occurs in old fields, especially on dry soils. Similar species include **SHARP-LEAVED GOLDENROD, ROUGH-LEAVED GOLDENROD,** and **ELM-LEAVED GOLDENROD.**

EARLY GOLDENROD has plume-like flower clusters and blooms very early in summer. Its upper leaves are oblong and untoothed, and its lower leaves are oval and toothed. All leaves have two tiny leaflets at the axil. This species is widespread in dry fields and woodland edges. Similar species include **SHARP-LEAVED GOLDENROD** and **BOOTT'S GOLDENROD.**

DOWNY GOLDENROD is wandlike, with oval, untoothed upper leaves and toothed basal leaves. It is common on dry sandy soils, especially on coastal plains. Similar species include **WANDLIKE GOLDENROD** and **ERECT GOLDENROD.**

ASTERS **PL. 23**

Asters number over 600 species worldwide, and approximately 150 occur in the United States. Most of the common species are native to this continent, and virtually all are perennials. Aster flowers are typically "daisy-like" but have more rays than daisies do. They range from white to pink to blue and violet. Most asters are bushy in shape, but a few species are tall. Seeds are similar to those of goldenrods, and are eaten by many bird species. Many asters are common to old fields but many are found in woodlands. There are more woodland asters than goldenrods.

The following three species are found in different habitats. The first, an old field species, displays considerably more flower clusters than the third, a woodland species (see pp. 326, 330).

BUSHY ASTER is one of many asters common in old fields, especially on sandy soils. As its name implies, it is many-branched, with numerous flowers. Flower color is white to blue. Tiny leaflets are found at the base of the leaves.

PURPLE-STEMMED ASTER is found mostly in wet areas, including swamps. It has purple stems, and both leaves and stems feel rough to the touch. Its leaves clasp the stem. Ray flowers are violet-blue.

WHITE WOOD-ASTER occurs in woodlands, especially woodland edges. Flower clusters are flat-topped and the flowers are white. Leaves are heart-shaped.

OTHER NATIVE COMPOSITES **PL. 27**

GOLDEN RAGWORT is common in wet meadows, mesic woodlands, and swamps throughout most of eastern North America. Look for it in late successional fields and young woodlands, but only in wet areas. It blooms early. Its yellow flower clusters look flat-topped. It is also recognized by its large, heart-shaped leaves.

NEW YORK IRONWEED is another native composite of mesic areas. Most common along streambanks, this species requires not only moist soil but high light levels. Flower clusters, which are abundantly grouped atop the stem, range from deep pink to lavender. They have hairlike tips. As is characteristic of many wet-area species, New York Ironweed can reach heights of up to 7 ft. (see p. 327). A similar species, **TALL IRONWEED,** has lavender flower clusters and reaches 10 ft. in height. It too is found only in mesic soils. Ironweeds take their name from their strong stems.

SNEEZEWEED is found in mesic soils and swamps, and grows to 5 ft. tall. Its distinctive flower clusters are yellow; the center is a rounded, greenish yellow "ball," with yellow ray flowers that bend backwards. The closely related **PURPLE-HEADED SNEEZEWEED,** common in old fields, has a deep purple center. Both species flower in mid- to late summer.

SWEET JOE-PYE-WEED is a tall (up to 6 ft.) plant of old fields and open woodlands. It can be recognized by its dense, flattened mass of pink-purple flower clusters. This common plant takes its name from the odor of vanilla that is evident when the leaves are crushed. Leaves are arranged in distinctive whorls of 3–4 and the stem is greenish with a glaucous (whitish) tinge. The closely related **SPOTTED JOE-PYE-WEED** is more northern and mountainous and has leaves in whorls of 4–5.

BONESET is readily identified by its opposite, wrinkled leaves, connected at the base (the Latin name—*perfoliatum*—refers to the stem "perforating" the leaves), and its flattened clusters of white flowerheads. The stem is hairy. Boneset grows in mesic areas, along woodland edges, and in moist thickets. Alleged to have strong medicinal properties, Boneset preparations have long been used to treat fevers and colds, and Native Americans used Boneset to treat arrow wounds.

BLACK-EYED SUSAN is well known for its large flower clusters, deep brownish in the middle, with bright yellow ray flowers. This species is now common in old fields in much of our area but was originally a prairie species.

WILD LETTUCE is a tall (up to 10 ft.) composite with widely spreading branches and small yellow upright flower clusters that somewhat resemble tiny dandelions. The leaves are very deeply lobed. A similar species, **PRICKLY LETTUCE,** has spiny leaves and is an alien species.

Other Native Perennial Herbs (Pl. 22)

FIREWEED is common in northern areas, especially on burned sites. This erect plant, which belongs to the evening-primrose family, is topped with an erect spike lined with four-petaled pink flowers. Below the flowers are slender pods. Leaves are alternate, narrow, pointed, and unlobed. The colorful flowers are pollinized by both bees and hummingbirds. Fireweed is a rapid invader following fires.

COMMON CINQUEFOIL is one of several common cinquefoils. Cinquefoils are named for their yellow or white, five-sepaled flowers, with 5 alternating bracts below. Superficially, many cinquefoil flowers resemble buttercups, though cinquefoils are actually members of the rose family. Of the 33 North American species, most are native. Some cinquefoils have erect stems, but many are prostrate. Common Cinquefoil has compound, five-part leaves that resemble strawberry leaves. The plant spreads by prostrate runners. This species is found in dry old fields with sandy or gravelly soil. The plant produces relatively few seeds for an old field species.

HORSE-NETTLE is a common perennial relative of Nightshade and

Fireweed. As the name implies, this species rapidly invades burned areas, both in the Northeast and in the West. Goldenrod in foreground.

Jimsonweed (see p. 169). All are members of the tomato family. Horse-nettle has pale blue, five-petaled flowers. Stems and leaves have prickles and leaves are alternate, toothed, and slightly lobed. Fruits resemble tiny yellow tomatoes. This species reproduces both by seeds and by spreading from underground stems call rhizomes. Many birds eat the fruit and hence disperse the seeds. Found in old fields, especially in sandy soil.

BUTTERFLYWEED is a member of the milkweed family, but it lacks milky sap and can be recognized by the dense clusters of bright orange flowers and alternate, elongate, unlobed, hairy leaves. The "root" is actually an underground stem or rhizome. This rhizome has been used medicinally as a diaphoretic, expectorant, emetic, and cathartic, and a common name for the species is Pleurisyroot. The medicinal properties of the rhizome are probably caused by powerful chemicals which protect the plant from insect damage (see p. 345).

INDIAN HEMP, or **DOGBANE,** is also a member of the milkweed family and has milky sap. It is identified by its clusters of white flowers at the tip of the erect stem. Pods are long and slender. It has opposite leaves, elongate and unlobed. Seeds are borne on feathery tufts and are dispersed by wind. This species also spreads by horizontal rootstocks and, like most milkweeds, is toxic to livestock (see p. 345).

POKEWEED is one of the largest non-woody plants in our area, growing up to 10 ft. tall in thick, shrublike clumps. Note the rows of white, five-petaled flowers which, by late summer, become deep purple-black berries with red juice. Each berry contains about 12 black seeds. Stems are very reddish and hollow; leaves are alternate, unlobed. The large taproot is quite poisonous. Many birds feed on the berries and thus aid in dispersal of the seeds. The most common birds that feed on Pokeweed are the Mourning Dove, American Robin, Gray Catbird, Northern Mockingbird, and Cedar Waxwing. Birds occasionally become intoxicated from ingesting fermented berries.

Alien Perennial Herbs (Pl. 29)

FIELD GARLIC is easily detected by its pungent odor. Flowers cluster atop the stem and are pink to white in tiny bulblets. Leaves are long, narrow, and hollow. Similar species are **WILD ONION** and **WILD GARLIC,** both native. The bulbs of all three species closely resemble a miniature onion in both appearance and odor.

WINTER CRESS, or **YELLOW ROCKET,** is an abundant alien member of the mustard family. Like most mustards, it blooms in early spring and summer, often covering early old fields. Though sometimes confused with goldenrods, the differences are many and goldenrods never bloom in spring. Winter Cress has bright yellow, four-petaled flowers on stalks. Seeds are in long slender pods. Leaves are alternate, deeply lobed at the base of the plant, and untoothed. Plants overwinter as rosettes, but are perennials and have a thick taproot. Found in rich alluvial soils. A similar species is the closely related **EARLY WINTER CRESS**, also an alien.

COMMON ST. JOHNSWORT is the most successful alien member of the St. Johnswort family. Though most St. Johnsworts are found in mesic soils, Common St. Johnswort is also common in dry fields and meadows. The bright yellow flowers have 5 petals with black spots on their margins. Stamens have noticeable red tips. Flowers are located on a spike and the entire plant can be 1–2½ ft. tall. Leaves are opposite, oblong, and unlobed. Reproduction occurs by the rootstock spreading as well as by seeds. Common St. Johnswort, known in the West as **KLAMATH WEED**, is poisonous to grazing animals. Cattle that eat it develop skin irritations, rashes, blisters, and hair loss.

BUTTER-AND-EGGS is an alien plant in the snapdragon family. It is easily identified by the spikes of yellow, trumpetlike flowers with orange lips. Grows up to 3 ft. tall. Leaves are narrow and lanceolate. Butter-and-eggs spreads by creeping rhizomes and seeds. The plant is mildly poisonous and is very common in dry fields and roadsides.

COMMON TALL BUTTERCUP is the most common alien buttercup. Approximately 36 buttercup species occur in our area and most are native. This tall species (up to 3 ft.) has typical buttercup flowers: yellow, with 4 petals in a cuplike arrangement. Leaves are very intricately cut (termed "dissected") and remind one of Common Ragweed leaves. The plant is somewhat poisonous to livestock. Roots are very dense and fibrous, without a taproot.

CURLED DOCK and **SHEEP SORREL** are abundant alien members of the buckwheat family. Curled Dock is very tall (4 ft.) and is easily identified by its stalk with clusters of brown, heart-shaped seeds remaining on the plant well after the growing season. Leaf margins are wavy. The deep taproot branches out in several directions. Sheep Sorrel is recognized by its unique arrowhead-shaped leaves. Leaves are often quite reddish. The tiny green flowers, located on a small stalk at the tip of the plant, eventually turn red-brown. Both species are pollinated by wind, a characteristic evidenced by their inconspicuous flowers.

COMMON PLANTAIN and **ENGLISH PLANTAIN** are abundant in fields, lawns, and roadsides. Common Plantain has tiny white flowers numbering in the hundreds, arranged along a prominent spike. Basal leaves are quite wide and very slightly toothed. English

Plantain has short, spearlike flower heads and lanceolate leaves. Both species have thick, fibrous roots and reproduce only by seeds.

STINGING NETTLE is covered by stinging hairs and is very irritating to touch. Nondescript greenish flowers are on stalks emanating from the leaf axils. Leaves are opposite, arrow-shaped, and toothed. This species spreads by both seeds and rootstocks and is common in fields with mesic (moist) soil.

BITTERSWEET NIGHTSHADE (Pl. 26), an alien member of the tomato family, is easily identified by its arrow-shaped leaf with 2 small basal lobes. Flowers are blue to pale violet with yellow anthers. Fruit turns red and resembles a miniature tomato. It may grow as a vine.

VIPER'S BUGLOSS is an alien plant common in old fields throughout our area. It is a very hairy, bristly plant, which is attractive enough to be considered a desirable flower by some, rather than a weed. The trumpetlike flowers are blue with long red stamens and grow from a single central stem. The plant is a member of the forget-me-not family.

Origin of Old Field Herbs

Since most of eastern North America is normally forest, why are there so many species of plants adapted to old fields? Part of the answer lies in the fact that many alien species occur. In New Jersey, for instance, between one-half and one-third of old field species are alien and most are herbaceous (green, not woody). But, what of the others? In a region normally made up of forest, how did the sun-loving species common in old fields evolve?

Four ideas have been suggested to account for the presence of old field species.

1. These species evolved on agricultural land abandoned by Indians.
2. These species migrated eastward from prairie regions along corridors such as railroads, canals, and wagon trails.
3. These species adapted to forest openings caused by storms, fire, etc.
4. These species evolved in marginal habitats, such as rocky outcrops; areas with very dry, wet, or nutritionally poor soil; and along streams or floodplains or other natural corridors.

There are elements of truth in each of the ideas stated above but the one that seems to fit best with at least several of the most common herbs is number 4. Common Ragweed grows on sand plains, oak openings, ridgetop prairies, and dry oak forest, all areas of rather marginal soil quality. Tall Goldenrod is found in midwestern prairies, gravel bars, and floodplains, and New England Aster grows on prairies in Michigan and Ohio and also is found often in dry soils that characterize oak openings. Many grasses and some trees, such as Eastern Red Cedar, are particularly characteristic of rocky outcrops.

Many field plants may have a difficult time colonizing forest openings, and it is unlikely that these species first evolved in such areas. Forest openings are often quickly closed by the rapid regrowth of woody species from stump and root sprouts. In addition, the thick leaf litter can block seed sprouting as well as seedbed establishment. Finally, many old field species have relatively poor seed dispersal and it is unlikely that they could effectively colonize scattered forest openings. An exception is the large composite family, in which wind dispersal of seeds is effective.

Legumes (Pl. 30)

The legume family is a large and important group with 13,000 species, exceeded in diversity only by the composites. This family includes many old-field colonizing species such as clovers, vetches, and locusts. Legumes may be herbs, vines, or woody trees. Most species have delicate compound leaves, attractive flowers, and seeds in pods. Many of the woody species have thorns. Familiar examples of agricultural legumes include peanuts, peas, beans, and soybeans.

WHITE SWEET CLOVER is an abundant alien clover found in old fields and roadsides throughout much of North America. Its white flowers are on long, upright, wandlike clusters and it is most easily recognized as a clover by its tripartite, cloverlike leaves. On rich soils, the plant can reach a height of 8 ft., though more commonly it is 2–4 ft. tall. **YELLOW SWEET CLOVER** is very similar (and also an alien) but has yellow flowers.

BUSH-CLOVER is, as its name implies, an almost shrublike species, which grows up to 3 ft. in height and has widely spreading branches topped by delicate purple flowers. There are many other common species of bush-clovers, all of which are native. Most are old field species, but some are found in woodlands, especially in gaps.

RABBIT'S-FOOT CLOVER is an alien legume easily recognized by its very fuzzy, gray-pink flowerheads and narrow leaflets. **RED CLOVER,** also an alien, is closely related. It can be recognized by its large, rounded, reddish purple flower and its leaves, which are marked by pale chevrons. Both clovers are abundant inhabitants of old fields. Several other clovers in this group (genus *Trifolium*) are common aliens throughout our area.

BLACK MEDICK is a common shrubby but prostrate legume with yellow flowers. It is named for its black seedpods. Also an alien, it is an old field species similar to the alien **HOP CLOVER**. Hop Clover has larger flowers and is always upright.

VETCHES are common ground-hugging, leguminous vines. Virtually all species in our area except one, **AMERICAN VETCH**, are aliens. **NARROW-LEAVED VETCH** is found throughout North America and **COW VETCH** is abundant in old fields from middle Atlantic states through the Northeast. Vetches have branches ending in tendrils, and the flowers are bluish purple.

One introduced species, **CROWN VETCH**, has pea-like, pinkish flowers. This alien is widely planted as a ground cover along interstate highways.

Other common legumes include Hog Peanut, Groundnut, Wild Indigo, Wild Senna, Butterfly Pea, Beach Pea, Southeastern Coralbean, and Birdsfoot Trefoil.

See also Black Locust (p. 197) and Kudzu-vine (p. 188).

Legume Natural History

Though legumes are found in virtually all terrestrial habitats, they are basically tropical in distribution. Mimosa, the common sensitive-plant whose leaves close when touched, grows in tropical old fields. Acacia and Mesquite frequent dry areas of the tropics and subtropics. Many common eastern North American species are aliens and many native species are most abundant on the Prairie Grasslands. These latter include such species as Leadplant, Woolly Locoweed, Locoweed, and Purple Prairie Clover.

One important reason why legumes succeed so well in old fields is their ability to capture gaseous nitrogen from the atmosphere and convert it into usable chemical form. Legumes

Fig. 22. Leadplant.

have the ability to provide for their nitrogen needs by literally manufacturing fertilizer. The capture of gaseous nitrogen is called nitrogen fixation and legumes are the most important of the nitrogen-fixing plants. Approximately 300 species of other plants, such as alders and Sweet Gale, also fix nitrogen, but the legumes contain by far the majority of nitrogen-fixers.

Nitrogen fixation is done within tumorlike growths called nodules, located on the roots. Bacteria of the genus *Rhizobium* are housed within the nodules and actually do the nitrogen capturing. The nitrogen fixed by bacterial labor is used by the plant. The bacteria are provided with food and shelter by the legume, making the relationship between bacteria and legume a case of *mutualism,* where both species need each other and benefit each from the other. The *Rhizobium* bacteria invade the legume via the tiny root hairs. The plant soon forms nodules where the bacteria have invaded. *Rhizobium* require an oxygen-free environment and the red pigment hemoglobin is manufactured to trap oxygen before it reaches the bacteria. The nodules are pink because of the pigment. Hemoglobin consists of both proteins and a complex chemical called heme. The legumes have genes that make the protein but the bacteria manufacture the heme, thus the two combine their biochemistry to produce hemoglobin and provide the bacteria with the required environment.

Legume flowers are pollinated by insects, mostly bees, and are thus showy and colorful. The typical flower consists of 5 petals of 3 different types. The uppermost petal is called the standard, or banner, petal. It is the largest and attracts the insect with color. The lower petal is actually 2 petals joined together to form a keel. The 2

Everlasting Pea. The flower of this species demonstrates the unique structure typical of the group, with five petals: an upper "standard," a lower "keel" (two joined petals), and two "laterals."

side, or lateral, petals complete the arrangement. The bee locates the flower by color and odor and must force its way in between the keel and standard. As it does so it comes in contact with both the pollen and pistil. For a further discussion of insect pollination, see p. 273.

GRASSES

To identify grasses, consult *Grasses: An Identification Guide* by Lauren Brown.

Members of the grass family are the most important crop plants used by humans. Corn, rice, wheat, barley, oats, sorghum, and millet are all grasses. Grasses are superbly adapted to open field environments and are important species during old field succession. They are especially successful in areas prone to drought and fire and they tolerate high levels of grazing by animals.

Grasses are wind-pollinated flowering plants. The flowers are usually inconspicuous. Characteristics unique to grasses include leaves with parallel veins attached at joints to mostly hollow stems; thick, fibrous root systems, which may make up 90% of the total weight of the plant; and a unique kind of photosynthesis (see below). The growing part of the stem is at the base rather than the tip. This is the reason why grasses grow back so quickly after being mowed or grazed. Many grasses form sod, which is due to rapid spreading by rhizomes.

Grasses form a major biome in midwestern North America, the Prairie (see p. 118). The line separating forest biome from prairie biome is anything but sharply defined and prairie tracts mingle with forest in Minnesota, Iowa, Wisconsin, Indiana, Kansas, Nebraska, Oklahoma, and Texas.

Two basic growth forms are apparent among grasses, the sod form and the bunch form. Sod grasses, such as Kentucky Bluegrass, spread by rhizomes and evenly carpet an area. Bunch grasses, such as the bluestems, are scattered in clumps.

The North American grasslands can be divided into three regions from east to west, the Tall, Medium, and Short grasslands. Sharp boundaries do not exist between them. Tall Grassland is characterized by Big Bluestem, Switch Grass, and India Grass. Short Grassland melds with desert biomes and is characterized by such species as Buffalo Grass and Blue Grama. Medium Grassland is a mixture between the two extremes and such species as Little Bluestem and June Grass are common.

Why are there grasslands? Ecologists believe moisture level to be a strong determinant. As the climate becomes wetter from west to east, the grasses become taller. However, in areas protected from fire, trees often invade, especially toward the east.

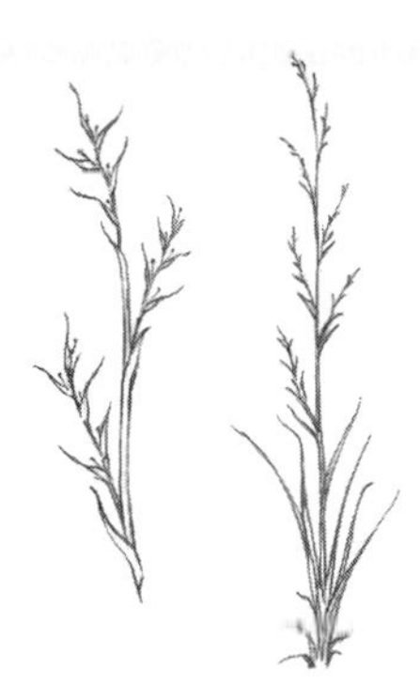

Fig. 23. Little Bluestem grass.

Therefore, grasses are probably dependent on fire to eliminate competing trees. Finally, grazing is also a factor. Grasses grow back quickly when grazed moderately but can be destroyed by overgrazing.

Grasses have a unique type of photosynthesis shared by a few other plants native to hot dry regions. Called C-4 photosynthesis, it allows grasses to tolerate very high light intensities and high temperature without any loss of their abilities to extract carbon from the carbon dioxide in the atmosphere. Grasses use water very efficiently, requiring only 250–350 grams of water to produce one gram of dry matter. This may seem high, but non-C-4 plants require between 400 and 1 000 grams of water.

Herbaceous and Woody Vines (Pl. 31)

COMMON VINES

POISON-IVY is a notorious vine. A member of the mostly tropical cashew family, this noxious vine is recognized by its compound leaves with 3 shiny leaflets that turn brilliant red in the autumn. Fruits are in clusters of white berries. Poison-ivy can grow as an erect shrub, as a prostrate ground cover, or as a vine. When growing as a vine it sends out aerial roots that cling to the object upon which it is climbing, usually a tree trunk. Its fruits are eaten by many birds, the principal dispersers of its seeds. Poison-ivy, as virtually everyone knows, produces a volatile oil on all of its tissues. This oil, though tolerable and of no harm to most animals, is highly irritating to most humans when in contact with the skin.

KUDZU-VINE is an alien legume (see p. 184). This Japanese vine may grow to heights of 60 ft. or more and can drape itself as a virtually solid blanket over woodlots, especially in the southeastern states. It was introduced in 1876 in the hope it would restore and bind soils (preventing erosion) and for its high feed value. Like many alien species, it has encountered few natural enemies or competitors and has prospered, becoming a nuisance in many areas. It is recognized by its hairy stem and violet, pea-like flowers. Fruits are in a large hairy pod. Leaves are broad, suggesting a maple-like shape. Principally confined to the South, Kudzu rarely is found farther north than Pennsylvania.

GRAPES are abundant in sunlit areas. Approximately 30 species occur in North America, and most are found in our area. Grapes occupy many habitats, including old fields, stream banks, and forest edges and gaps. They have generally large leaves on long petioles. Many species have lobed or heart-shaped leaves. Grape bark peels easily and is heavily utilized by songbirds as nesting material. Flowers are small and greenish, a characteristic suggesting wind pollination. However, the flowers are fragrant, open early in spring, and are actually pollinated by insects. The fruits are quite sweet and eaten by many bird species: Ruffed Grouse, Wild Turkey, Ring-necked Pheasant, Northern Cardinal, thrushes, mimic thrushes, Cedar Waxwing, and even the Pileated and Red-headed woodpeckers. Mammals such as the Black Bear, Raccoon, foxes, squirrels, and skunks are also grape-eaters. See also p. 335.

VIRGINIA CREEPER is an abundant member of the grape family. Identify it by its compound leaves of five leaflets arranged circularly. Leaflets turn bright red in fall and the berries are deep blue-black, in clusters. Flowers are in small green clusters that bloom in late summer and are pollinated principally by bees. This vine climbs both by tendrils and by aerial roots with adhesive pads. The berries are eaten in quantity by birds of many species. Pileated Woodpeckers are especially fond of Virginia Creeper berries. See also p. 334.

TRUMPET-CREEPER is a striking member of the mostly tropical begonia family. This vine is found from Pennsylvania to Florida. Its trumpetlike, brilliant orange-red flowers are unmistakable (see Pl. 31) and attract numerous bees, ants, and hummingbirds. Leaves are compound, with 7–11 pointed, toothed leaflets. The vine climbs by aerial roots.

BITTERSWEET is noted for the way it tightly entwines around its host plant, often forcing the host stem to widen around the vine. Its oval leaves are finely toothed and the flowers are in greenish terminal clusters. Male and female flowers occur on separate plants. Fruits are orange-yellow and berries are red. Bittersweet, as the name implies, is not heavily fed upon by wildlife. The native species is currently being outcompeted in the northeastern states by **ASIATIC BITTERSWEET.**

GREENBRIERS are members of the lily family. Noted for their very thorny stems and large bright shiny green leaves with parallel veins, these vines grow by tendrils. Like Bittersweet, Greenbrier produces small greenish flowers in clusters, with the male and female flowers on separate plants. Flies are the principal pollinators and the fruits are blue-black berries that are an important winter food for White-tailed Deer, grouse, mimic thrushes, Robins, Black Bears, and Raccoons. All but one species of Greenbrier are deciduous and the base of the leaf stalk remains attached to the stem after the leaf drops.

HEDGE-BINDWEED, like several other common members of the morning-glory family, is important during old field succession. It spreads by rootstocks and twining and is easily identified by the large, funnel-shaped flowers, pink with white stripes. Leaves are triangular. Hedge-bindweed is often abundant in early old fields but is gone by 30 years post-abandonment. It is initially successful because it poisons other species which grow nearby, a characteristic called *allelopathy* (see p. 171). However, the buildup of its toxins in the soil eventually poisons Hedge-bindweed itself (see Horseweed, p. 167).

Other common vines include gooseberries and Wisteria.

Natural History of Vines

A vine is a growth form, not a family of plants. Vines succeed by climbing; they use other plants or objects as a means of support as they grow outward and/or upward. Many species representing several different plant families are abundant in old fields. Others are adapted to spread in forest gaps and other areas where light is abundant, such as along river banks and forest edges. Vines grow quickly, using relatively little energy for woody tissue. Instead they cling to or entwine around other plants. Vines grow by any of several means. Some, such as Bittersweet Nightshade (p. 182), drape themselves over other plants. Others, such as grapes and greenbriers, climb by minute curling stems called tendrils. Still others, such as Honeysuckle and Bittersweet, entwine themselves around other plants.

Shrubs and Trees

Many shrubs thrive in sunlit areas and are among the first woody species to become abundant during succession. Shade produced by patches of shrubs contributes to the decline of both herbs and grasses, which require uninterrupted sunlight. In most cases trees eventually overtop the shade-intolerant shrubs and replace them.

SELECTED COMMON SHRUBS PLS. 32, 33

NORTHERN BAYBERRY is abundant in the Northeast on poor sandy soils, particularly along coastal areas. A spreading deciduous shrub, this many-branched plant rarely exceeds 6 ft. Noted for its fragrance, it has grayish, thick twigs and small waxy leaves spotted with minute dots of resin. The female flowers become clusters of hard whitish berries that feel oily and have a strong pungent odor. These berries form an important winter food source for thrushes and Yellow-rumped Warblers (which were formerly named Myrtle Warblers). Tree Swallows eat bayberries during their fall migration (see p. 338). Male flowers are on short terminal catkins. Although most plants have either male or female flowers, some have both. A few even have bisexual flowers, where both pollen and egg are present. To make matters more complex, some individual plants change sex, producing male flowers one year, females another. Northern Bayberry, though not a legume, is a nitrogen-fixer (see p. 185), a characteristic which provides an advantage in nutrient-poor sandy soils. In the Southeast, similar species include **EVERGREEN BAYBERRY** and **SOUTHERN BAYBERRY**, which is also called **WAX-MYRTLE.**

SWEET-FERN is an aromatic deciduous shrub found in dry sandy soils as well as fertile pastures. It is named for the fragrance of its crushed leaves and for its superficial fernlike appearance—the waxy leaves are oblong with multiple lobes, suggesting a fern frond. Each leaf axil has a hornlike pair of tiny structures called stipules. Male catkins are brown, located at branch tips. Female catkins are found below. The fruit is nutlike, in a structure resembling a bur.

SUMACS are, like Poison-ivy, members of the cashew family. Moderately sized spreading shrubs or small trees, sumacs are recognized by their large, alternate, compound leaves with 11 oval leaflets. Flowers are in dense, greenish white clusters that, in all but Poison-sumac, become fuzzy red berries in fall (Poison-sumac has white berries and is found in swampy areas, not old fields). The flowers are bee-pollinated and plants produce either male or female flowers. Male plants outnumber females in early and late succession, but females prevail during mid-succession. Sumac twigs have hollow pith, a possible adaptation for rapid growth since hollow twigs require less energy to make. Bundle scars are obvious on the twigs and the sap is milky. Sumac berries (see Pl. 50) remain on the plant in winter, and often remain as long as several years. They are not favored by birds but may be eaten when other foods are scarce. Deer also feed on sumac (see p. 336).

STEEPLEBUSH and MEADOWSWEET, both also known as spireas, are members of the rose family. These are small woody plants with very slender stems. The central stem ends in a large torchlike cluster of small, bee-pollinated flowers. The uppermost flowers are male and the lower ones are female. The plant blooms from the top down. In Meadowsweet, flowers are white with a pink center. In Steeplebush, the flowers are bright pink. Leaves are alternate, oval, unlobed, and slightly toothed.

There are approximately 25 species of shrubs in a group whose members are called variously JUNEBERRIES, SHADBUSH, or SERVICEBERRY. Like the spireas (above), they belong to the rose family. Individual species are highly variable and hybrids often occur. Most grow as shrubs but a few grow as small trees. The reddish purple fruits resemble small apples and are eaten by many birds. Twigs are browsed by deer, moose, beaver, and cottontails. Bark is gray and leaves are alternate and simple, widest at or below the middle. Leaves are often but not always toothed. Flowers are white and contain both sexes. The long, pointed buds are distinctive for their deep pink color. These shrubs flower and fruit earlier than most shrubs. They often spread by underground stems.

HAWTHORNS make up another complex group with many species and much hybridization. Identification of individual species can be tricky. Many occur in old fields, but others can be found in woodlands and swamps. They grow as dense thorny shrubs or small trees. Fruits are apple-like and yellow to red (see Pl. 50); they form an excellent winter food for many animals. Leaves vary depending on species: some are lobed, some unlobed, but all are simple and toothed.

Many species of WILD ROSE are found in old fields. These thorny plants have compound leaves with toothed leaflets. The distinctive five-petaled, pink-red flowers are pollinated by bees and mature into a fragrant cup, called a hip, in which the seeds are encased. The alien species MULTIFLORA ROSE, originally introduced as a hedge plant, has spread rapidly and provides excellent nest sites and cover for birds, especially the Mockingbird. In northern areas of its range, the Mockingbird is very commonly associated with Multiflora Rose.

COMMON JUNIPER is a distinctive species of northeastern and Appalachian hilltops, dry fields, and ridges. A wide, spreading shrub, it forms dense thickets. A member of the pine family, it is evergreen and covered by prickly needles, each with a whitish line down the middle. Needles are in bundles of three. Male and female flowers are on separate plants and pollen is distributed only by wind. Tiny, fleshy blue cones are very berry-like and are covered by a gray waxy coating. They are fed upon heavily by birds, which disperse the seeds. Gin is made from juniper cones. Common Juniper, though usually a shrub, can grow into a small tree. See Eastern Red Cedar (p. 194), with which it may occasionally be confused.

BRACKEN FERN is not a shrub but rather a true fern. It tolerates a very broad range of ecological conditions and is commonly found in dry open fields as well as shady moist woodlands. It is particularly abundant in burned-over areas. Its 3 triangular fronds originate from a single place and it spreads rapidly by rhizomes.

Other common shrubs of old fields include various viburnums, Elderberry, Lyonia, Holly, and Barberry. In moist meadows, alders, Buttonbush, and Pepperbush commonly occur.

HEATHS

PL. 33

Among the most important of the shrubs are members of the heath family. Over 2000 species, comprising approximately 80 genera, occur globally, mostly in temperate and alpine areas where acid soils and cool temperatures prevail. Growth forms include not only shrubs but also trees, vines, and herbs. Most heaths are evergreen (azaleas are examples) but many are deciduous. Leaves are simple and alternate and tend to be thick, leathery, waxy, and unlobed. Flowers may be either single large blossoms or clusters of small blossoms. The rhododendrons are the most diverse group, a colorful group making up nearly half of the heath species.

Heaths common to our area include several rhododendron species, blueberries, cranberries, huckleberries, azaleas, and Mountain Laurel.

BLUEBERRIES are among the most common heaths. Spreading shrubs, occasionally tall, they are found in both old fields and forest understories, especially where light penetrates well and where soils are decidedly acidic. Leaves are unlobed and leathery and the whitish, bell-like flowers become the familiar and tasty blueberries. Blueberries are an important food for wildlife. Many species of birds and mammals feed on the berries. One member of this group, **SPARKLEBERRY,** is a tree, not a shrub.

HUCKLEBERRIES closely resemble blueberries but have darker and smoother bark and yellowish resin dots on the leaves (lacking on blueberries). Common to open areas of sandy acidic soils, some huckleberries spread asexually by rootstocks. Flowers vary in color from white to pink; like those of blueberries, they are bell-like and hang downward in clusters. Bees and small butterflies are the principal pollinators. Each fruit has exactly 10 seeds; blueberries have over 100 seeds per fruit.

MOUNTAIN LAUREL grows primarily in woodlands, not in old fields. It is included here because it is an important heath in many forests throughout our region. Usually a shrub, it can grow to tree size under optimal conditions. Blooming from April through June, the flowers are uniquely adapted for pollination by bees. The stamens bend downward and are held in tension by the petals. A bee will inadvertently release the springlike stamen and get itself coated with pollen. As with most heaths, branches are stout and leaves are oval, dark green above and yellowish below. Thickets of Mountain Laurel make excellent cover. Deer consume the leaves, though the leaves are poisonous to humans and goats. The Worm-eating Warbler is common in woodland heath thickets of Mountain Laurel and rhododendron.

TREES OF OLD FIELD SUCCESSION **PL. 34**

Many species of trees are abundant during old field succession. They demand sunlight and grow rapidly, and many are eventually replaced by the more shade-tolerant and slower growing species. Some, however, are themselves shade tolerant and quite capable of remaining indefinitely. The natural histories of some of the most common and widely distributed species are included here.

SUCCESSIONAL TREES OF NORTHERN OLD FIELDS

GENERAL SITES

White Pine
Red Maple
Eastern Red Cedar
Gray Birch
Quaking and Bigtooth aspens
Choke, Pin, and Black cherries
Black Locust
White Ash
Alder Buckthorn
Gray-stemmed Dogwood
Sassafras

SAND PLAINS*

Pitch Pine
Jack Pine
Bear Oak
Post Oak
American Holly

*(successional only when protected from naturally occurring fires—see pp. 71, 92, 94–95).

GENERAL SITES
Eastern Red Cedar
Red Maple
Eastern Sycamore
American Elm
Black Locust
Ashes
Cherries
Northern Catalpa
Tuliptree
Sweetgum
Persimmon
Winged Elm
Redbay

SAND PLAINS*
Longleaf Pine
Shortleaf Pine
Loblolly Pine
Slash Pine
various oaks

EASTERN RED CEDAR is an abundant tree in abandoned pastures, outcrops, and calcareous (lime-rich) soils throughout our region except the northern boreal area and southern Florida. This cone-shaped tree grows to a height of 50 ft. An evergreen member of the pine family, its branches are covered by tiny overlapping scales that form prickly needles. As the needles age they become less prickly and browner. Sexes are usually on separate trees and the tiny, berry-like cones are waxy and pungent with the odor of gin. Eastern Red Cedar can tolerate a wide range of soil moisture but its relatively thin, peeling bark is quite susceptible to fire. It prefers soil high in calcium and is an indicator of limestone and shale. Red Cedar decreases the acidity of soils and concentrates calcium, adding it to the upper soil. Earthworms, which require soils high in calcium, are abundant in fields with Red Cedar. Eastern Red Cedar is a preferred food for thrushes, American Robins, and Cedar Waxwings, each important in disseminating the seeds. Seed dispersal by birds is so effective that Red Cedar can quickly invade fields and form even-aged stands. This species requires high levels of light and is eventually shaded out, though under favorable conditions individuals can live for 200–350 years. Red Cedar is often parasitized by large masses of gelatinous orange tissue, the Cedar-apple Rust (a fungus).

SOUTHERN RED CEDAR is a common coastal plain, floodplain, and swamp species from North Carolina through Florida.

*(successional only when protected from naturally occurring fires—see pp. 71, 92, 94–95).

PIN or **FIRE CHERRY** is an important successional tree in old fields in northern states and throughout the Appalachians. It is particularly important in Northern Hardwood and Boreal forests. Cherries of all species have alternate lanceolate leaves with small teeth. The bark tends to be smooth when trees are young but roughens with age. It is reddish brown and has very prominent horizontal lenticels. Pin Cherry flowers are in rounded clusters. Each flower has 5 white petals with yellow stamens and red anthers.

Pin Cherry is well dispersed by birds. Waxwings, American Robins, thrushes, and grosbeaks all feed on the fleshy fruits and thus distribute the seeds. Large numbers of viable seeds accumulate in the soil and germinate when a light gap or other opening appears. In two sites in New Hampshire, there were estimated to be 200,000 and 140,000 Pin Cherry seeds per acre, respectively. Pin Cherry grows very quickly and lives only 25–30 years. As it grows, the tree accumulates nitrogen, potassium, magnesium, and other important nutrients; thus it plays an important role in restoring and maintaining soil fertility following disturbance. The key to Pin Cherry's success is that, though it is short-lived, its seeds remain viable in the soil for long periods. It has been called a "fugitive species," meaning it survives by dispersing widely and growing and reproducing quickly when conditions permit. In this regard, it is similar to Common Ragweed.

BLACK CHERRY shares a similar natural history with the Pin Cherry and is widely distributed throughout our region, though particularly abundant in southern areas of the Northern Hardwood Forest.

ASPENS (sometimes called poplars) are abundant trees in both the northern states and the Far West. Though typically successional, aspen stands can persist indefinitely (see below). Two species occur in our region, the **QUAKING ASPEN** and the **BIG-TOOTH ASPEN.** Big-tooth Aspen is particularly common on sandy soil and fire-disturbed sites. It is found throughout the Appalachians as well as in northern states. Aspen leaves are green above and silvery below and are triangular in shape, on long petioles (which make the leaves "tremble" in the slightest breeze). Big-tooth leaves have large teeth. Bark varies but is usually very smooth and pale greenish or, occasionally, cream-colored. Male and female flowers grow as catkins on separate trees. Catkins mature before leaves and pollination is by wind. Aspens may live for 200 years or more but such longevity is rare. More usually, aspens are replaced or destroyed by fire before such an age. They reproduce when only 15–20 years old and good seed crops are produced about every 4–5 years. The tiny (3 million to a pound) seeds are wind dispersed and, unlike those of Pin Cherry (see above), do not remain viable for very long.

The key to the aspens' success is the ability to grow very quickly and spread by root sprouts. Quaking Aspen is 2 ft. tall when only 2 years old and has an extensive root system, with lateral roots radiating. (Compare with the growth pattern of Longleaf Pine, another important successional species—see p. 94.) Roots sprout other stems and a clone quickly forms, usually with one "dominant" stem in the center. Some aspen clones in Minnesota contain nearly 50,000 trees and occupy 200 acres. One clone has been aged at approximately 8000 years, which means it began soon after the retreat of the glaciers. Aspen clones may, therefore, be the oldest organisms on the planet (Bristlecone Pines are "only" 4600 years old). Big-toothed Aspen spreads not only by root sprouts but by stump sprouts as well. Aspens are most common on soil high in calcium and are totally intolerant of shade; they are quickly replaced by shade-tolerant spruces or other species.

Aspens are very important trees to wildlife. Beavers use aspen stems extensively for dams and lodges and they eat the bark. Deer, moose, and porcupine also feed on aspen bark, and many birds feed on the catkins.

Two species of birch, **GRAY BIRCH** (Pl. 34) and **PAPER BIRCH** (Pl. 4), are abundant successional species in the northern states. Both are small trees, 50–70 ft. tall, with smooth, thin bark marked with triangular black patches. Paper Birch, also called White Birch, has peeling white bark. Gray Birch has grayer bark. Leaves of both are alternate, triangular, toothed, on short petioles (stalks). Male and female flowers occur on the same plant and are catkins. Male catkins are long and dangle at branch tips. Female catkins are shorter, thicker, and erect. Female catkins become cones containing seeds. Gray Birch often grows in clumps, each with several thin stems.

Gray Birch is short-lived, rarely exceeding 50 years of age, and flowering at about 10 years of age. It is found in both wet and dry sandy or gravelly soils. Paper Birch is a very quick colonizing species, especially in fire-damaged areas. It is most common in northern states and is longer-lived than Gray Birch.

RED MAPLE (Pl. 10) is an abundant tree throughout eastern North America. A highly adaptable species, it is often a subcanopy species in mature forests and can be dominant in the canopy on very wet or very dry sites. Moist rich soils are optimal for its growth. Red Maple grows rapidly during the first 20–30 years and can live well over 100. Growing up to 90 ft. tall, it has three-lobed, triangular, toothed leaves, which, like all maple leaves, are

Black Locust. This species is among those that are early woody invaders of old fields.

opposite. Leaves turn brilliant red in fall, especially on acidic soils. Bark is gray and smooth on young trees but blackish and ridged on older trees. Red flowers appear early in spring, well before the leaves open. Flowers may be male, female, or bisexual, and all three types may occur on the same tree. Red Maples are subject to damage by both storms and insects.

BLACK LOCUST (Pl. 34) is a woody legume very common in old fields ranging from the Northeast through the Appalachians, Ozarks, Missouri, Arkansas, and eastern Oklahoma. It is commonest on soils rich in limestone. An extremely rapidly growing successional species, it can flower as early as 6 years of age, but normally flowers at 10–12 years. Once a Black Locust becomes established, it can spread rapidly by root sprouts, forming clones as aspens do. Flowers hang in large fragrant clusters and resemble the flowers of peas. They are white with yellow on the upper lip petal. Bees are the major pollinators. Seeds are contained in brown leathery pods that hang from the branches. Leaves are compound and alternate, each leaf consisting of 7–19 soft, oval, untoothed leaflets. Bark is very deeply furrowed and tends to peel easily. Branches are reddish and have sharp spines.

Black Locust, like most legumes, fixes gaseous nitrogen in nodules on its roots (see p. 185). Nitrogen fixation increases until mid-succession, then lessens. As the locusts deteriorate from insect damage and competition from other plants, the nitrogen is released and added to the soil.

Three species of ash, **WHITE ASH, GREEN ASH,** and **BLACK ASH,** are common successional trees. White Ash is most common on upland sites in old fields, Green Ash frequents floodplains and streamsides, and Black Ash is also most common on moist sites. Tall straight trees, ashes grow over 100 ft. tall. They grow rapidly and their seeds are spread widely by wind as well as by birds and mammals. Ashes have compound leaves.

TULIPTREE (Pl. 13), also called Yellow Poplar, is most common on rich soils of the southeastern and south-central states. Though common in old fields, it usually persists in mature forests as well. A member of the tropical magnolia family, it is named for its flowers, which are large and cup-shaped, suggesting a tulip, and conspicuously yellow-green, with an orange band at the base of the petals. Flowers develop into dry, cone-shaped fruits, tapered at the far end. Tuliptrees can grow to nearly 200 ft. tall but normally are 100–165 ft. tall. Trunks are straight with grayish, deeply fissured bark. Leaves suggest those of a maple but are alternate, not opposite. They are broad and untoothed, with 2 lobes at the tip and 2–4 lobes on the sides. Leaves are bright green above, pale green below.

The colorful flowers are pollinated by bees, and many birds and mammals eat and disperse the seeds. Tuliptree is particularly common in the Appalachian Cove Forest (p. 98), the Oak-Hickory Forest (p. 81), and the Beech-Maple (p. 77).

SWEETGUM (Pl. 15), like Tuliptree (above), is a pioneer species that often persists in the mature canopy, especially in very mesic areas in the South. It can be recognized by its star-shaped, five-lobed, alternate leaves with pointed lobes. Sweetgum grows straight, as tall as 130 ft. The crown tends to be rather cone-shaped. Bark is grayish or brown and ridged. Male and female flowers occur on separate branches. Male flowers are in small, dense globular clusters on an upright stalk. Female flowers are also densely clustered but on a hanging stalk. Fruits are very distinctive spiny balls, suggesting the fruits of a Sycamore. They often remain on the tree throughout the winter. Each seed has a small wing but the seeds are dispersed mostly by birds, not by wind.

SHRUBS ALONG POWERLINES

OBSERVATIONS: In many old fields, dense shrub cover predominates over trees, especially where the field is crossed by a powerline right-of-way. Powerline rights-of-way are usually characterized by a thick growth of low spreading shrubs. Species such as huckleberries, blueberries, Sweet-fern, Greenbrier, alders (wet sites), Meadowsweet, Arrowwood, and Common Juniper may occur in dense single stands or in various combinations. A careful look at these shrub communities will reveal few, if any, tree seedlings or saplings. The idea that trees will eventually replace shrubs as they do in "normal" old-field succession will likely prove untrue in

Powerline. Frequent cutting and use of herbicides create permanent shrub communities under powerlines, as old field succession is arrested. Photo shows Common Juniper, Eastern Red Cedar, and other successional species.

these shrub-dominated communities. These communities are examples of "arrested succession." The shrubs are there to stay.

EXPLANATION: Why do the shrubs become sufficiently dense to resist invasion by trees? In many cases the answer is that human manipulation is responsible. By selectively using herbicides to kill trees at the roots or directly removing developing trees, the people who maintain the rights-of-way create conditions that allow shrub species to continue spreading, developing an ever-denser cover. Once a dense shrub cover is established, the shrubs are able to outcompete invading tree seedlings. Dense cover can reduce the level of light at the soil surface sufficiently to prevent seedling growth. Shrubs may outcompete trees for moisture and/or nutrients as well. Some evidence also indicates that the litter from huckleberries, lowbush blueberries, and Greenbrier may be somewhat toxic to tree seedlings. It is desirable to slow down vegetation development (succession) on powerline rights-of-way so that access for servicing is not interrupted. Though shrub dominance often can be attributed to human efforts, it can occur naturally. Not all old fields become forests; some become shrub communities.

Riverine Zonation

Rivers are dynamic. A flowing body of water causes erosion in some areas and deposits silt and sand in others. When sediments are deposited as the energy of a river dissipates, a silt or sand bar is often created. Such bars, called *point bars*, are colonized by plants germinated from seeds transported by the river and wind. As the process of plant invasion and bar stabilization continues, it

Riverine Islands. The dynamics of river flow form point bars and sediment islands, creating habitats for successional species.

creates a series of vegetation zones. Point bars extend from curves in rivers and have a characteristic vegetation-zonation pattern.

Black Willow and various other trees and shrubs are pioneer invaders. They grow closest to the river and are followed by zones of Cottonwood and Silver Maple.

The most common invader species on point bars are Buttonbush, Swamp Dogwood, Smooth Alder, River Birch, and willows, especially Black Willow.

Eventually, riverine tree species characteristic of mature alluvial communities may dominate the bar. Sycamore, River Birch, Box-elder, Silver Maple, and Swamp White Oak become common. For more on the mature Riverine Forests, see pp. 85 (northern) and 106 (southern).

Vegetation Development Without a Soil Base

Soil provides the physical substrate in which plants take root and obtain minerals and organic matter. Embedded in the soil are the dormant seeds of the invader plants characteristic of early colonization. However, many sites, such as sand dunes and rocky outcrops, lack an initial soil base. Plants must take root in mineral-poor sand or on bare rock. Because soil is lacking, vegetation development is usually very slow compared with that in old fields.

Sand Dune Zonation

Dunes occur on the sea coast, along shores of lakes like the Great Lakes, and, occasionally, along river banks. Wind shifts sand dunes, leveling dunes in some areas and building new dunes in others. The dynamic instability brought about by wind action, especially during winter storms, makes dunes a challenging environment for colonizing plant species.

Shifting sand is but one of several significant problems that sand dunes present to plants. Wind also causes high rates of evaporation and plants may become easily dried out. Desiccation is worsened by the fact that sand holds very little water, thus dunes are essentially like cold deserts. Wind can also damage plants directly through sand abrasion, stem breakage, or outright upheaval (windthrow). On coastal dunes plants are battered by surf spray with high salt concentrations.

The plant community of a dune changes with degree of exposure. Highly exposed sites have far fewer species than well-protected sites. Most species occur in areas where protection from wind is maximized. As you traverse the dune from the water's edge inward and away from the highly exposed areas you will pass through the following ecological zones:

1. Beach
2. Foredunes
3. Interdune meadow, or swale
4. Backdunes

As you move from the beach to the backdunes, plant species richness increases and the soil changes from unstable shifting sand to more loamy sand with a thin upper layer of organic matter and minerals. Some of the same herbs and grasses found in early old fields colonize the foredunes. Those of the backdunes are mostly shrubs and trees, many of them shade-intolerant and fast-growing.

Note: Dunes are fragile. Trampling can significantly disturb the sand and destroy vegetation that has taken many years to become established. When exploring dunes, please remain on trails and never collect plant specimens.

Beach

The beach is often bare of rooted plants. It is a highly dynamic area where plants receive maximum exposure to wind and water. Very few individual plants survive for long, thus the species that occur are constantly recolonizing. Look for herbs such as Sea Rocket, Cocklebur, and Bugweed scattered well above waterline. In addition, American Beachgrass occurs in uniform small stands. Beachgrass is the most important pioneer plant (see next section). Other pioneer grasses include Sand Reed Grass and Wild Rye.

Foredunes

The foredunes result from the accumulation of sand thrown up by water and wind. Although they are intrinsically dynamic and unstable, the foredunes are nonetheless colonized by herbs, grasses, shrubs, and two tree species.

The most important colonizing species, often representing 100% of the vegetation cover, is American Beachgrass. This species is an important ecological stabilizer of the dune. Spreading by rhizomes just below the sand, the grass clones itself rapidly, with a single genet producing hundreds of ramets, all interconnected by rhizomes. The effect of the rhizome network is to stabilize the sand and reduce its tendency to shift. For this reason, dune-stabilization programs largely rely on the systematic planting and cultivation of dense stands of Beachgrass.

American Beachgrass is well adapted to withstand the rigors of dune exposure. Its blades tend to curl inward, forming a tube covering the stomata, the openings through which water is evaporated. Thus, evaporation is minimized.

Another common foredune species is Beach Pea, which is a spreading legume. Its ability to fix nitrogren probably provides a strong advantage in mineral-poor sand.

Along coastal dunes, Dusty Miller and Seaside Goldenrod are both very common. Each is adapted to the desertlike conditions found on dunes. Dusty Miller leaves are coated with fine hairs that reduce the effects of wind and aid in minimizing evaporative water loss. Seaside Goldenrod is succulent, storing much moisture in its leaves.

Foredune. This series of sand dunes on Plum Island, Massachusetts, shows characteristic patchiness, as successional species and American Beachgrass stabilize the shifting sands.

Common foredune shrubs include Poison-ivy, Northern Bayberry, various roses, Sand Cherry, Dune Willow, Beach Plum, and sumacs.

The only trees to persist on foredunes are Pitch Pine (see p. 90) and Cottonwood. Cottonwoods develop very deep root systems, enabling them to tap into water well below the surface of the dune. They are also adapted to shifting sand because they can avoid being buried by sprouting new roots from stems nearest the ground. As sand builds up around a Cottonwood, it merely sends down new roots from a branch and reestablishes itself at a new level.

Blowouts

Despite the stabilization provided by Beachgrass, Cottonwoods, and various other dune colonizers, wind will occasionally overpower a section of a foredune and cause a blowout. Blowouts attest to the intrinsic instability of sand dunes. Blowouts are recolonized by pioneer species.

Interdune Meadows (Swales)

Interdunes, located behind and below the foredunes, enjoy a measure of protection from wind and waves. Because of this protection, more species occur. All of the foredune species are present and plenty more, including species abundant in old field succession, especially the goldenrods, asters, and milkweeds. Willows are common, and old field species such as Common Juniper, Eastern Red Cedar, Quaking Aspen, and Red Maple are found. One herb, False Heather, is characteristic of interdune sites. It is a prostrate plant that grows in dense mats.

Interdune sites are often low enough to allow accumulation of fresh water, and thus wetland species such as rushes, cattails, willows, and alders are found.

Backdunes

The backdunes are sufficiently protected to permit a humus layer to develop, making the soil a sandy loam. Small forests of cherries, aspens, oaks, Pitch Pines, White Pines, and maples are interrupted by thickets of bayberry, heaths, Witch-hazel, Shadbush, blueberries, Poison-ivy, and sumacs. Many of these species have sprouted from seeds brought in by birds during their migrations. Many of the common species found in old fields occur, and species richness is high.

Dune Succession Along Lake Michigan Shoreline

These photos illustrate the pattern of dune succession along the eastern shore of Lake Michigan, in Indiana.

Beach bordering lake is sparsely populated due to unstable environment, with great exposure to wind and waves. One line of Beachgrass (a hardy, colonizing species) and a single Cottonwood manage to survive.

Just behind the foredunes, clumps of Beachgrass mix with herbaceous and woody species such as Poison-ivy. Cottonwood grows more densely here.

Winds shift sand, making dunes intrinsically unstable. This photo shows a recent blowout, where wind has cleared an area of vegetation.

Backdunes, protected, are more stable, and thus support a forest of oaks and other tree species. Still, blowouts occasionally occur, as this photo attests.

Rocky outcrops, mostly of granite, are widely scattered throughout eastern North America. Outcrops pose immense problems for colonizing plant species. Rock exposure usually ensures severe wind stress and the lack of a soil base requires plants to colonize bare rock.

Rocky outcrops have scattered mats of plants, resembling small vegetation islands on a sea of exposed rock. Plants colonize cracks, crevices, and any area affording the least bit of protection from the elements or where soil particles can become trapped. With time, some of these scattered island-mats increase both in breadth and soil depth as colonization continues and decomposition enhances soil production. Plant species richness and biomass increase and woody species sometimes invade.

As you inspect an outcrop area look for examples of each of the following kinds of vegetation patches:

1. Bare-Rock Lichen
2. Mossy Mat
3. Moss-Herbaceous Mat
4. Woody Mat
5. Mixed Species Mat

Bare-Rock Lichen Patches

The first of the bare-rock colonizers are lichens, blown in as spores in the air, along with a few mosses and several herb species. Lichens are remarkable in their abilities to grow and survive where no soil exists. Pioneer outcrop species are called crustose lichens. They hug the rockface tightly, appearing like blobs of dried paint on the rock.

A lichen represents an evolutionarily ancient mutualism between an alga and a fungus. Two very different types of plants (one of which, the alga, photosynthesizes, and the other of which, the fungus, does not) grow in such an intimate arrangement as to appear to be a single type of plant. Both the alga and the fungus require the other's presence for survival, hence the term *mutualism.* Each benefits from and needs the other. The alga captures sunlight and photosynthesizes, thus providing energy-rich food molecules to the fungus, and the fungus holds tightly to the rockface.

Lichens possess a remarkable ability to tolerate prolonged dryness, a great advantage on exposed sites. Crustose lichens grow outward in rings, like slow-motion ripples in a pool, continually expanding to cover an ever-widening area. As it expands, the lichen builds up organic matter and traps soil particles, occasionally making it possible for other plants such as foliose (upright) lichens, mosses, and some herbs to invade.

A common herb is Stonecrop, a winter annual. Stonecrop is a creeping plant that forms mosslike mats.

Mossy Mat

Mossy mats have dense moss cover overgrowing the still-present lichens. Mosses take advantage of moisture trapped by the growing amount of organic matter from both the lichens and herbs. Moss growth adds organic material, building a soil base, enhancing the water retention of the mat.

Moss-Herbaceous Mat

Once a moss mat is well established, grasses, sedges, horsetails, ferns, and herbs sometimes invade, having arrived by windborne spores or seeds. The mat begins to have a vertical complexity as soil depth, soil moisture, and organic matter accumulate further and plant richness increases.

Woody Mat

If a sufficient depth of soil is built up, the more extensive root systems of woody species can be supported, and tree species such as Red Cedar and Red Maple grow on the mat along with various shrubs. Depending upon factors such as degree of exposure, soil depth, and water, woody plants may grow to normal size or be stunted. Herbaceous species can be shaded out by the woody species.

Mixed species Mat

Old mats sometimes support such species as White Pine, Eastern Hemlock, rhododendrons, and other forest species. If this point is reached, you could be standing on the rocky outcrop and not know it. The vegetation cover will be complete and will have largely converged with the typical plant community of the surrounding area.

Rate of Vegetation Change on the Mats

As in dunes, exposure is often severe enough that only lichens and a scattering of moss are able to persevere. Even when the area is sufficiently protected for moss-herb mats to form, many centuries may pass without the formation of mixed species mats. The thin soil base is delicate and subject to damage by storms, drought, and fire. Rocky outcrops have very slow rates of vegetation development.

Overview of a granite outcrop in Georgia, showing mats of vegetation. Dead trees illustrate the severe and unstable nature of this exposed environment.

Species such as lichens and Stonecrop grow directly on the bare rock face and do not require a soil base. They are the initial colonizers.

In more protected crevices, where soil particles can be trapped, mosses and annuals invade the lichen-Stonecrop mat. The mosses and lichens help trap more soil.

Photos on pp. 208–9 courtesy of D. J. Shure.

If soil builds up, the mats become increasingly stabilized and perennials invade, continuing to thicken the soil base.

Stable mats with an adequate soil base are invaded by bunch grasses and shrubs.

Trees can eventually invade vegetation mats on rocky outcrops.

LANDSCAPE PATCHES, FOREST FRAGMENTATION

Eastern North America is a patchwork of varying landscapes. Rather than large uninterrupted tracts of forest, there is instead a mosaic of agricultural, suburban, and urban landscapes imposed on the natural ecology of any area. One excellent way to get an overview of this mosaic is to fly for an hour or so in a small airplane, but the basic landscape types can be seen from ground level as well. Each of the various components of this mosaic is a *landscape patch,* a distinct unit with particular ecological characteristics. Look around at the various landscape patches in your area. This section will be your guide to recognizing landscape patch types.

Forest Islands

When is a forest a forest and when is it a woodlot? We tend to associate the term "forest" with a large tract and the term "woodlot" with a small tract. As suburbia and agriculture have each claimed ever-increasing amounts of land, forests in many areas have become fragmented, reduced to a series of scattered *forest islands.* Unbroken forest tracts, with few exceptions, are uncommon. Instead, you see "archipelagos" of forest islands surrounded by "seas" of agricultural areas, old fields, housing developments, and industrial parks. Some forest islands are relatively large, from 200 acres to several square miles, but many are quite small, a few

Forest Island. Because of its isolation from other forest and its small area, this forest island will support relatively few species.

acres or less. The size of the forest island is very important. The home ranges and territories of animals are size-dependent. For instance, most large hawks and owls require at least a square mile to sustain a territory. These species will be eliminated from small forest islands. Other bird species are classified as forest-interior species, because they must have territories completely within a forest, not along a forest edge (see below). You can demonstrate the effect of area on bird diversity by taking a census of the nesting and/or wintering birds of several different-sized forest islands in your area. The effect of forest fragmentation on bird communities is discussed further on pp. 214–19.

Other Patch Types

Patches are communities or species assemblages surrounded by other dissimilar assemblages. A forest island, surrounded by alfalfa fields, is a type of patch. A patch is a small but recognizable landscape, with its own identity and characteristics. Patches fall into the following categories:

REMNANT PATCH: A forest island within an agricultural or suburban area is a remnant patch. The forest is cut away and removed save for a small area that remains, perhaps to be a picnic area, a park, or simply because it has by chance been spared. Remnant patches are useful in showing what the mature forest community would likely be in a given area.

SPOT-DISTURBANCE PATCH: These are temporary patches, created by fire, weather, or animals. A fire in a Pine-Oak Forest opens up a section of the tract. A late-summer hurricane blows down trees in a forest, creating many forest gaps where light-loving species will thrive. A Gypsy Moth invasion defoliates several hundred acres of Oak-Hickory Forest. Any area undergoing vegetation change is a spot-disturbance patch.

EPHEMERAL RESOURCE PATCH: This patch type is similar to a spot-disturbance patch, but it is caused by short-term resource fluctuations. For example, flocks of blackbirds may roost in a Red Maple swamp for several days while migrating north. Thousands of Tree Swallows may gather to feed on Northern Bayberry during autumn migration. Ephemeral patches are of short duration, and depend on short-lived fluctuations in resource levels.

UNIQUE COMMUNITY PATCH: The heath balds of the Appalachians, the alpine tundra communities of the New England mountains, and the areas supporting Kirtland's Warbler populations in young Michigan Jack Pine Forests are examples of patches that have value for their unique ecological characteristics. It is very important that they be recognized and preserved.

INTRODUCED PATCH: Introduced patches are created by and depend on humans. Examples are pine plantations, cornfields, orchards, and golf courses. Without continued maintenance by humans, these patches would usually not persist and would be replaced by native ecological communities.

Corridors and Peninsulas

Corridors are connections between or paths through patches. As in patches, several types of corridors are recognized.

LINE CORRIDOR: A road, hedgerow, property boundary, or even a drainage ditch are all examples of line corridors. Line corridors are narrow, but they do provide habitat for certain species. A drainage ditch through agricultural fields can support cattails in which Red-winged Blackbirds can nest, shrubs in which Song Sparrows can nest, as well as habitat for various frog species. Hedgerows support many plant and animal species that would otherwise be confined to old fields or forest edges.

STRIP CORRIDOR: Sometimes two forest islands will be connected by a relatively narrow strip of forest between them. This is a strip corridor. Strip corridors are very important because they permit dispersal of species from one forest patch to another. They are the "highways" by which species can move without having to leave their preferred habitat type.

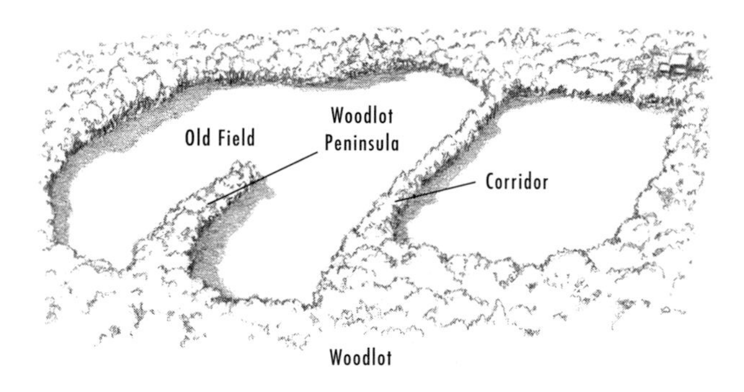

Fig. 24. Landscape features.

STREAM CORRIDOR: Streams and rivers are bordered by hydric (water-adapted) communities (see p. 29). Stream corridors typically border larger water courses, forming small tributaries that feed into the larger water course. The presence of stream corridors adds diversity by providing habitat for hydric and mesic species. They also act as buffers between the land and the river, helping to control runoff and mineral loss and serving to reduce flooding.

PENINSULA: A peninsula is a narrow portion of an otherwise large patch that extends from the patch into surrounding landscapes. A strip of Oak-Hickory Forest protruding into a wheat field or separating two industrial parks is a peninsula. Though peninsulas are comprised of the same species that characterize the larger patch, overall species diversity commonly decreases toward the tip of the peninsula.

The Ecology of Patches and Corridors and the Effect of Forest Fragmentation

By surveying a few forest islands of varying size you will soon begin to realize that area and species richness are related. In general, the larger the area, the more species occur. You simply will not find as many different kinds of plants and animals in small patches as in large patches. The fragmentation of forests into smaller and smaller forest islands is contributing to the decline of certain bird species (see following section), because the forest islands are not large enough for some species to establish a territory.

Forest fragmentation also isolates tracts of forest, making it less possible for species to disperse or recolonize because of unsuitable landscapes between the forest fragments. This is why corridors, particularly strip corridors, are important. Strip corridors act as "friendly highways" of habitat connecting two otherwise separate fragments. If a forest is to be cut to make room for houses or agriculture, it is best to leave a connecting strip corridor between forest fragments. Doing so partially offsets the effects of reduced forest area.

Frequency of disturbance also contributes to the landscape diversity of an area. Disturbance presents opportunities for many species that would otherwise fare poorly. A patchwork of old fields of different ages adds significantly to the diversity of an area. Associated with this concept is edge effect (see ecotones, p. 35) and the presence of hedgerows and other "mini-habitats." Forest edges support a mixture of forest and old field species, as do hedgerows.

Birds of Old Fields, Edges, and the Forest Interior (Pl. 24)

Just as old fields are recognizable by their component plant species, so too do they contain distinct bird communities. Since European settlement, many species of old field birds have probably increased significantly in population. Depending on the age of the old field, one of three groups of birds will be present.

Birds of Early Old Fields

The following four species were formerly abundant residents of eastern grassy old fields in early stages of vegetation development. With the large-scale decline in these types of habitats, in part due to human encroachment for development and in part due to regrowth of forest, these species have each undergone major population declines in many places throughout the East.

EASTERN MEADOWLARK

(9.75 in.) Two, and possibly three, species of meadowlarks occur: the Eastern, the Western, and a southwestern variety that may be a separate species. They all look very much alike: chunky, mottled brown birds with bright yellow breasts, a black V separating throat from breast, and white outer tail feathers, conspicuous in flight. The Eastern Meadowlark is most easily identified by song, a melodious, whistled *ee-ohh-lee-oh-dah-lee,* quickening and rising at the end. Birds frequently perch on shrubs and fence posts when singing. Meadowlarks walk on the ground in search of arthropods and worms. Nests are well hidden, on the ground, covered by grass.

VESPER SPARROW

(6 in.) Like the meadowlarks, this streak-breasted largish sparrow has conspicuous white outer tail feathers. When seen closely, it has a thin eye-ring and a russet shoulder patch on its wing. It breeds mostly in northern states and migrates to winter in southern states. Nest is hidden on the ground, usually at the base of shrubs or grassy clumps.

GRASSHOPPER SPARROW

(5 in.) This small chunky sparrow has a flattened head and pale, unstreaked breast. It is named for its brief, buzzy, insectlike song, often given while the bird is perched atop a grass clump, small shrub, or fence post. Often nests in hayfields and other grassy, agricultural land, habitats that are in serious decline throughout much of its northeastern breeding areas. Winters in Deep South.

UPLAND SANDPIPER

(12 in.) Though sandpipers are usually (as the name implies) associated with beaches and marshes, this species has adapted to prairie and grassland habitats and is almost never encountered along the ocean shore. It often perches on fence posts, raising its long, sharply pointed wings as it alights. Its song is a haunting, loud whistle, often given in flight, as wings are held stiffly, almost quivering. Breeds only in northern states in East and Midwest, wintering in southern South America. Nest is on the ground, well covered by dense grasses. In the East, the species is often found around airports, where short prairielike grass suits its ecological needs.

Birds of Mid-successional Old Fields

As ecological succession unfolds, open, grassy fields and meadows are invaded by woody shrubs and trees, creating new habitats suitable for an array of bird species not found either in open fields or closed woodlots. Many of these species have thrived in the past century as a result of an abundance of suitable habitat. Like the species noted above, however, some of these species are now declining in parts of their ranges as forest reclaims old fields.

FIELD SPARROW

(5 in.) As shrubs and small trees replace grasses in old fields, Grasshopper Sparrows disappear and Field Sparrows enter. The Field Sparrow is best identified by its pink bill, russet cap, and unstreaked breast. Its plaintive song is a pleasant, whistled trill, usually rising at the end. The species is a permanent resident over much of the East and has adapted well to powerlines. Field Sparrows feed on many kinds of animals (arthropods, worms) as well as seeds. Nests may be placed on the ground or in a small bush or tree.

EASTERN KINGBIRD

(8 in.) This flycatcher of open country is a common summer resident throughout all of the eastern and central states as well as southern Canada. It is usually seen perched on a wire, fence post, or treetop, where it sits awaiting potential insect prey to snatch in midair. Noisy but nonmusical, the chattering, buzzing song is often given in flight, when the tail is spread, revealing the definitive white outer tail band. Kingbirds are aggressive and routinely harass crows and hawks in flight. The species flocks in fall and migrates to winter in South America, where they feed mostly on fruits.

GRAY CATBIRD

(8.25 in.) This slender gray relative of the Northern Mockingbird shares some of its cousin's ability to mimic other birds. But the catbird's unique vocal attribute is its distinctive catlike mewing call, often given when the bird is well hidden in dense shrubs. Catbirds remain common throughout most of their breeding range, which extends westward to northern Idaho and parts of British Columbia. They are persistently vocal and often perch atop shrubs while singing. Their diet includes many kinds of small animals as well as much fruit. Seeds of some species of old field shrubs and trees are disseminated by catbirds. The species is migratory, some wintering in the Deep South, but most in Central America.

BROWN THRASHER

(11.25 in.) A close relative of the Gray Catbird and Northern Mockingbird, the Brown Thrasher is rich russet brown above with a heavily streaked pale breast and two white wing bars. Thrashers are persistent singers, their complex whistles and squeaks often given while the bird is perched atop a tall tree. Thrashers often forage on the ground, tossing leaves as they search for animal prey. Like catbirds, they feed heavily on fruit and are likely to be important seed dispersers. Thrashers nest throughout the East, often near dwellings, where they forage on lawns. They are permanent residents throughout the South, and northern birds migrate to winter in southern states.

INDIGO BUNTING

(5.5 in.) Reasonably common in northern states, this colorful species remains abundant in suitable habitats throughout much of the South and Mid-Atlantic states. It thrives in ecotones between old field and woodland and has benefited from creation of powerlines, which produce suitable edge habitat. Males are deep blue, especially on the head, while females are uniform rich brown. Song is a melodious whistle, distinctive because each part of it is repeated. Winters in Central America.

PAINTED BUNTING

(5.25 in.) The male Painted Bunting ranks among the most striking of North American passerines, an extraordinary assemblage of brilliant colors. Females are uniform green. Once relatively widespread throughout southern states, the species is in decline in many areas, though it remains common in much of Texas, Louisiana, and Arkansas. The preferred habitat is successional areas in which there is dense undergrowth and thickets. Will come to bird feeders. Winters in southern Florida, the Caribbean, and Central America.

BLUE GROSBEAK

(7 in.) At first sight, this species resembles a larger version of the Indigo Bunting, with which it often shares habitat, especially in the South. It is distinctive for its larger body size, much thicker bill, and chestnut wing bars. Buntings and grosbeaks consume many kinds of seeds, though during breeding season they also feed heavily on worms and arthropod prey, as nestlings require much protein for growth. Essentially a southern species, Blue Grosbeaks extend as far west as Arizona and parts of California. Most recently, the species seems to be expanding its range northward. Winters mostly in Central America.

COMMON YELLOWTHROAT

(5 in.) This masked member of the colorful wood warbler family is found in brushy fields, wet meadows, and freshwater marshes. Though still common, the species has been reduced in some places by draining of marshland. Male is unmistakable, with sharply defined black facial mask and bright yellow breast. The female, like the male, is olive on the back with a yellow throat, but lacks the mask. Male Common Yellowthroat sings a staccato *witchity, witchity,* either from cover or perched atop a reed or cattail. Found throughout much of North America; winters mostly in Central America.

YELLOW-BREASTED CHAT

(7 in.) By far the largest member of the wood warbler family, the chat is a secretive bird of overgrown successional fields with dense shrub cover. It is easy to identify, with bold white eye "spectacles" and a bright orangy yellow throat and breast. Chats feed heavily on fruits such as blackberries, elderberries, and various wild grapes. Because of habitat loss, the species is declining in many places throughout the East. Chat songs are noisy, ventriloquial, and varied. Courting males often sing while flying butterflylike, legs drooping beneath them. Winters mostly in Central America.

NORTHERN BOBWHITE

(10 in.) The familiar eastern quail, named for its demonstrative, whistled, *bob-white . . . oh bob-white!,* is in decline in much of its former range because of habitat loss. It is gone from many formerly inhabited areas in the Northeast, though it remains common throughout much of the South. Ideal habitat is a succes-

sional field near hedgerow or wooded edge. Thrives in ecotones between forest and old field. The species is generally a permanent resident throughout its range. Generally a flocking species, groups, often numbering a dozen or more, are called "coveys." Though colorful, bobwhites are cryptically colored against their normal background. Males and females differ in that throat and eye-stripe are white in males, tan in females. Feeds on a wide variety of animals and plant foods.

Forest Interior Species

The majority of land bird species in eastern North America require a wooded habitat. Obviously, birds such as woodpeckers, nuthatches, and creepers, all of which forage directly on bark, require trees. So too do species such as Wild Turkeys and Blue Jays, which are dependent on nut crops. Many insectivorous species, the titmice, warblers, orioles, and tanagers, feed among the foliage of forests.

What has been recognized recently is that certain forest species survive well and even thrive along forest edges, while other species are strictly confined to the forest interior. You will not find these species in small forest islands or along forest edges—you must seek them well into a large forest. Finally, there are species that occur along edges but which support much larger populations in interior forests. Species such as the White-breasted Nuthatch, Veery, Ovenbird, and Scarlet Tanager fall into this category. The strict requirements of certain forest interior species make this group very vulnerable to forest fragmentation.

FOREST INTERIOR-EDGE SPECIES

Yellow-billed Cuckoo
Red-bellied Woodpecker
Downy Woodpecker
Great Crested Flycatcher
Blue Jay
Black-capped Chickadee
Tufted Titmouse
Carolina Wren
Blue-gray Gnatcatcher
Wood Thrush
Gray Catbird
Common Flicker
Eastern Wood-pewee
Eastern Phoebe
White-eyed Vireo
Yellow-throated Vireo
Red-eyed Vireo
Common Yellowthroat
Northern Cardinal
Rose-breasted Grosbeak
Rufous-sided Towhee
American Goldfinch

Broad-winged Hawk
Ruffed Grouse*
Barred Owl
Hairy Woodpecker*
Pileated Woodpecker*
Acadian Flycatcher
Red-breasted Nuthatch*
White-breasted Nuthatch*
Brown Creeper*
Veery*
Hermit Thrush
Black-throated Green Warbler
Cerulean Warbler
Black-and-white Warbler
American Redstart
Ovenbird*
Louisiana Waterthrush
Hooded Warbler
Canada Warbler
Scarlet Tanager*

Birds recognize and prefer certain types of habitat. Every birder knows that you associate certain species with certain habitats. Don't bother looking for Field Sparrows in an Oak-Hickory Forest and forget about finding Hermit Thrushes hopping between ragweed stalks. The factors by which birds recognize and orient themselves to their chosen habitats are poorly known. What is becoming more apparent, however, is the degree of sensitivity that certain species exhibit toward edge effect. While some species thrive in late old fields and forest edges, some decline radically in such areas. One Connecticut study has shown that Hermit Thrush, Black-throated Green Warbler, and Cerulean Warbler were absent from all forests with tracts smaller than about 460 acres. Forest fragmentation can devastate these populations.

Other species, such as the Brown Creeper and White-breasted Nuthatch, though true forest species, have adapted to suburbia as long as trees are around and there is access to forest. Even the large Pileated Woodpecker will occasionally come to bird feeders for suet. These species are much less abundant along forest edges than in the forest interior, however.

Still other species seem like true interior-edge dwellers. Many of these, such as the Gray Catbird, Northern Cardinal, and Rufous-sided Towhee, are actually more closely associated with old fields, though they must have woody vegetation.

*Found also along edges but in lower populations than in interior forest.

PLATE 20

HERBS OF OLD FIELDS — ADAPTATIONS (PP. 168, 171, 173)

The species illustrated show an array of adaptations common to old field herbs. See text for details. See also Pls. 40 and 41.

COMMON DANDELION

An abundant alien; readily invades lawns as well as old fields. Perennial—survives the winter by storing energy in a long, thick taproot, from which sprout multiple flower stalks. A member of the composite family: flowerhead is a dense cluster of bright yellow flowers that attract insect pollinators. Seeds are dispersed by wind, drifting with the help of tiny, parachute-like hairs. Leaves are usually deeply lobed and spread as a rosette over the ground.

COMMON MULLEIN

Biennial—survives the winter as a rosette of leaves. Long taproot firmly anchors plant and stores energy. Leaves large and velvetlike, covered by tiny hairs that help protect the leaf from drying out and from insect damage. Flowers throughout summer, with individual flowers on a long spike that can reach 6 ft. tall, making them easy for pollinating insects to locate. Seeds line stalk and, once in soil, maintain viability for a hundred years or more. Alien; snapdragon family.

BEGGAR-TICKS or STICKTIGHT

Seeds are in achenes, each of which has 2 barbed awns that cling to fur and make it easy for seeds to be dispersed by animals. Plate shows mature flowerhead with cluster of achenes and 2 individual achenes. Native; composite family. See also Pl. 25.

SHEPHERD'S PURSE

An annual or winter annual (may survive a single winter in rosette form). Taproot thin but highly branched. Basal leaves form a flattened rosette. Much energy put into reproduction. Each stalk contains many small white flowers that become seedpods. Each pod contains a few or many tiny seeds. Alien; mustard family.

COMMON YARROW

Perennial; spread not only by seed but also by rootstocks. A composite; flowerheads are large, flattened clusters of tiny white flowers, affording a wide surface for pollinators to traverse. Flowerhead tops plant, easily visible to pollinators. Leaves fernlike, a shape that provides extensive surface area for photosynthesis but requires less actual energy to make the leaf. Alien.

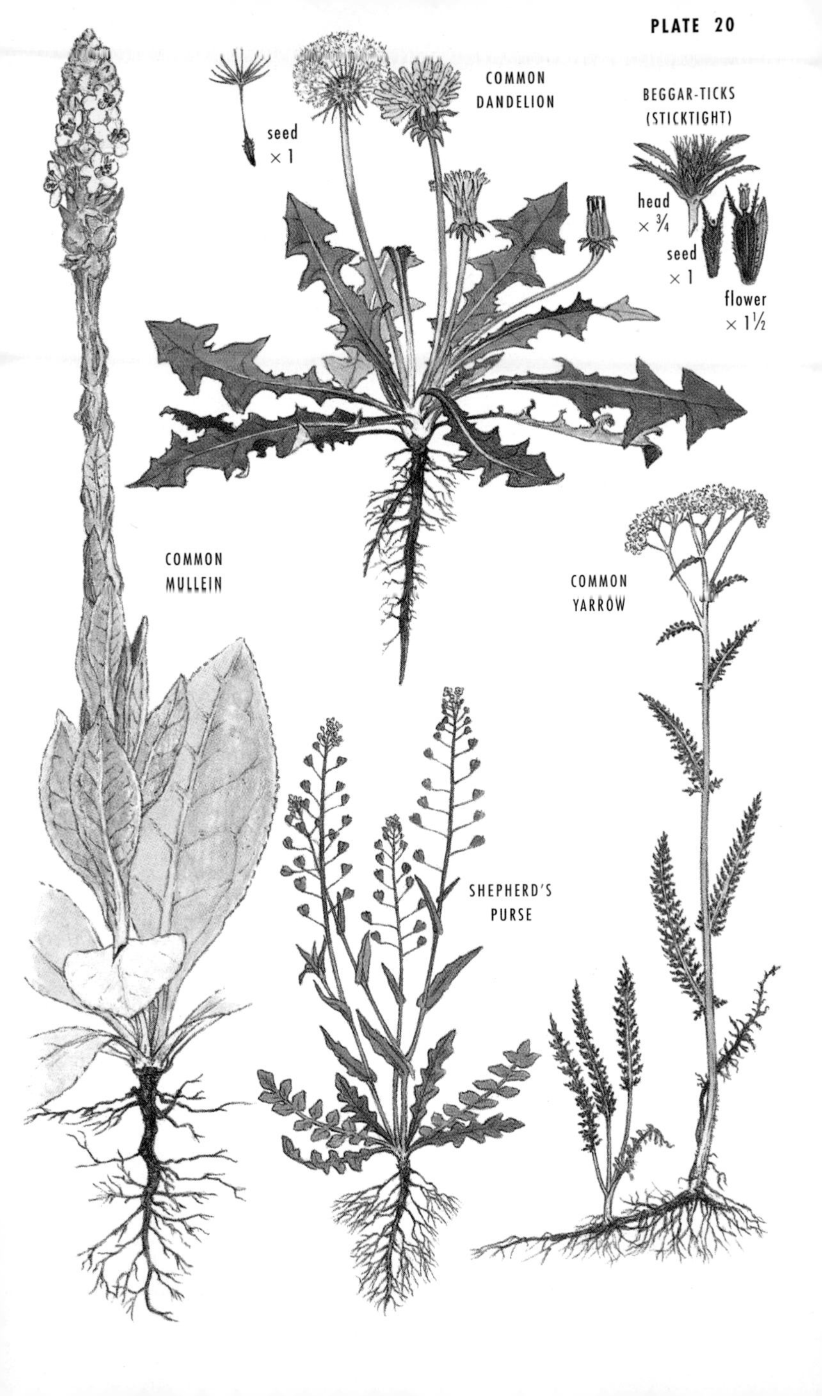
COMMON
DANDELION
seed
× 1
BEGGAR-TICKS
(STICKTIGHT)
head
× ¾
seed
× 1
flower
× 1½
COMMON
MULLEIN
COMMON
YARROW
SHEPHERD'S
PURSE

HERBS OF OLD FIELDS — BIENNIALS (P. 171)

Biennials are less abundantly represented than annuals and perennials, but the species shown here are quite common.

MOTH MULLEIN

Long stalk of 5-petaled yellow flowers with reddish centers. Flowers occasionally white. Leaves oval, slightly toothed. Stem up to 3 ft. tall. Blooms late spring through early fall. Alien; snapdragon family.

COMMON EVENING-PRIMROSE

Pale yellow, 4-petaled flowers that bloom in late afternoon. Green seedpods. Leaves lance-shaped, slightly toothed. Stem to 5 ft. tall. Blooms early summer through late summer. Native.

QUEEN ANNE'S LACE or WILD CARROT

Tall stalk with a flat, dishlike cluster of tiny white flowers (with 1 tiny purplish flower in the center). Leaves fernlike. Stem hairy, to 3 ft. tall. Blooms spring through fall. Taproot has the odor of carrot. Alien; parsley family.

SPOTTED KNAPWEED

Resembles thistle but not prickly. Flowers on wide-spreading branches; can be pink, reddish, or white. Leaves deeply lobed, with smooth margins. Stem to 4 ft. tall. Blooms throughout summer. Alien; composite family.

BULL THISTLE

Large, reddish purple flowerheads. Leaves very stiff and prickly, with sharp, spiny lobes. Stem to 6 ft. tall, very spiny. Blooms throughout summer. Alien; composite family.

COMMON BURDOCK

Resembles a thistle, with reddish flower rays (tipped with white) and a bulbous, bristly, green base (bracts). Leaves widely oval, smooth along margins. Stem to 5 ft. tall. Blooms midsummer through fall. Alien; composite family.

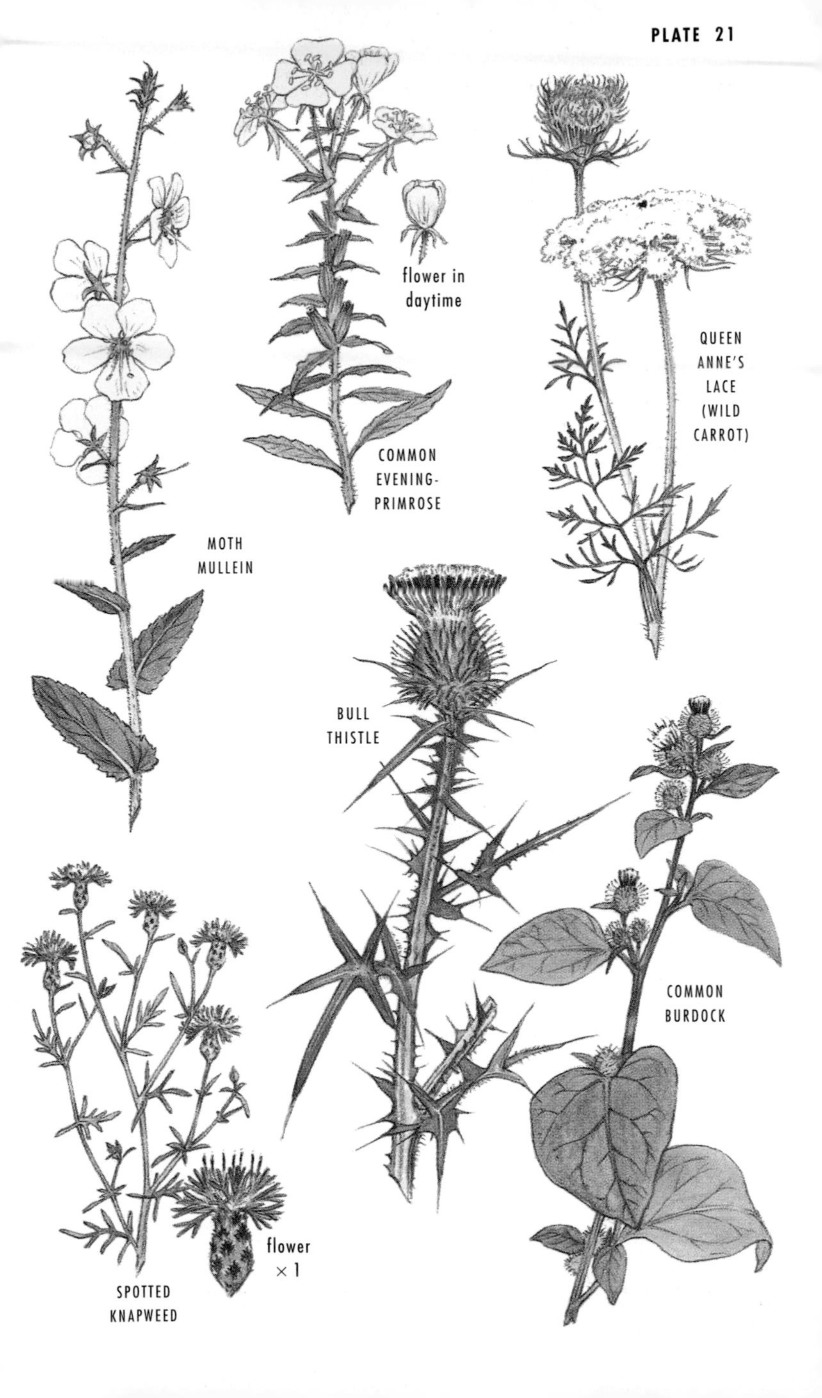
flower in
daytime
COMMON
EVENING-
PRIMROSE
QUEEN
ANNE'S
LACE
(WILD
CARROT)
MOTH
MULLEIN
BULL
THISTLE
COMMON
BURDOCK
flower
× 1
SPOTTED
KNAPWEED

HERBS OF OLD FIELDS — NATIVE PERENNIALS (P. 179)

This plate includes a few of the most abundant and widespread of the native perennial herbs.

FIREWEED

Tall (to 7 ft.), upright plant with stalk of red, 4-petaled flowers. Leaves lance-shaped, with slightly wavy margins. Invades fire-swept areas. Blooms mid- through late summer. Evening-primrose family.

COMMON CINQUEFOIL

One of many cinquefoil species (named because leaflets and flower petals come in fives), some alien, some native. This species spreads as a vine, and resembles a yellow-flowered strawberry. Leaf palmate, leaflets notched. Blooms early spring through early summer. Rose family.

INTERMEDIATE DOGBANE

Pale pink (often white), tubular flowers. Opposite leaves, oval in shape, untoothed. seedpods elongate, in pairs attached at base. Stem to 4 ft. tall. Blooms throughout summer. Dogbane family.

HORSE-NETTLE

White, soft, 5-petaled flowers with yellow, elongate anthers. Berries bright orange. Leaves wide, with large teeth. Stems prickly. Blooms spring through late summer. Tomato family.

BUTTERFLYWEED

Bright orange, flat-topped sprays of flowers. Elongate green pods. Leaves oblong, untoothed. Stem hairy, to 2 ft. tall. Blooms throughout summer. Milkweed family.

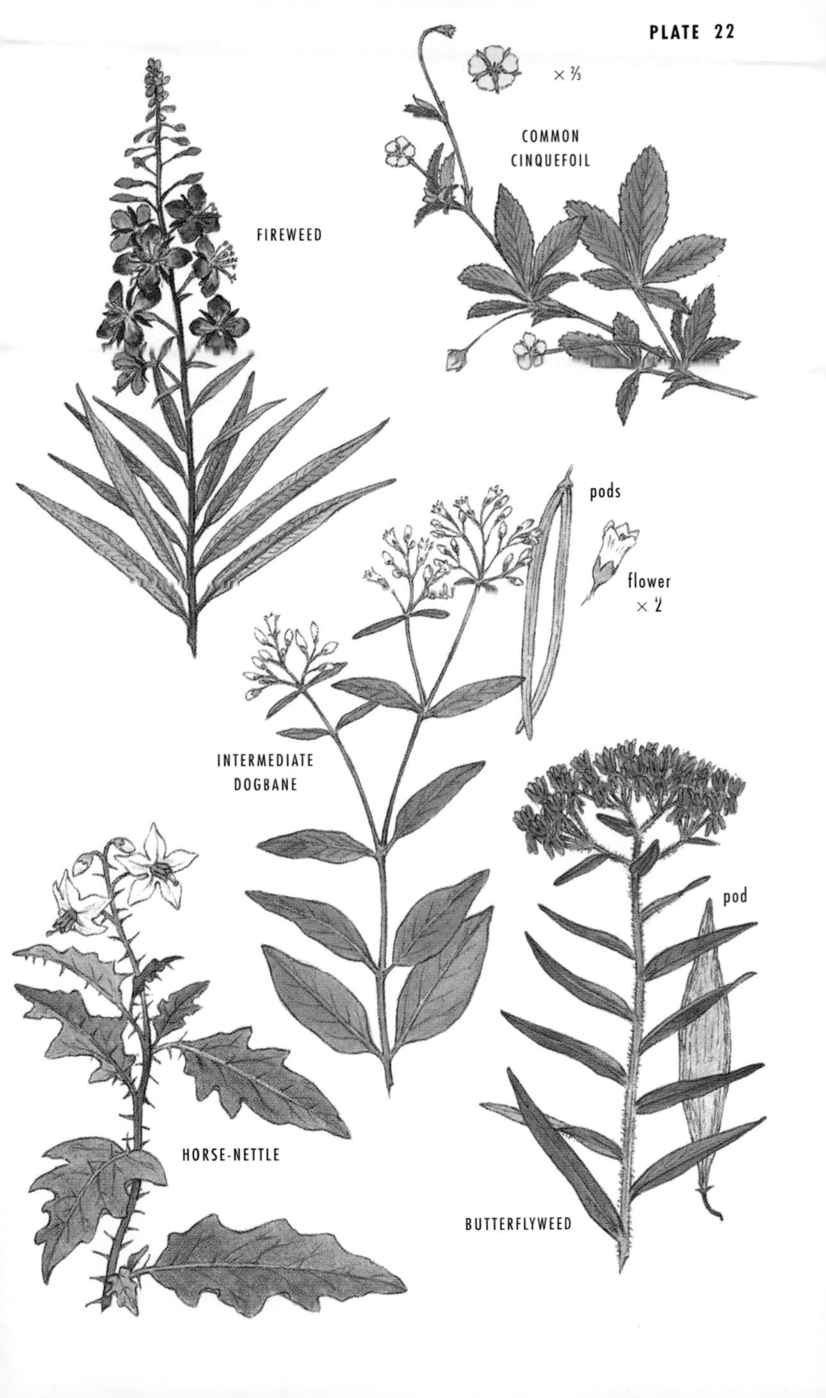
× ⅔
COMMON
CINQUEFOIL
FIREWEED
pods
flower
× 2
INTERMEDIATE
DOGBANE
pod
HORSE-NETTLE
BUTTERFLYWEED

PLATE 23

SELECTED GOLDENRODS AND ASTERS (PP. 176, 177)

Goldenrods and asters flower from midsummer through fall. Most species inhabit old fields and meadows, though some occur in open woodlands and forests. All are members of the composite family, and all are native. See also Pls. 41 and 43.

EARLY GOLDENROD

A plume-like goldenrod, to 4 ft., with a pair of tiny leaflets at the axil (base) of each leaf. Basal leaves oval, toothed.

TALL GOLDENROD

To 6 ft.; plume-like, with a hairy stem and lance-shaped leaves. Most leaves toothed, some smooth-margined. Similar to **Canada Goldenrod** (see Pl. 41).

ROUGH-STEMMED GOLDENROD

Plume-like; to 7 ft. tall but usually shorter. Very hairy leaves and stem. Leaves oval, sharply toothed.

LANCE-LEAVED GOLDENROD

Flat-topped goldenrod, to 4 ft. tall. Leaves lance-shaped, untoothed, but edges rough.

DOWNY GOLDENROD

Wandlike goldenrod, to 3 ft. tall. Most common on dry sandy soils. Leaves elliptical, slightly toothed. Stem purplish.

WHITE WOOD-ASTER

Flat-topped spray of flowerheads with relatively few white rays, yellow centers. Leaves arrowhead-shaped. Upper leaves with smooth margins, lower leaves toothed. Found in open woodlands and near edges.

BUSHY ASTER

Dense clusters of flowerheads; rays white or pale blue, centers yellow. Leaves slender, elongate, untoothed; some leaves have a pair of leaflets at base. Frequents sandy soils, old fields.

PURPLE-STEMMED ASTER

Stem purple, hairy. Rays violet to blue. Leaves clasp stem and are hairy, usually toothed. Found in moist areas, meadows. To 7 ft.

typical
flower
EARLY
GOLDENROD
TALL
GOLDENROD
ROUGH-STEMMED
GOLDENROD
LANCE-LEAVED
GOLDENROD
BUSHY
ASTER
DOWNY
GOLDENROD
WHITE
WOOD-ASTER
PURPLE-
STEMMED
ASTER

PLATE 24

BIRDS OF OLD FIELDS (P. 214)

Early Old Fields

EASTERN MEADOWLARK

A robin-sized, chunky bird. Yellow breast with a black V. White outer tail feathers. Rapid wingbeats alternating with short glides.

UPLAND SANDPIPER

A large (12 in.), mottled brown sandpiper of the grasslands. Longish neck with dovelike head. Often perches on fence posts.

GRASSHOPPER SPARROW

A chunky sparrow with a striped crown and unstreaked breast. Sharp tail. Song an insectlike buzz.

VESPER SPARROW

A largish sparrow with white outer tail feathers and an eye-ring. Heavily streaked. Rufous wing patch not easily visible.

Mid-age Old Fields

EASTERN KINGBIRD

An 8-in. flycatcher, often perched on wires or in treetops. Dark above, with a white band at tail tip. Voice a rapid series of *kitters.*

BROWN THRASHER

Slim, robin-sized; chestnut above, with brown streaks on breast. Long tail. Yellow eyes.

GRAY CATBIRD

Well named—mews like a cat. An all-gray, robin-sized bird with a black cap and rufous patch under tail. See also Pl. 39.

COMMON YELLOWTHROAT

A small yellow bird. Male has a black "mask"; female yellowish, lacks mask. Song a rapid, static *witchity, witchity, witchity.*

YELLOW-BREASTED CHAT

The largest warbler (7 in.). White "spectacles," bright yellow breast. Song a complex series of loud whistles, some harsh.

BLUE GROSBEAK

Cardinal-sized. Male deep blue with wide brown wing bars; female brown. Male often sings while perched on a telephone wire.

INDIGO BUNTING

Sparrow-sized. Male deep blue with no wing bars; female uniformly rich brown. Song a series of two-note whistles.

PAINTED BUNTING

Sparrow-sized. Male very colorful; female dull green. Warbling song, often delivered from wire or treetop. South only.

FIELD SPARROW

A pink-billed sparrow with an unstreaked breast and a rufous crown. Song a pleasant trill.

NORTHERN BOBWHITE (QUAIL)

A small, chickenlike bird with rufous streaking. Facial stripes and throat white on male, tan on female. Voice a strident *bob-white!*

EASTERN
MEADOWLARK
GRASSHOPPER
SPARROW
VESPER
SPARROW
UPLAND
SANDPIPER
EASTERN
KINGBIRD
BROWN
THRASHER
GRAY
CATBIRD
male
COMMON
YELLOWTHROAT
BLUE
GROSBEAK
male
INDIGO
BUNTING
male
male
PAINTED
BUNTING
FIELD
SPARROW
NORTHERN
BOBWHITE
(QUAIL)
YELLOW-
BREASTED CHAT

PLATE 25

HERBS OF OLD FIELDS — NATIVE ANNUALS AND ONE ALIEN ANNUAL (P. 166)

POOR-MAN'S-PEPPER

Also called Peppergrass. Spikes of tiny white flowers that become small pods. Each pod contains a peppery-tasting seed. Blooms early summer through autumn. Leaves elongate, toothed, on short stalks.

DAISY FLEABANE

Resembles small daisy or aster. Thin white petals surrounding a yellow center. Blooms from spring throughout the summer. Leaves toothed. Leaves and stem hairy.

HORSEWEED

A tall (up to 7 ft.) herb with many small, greenish yellow flowers with tiny whitish rays. Flowers along stalks. Thin, lance-shaped leaves. Blooms midsummer through fall.

COMMON RAGWEED

To 5 ft. Tiny yellow-green flowers on long spikes. Leaves "dissected," with deep, elaborate lobing. Stem hairy. Flowers from mid- to late summer and into fall.

GREEN AMARANTH or PIGWEED

The only alien (non-native) species on this plate. Dense greenish flowers at bases of upper leaves, becoming a spike. Leaves oval, on stalks, with smooth margins. Blooms late summer and fall.

PENNSYLVANIA SMARTWEED

One of many species of smartweed, some of which are alien. Small spikes of densely clustered, reddish pink flowers. Leaves attach to stem by a knotted "sheath." Leaves elongate, untoothed. Upper branches and stem hairy. Flowers spring through fall.

BEGGAR-TICKS or STICKTIGHT

To 4 ft. Bright yellow, daisylike flowers. Leaves compound, with 5 coarsely toothed leaflets. Seeds in structures called achenes, each with 2 barbed awns (see Pl. 20) that stick to fur and clothing. Flowers midsummer through fall.

SPOTTED SPURGE

To 3 ft., often with many branches. Small, inconspicuous flower clusters. Leaves opposite, oval, slightly toothed. Stem, when broken, exudes milky sap.

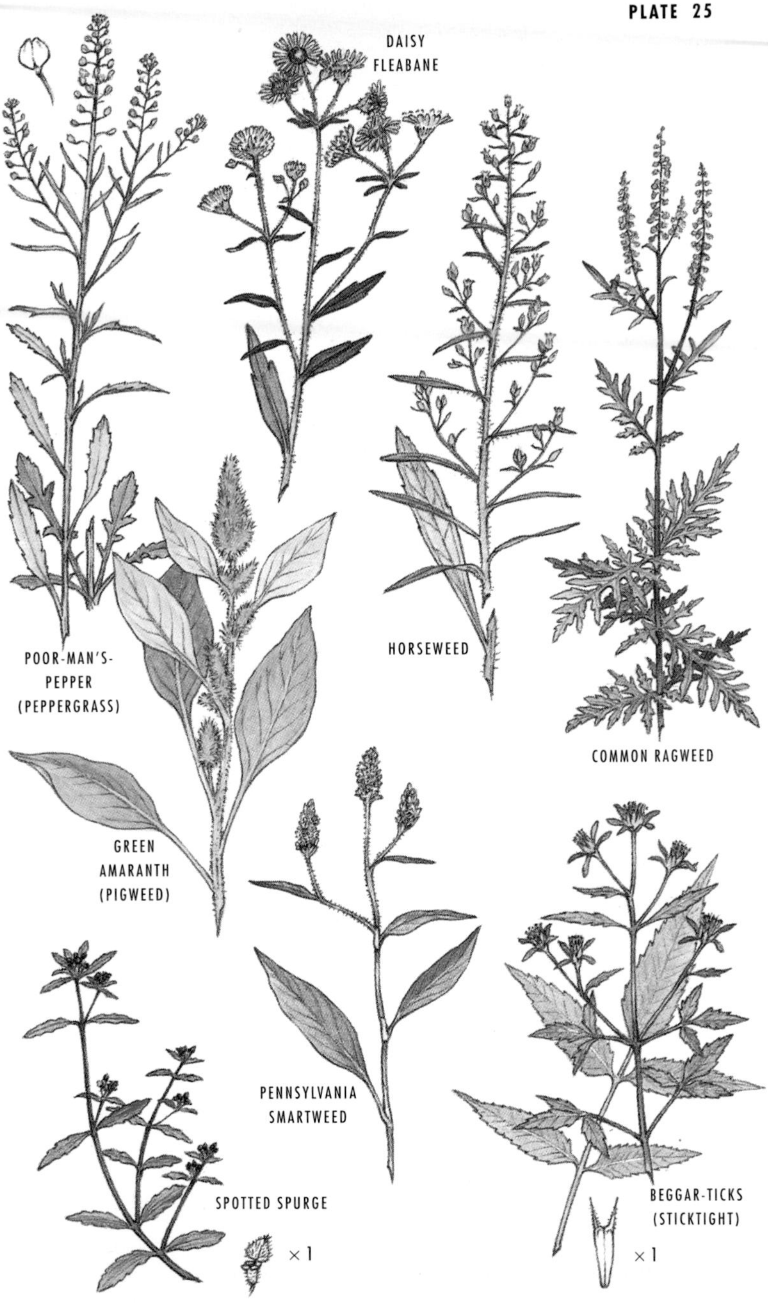

DAISY FLEABANE

POOR-MAN'S-PEPPER (PEPPERGRASS)

HORSEWEED

COMMON RAGWEED

GREEN AMARANTH (PIGWEED)

PENNSYLVANIA SMARTWEED

SPOTTED SPURGE

× 1

BEGGAR-TICKS (STICKTIGHT)

× 1

HERBS OF OLD FIELDS — ALIEN ANNUALS (P. 168)

FIELD BINDWEED

Closely resembles a morning glory. A creeping vine with arrowhead-shaped and smooth-margined leaves. Flowers are very small. Blooms throughout summer.

JIMSONWEED

Huge, trumpet-shaped, white or pale violet flowers. Seedpods green, very spiny. Leaves with varying-sized teeth. Grows as a dense vine. Blooms throughout summer. ***Poisonous.***

BITTERSWEET NIGHTSHADE

Hairy-stemmed vine or small shrub. Small, 5-petaled purple flowers have a yellow center projecting from flower. Petals point backwards. Berries turn from green to deep red. Leaves oval, untoothed. Blooms spring through summer. ***Poisonous.***

COMMON LAMB'S-QUARTERS or PIGWEED

Upright plant (to 3 ft.) with inconspicuous greenish flowers; flower stalk at leaf base (axil). Flowers turn reddish. Leaves arrowhead-shaped, with large, shallow notches. Blooms summer through fall.

FIELD PENNYCRESS

Upright, small (to 18 in.), with tiny, pale flowers that become little rounded pods. Leaves wavy, notched, clasp stem. Blooms early spring through summer.

SHEPHERD'S PURSE

Similar to Pennycress but with more heart-shaped seedpods and basal leaves that are deeply notched, resembling dandelion leaves. Blooms early spring through summer.

CHARLOCK or WILD MUSTARD

Pale yellow, 4-petaled flowers. Green, hairless seedpods. Leaves hairy, with irregular, deep lobes and teeth. Stem hairy. Blooms throughout summer, fall.

COMMON CHICKWEED

Small (to 16 in. tall) plant with opposite, oval, untoothed leaves. Often spreads vine-like. Flowers small, white, with single notch in each of 5 petals. Blooms spring through fall.

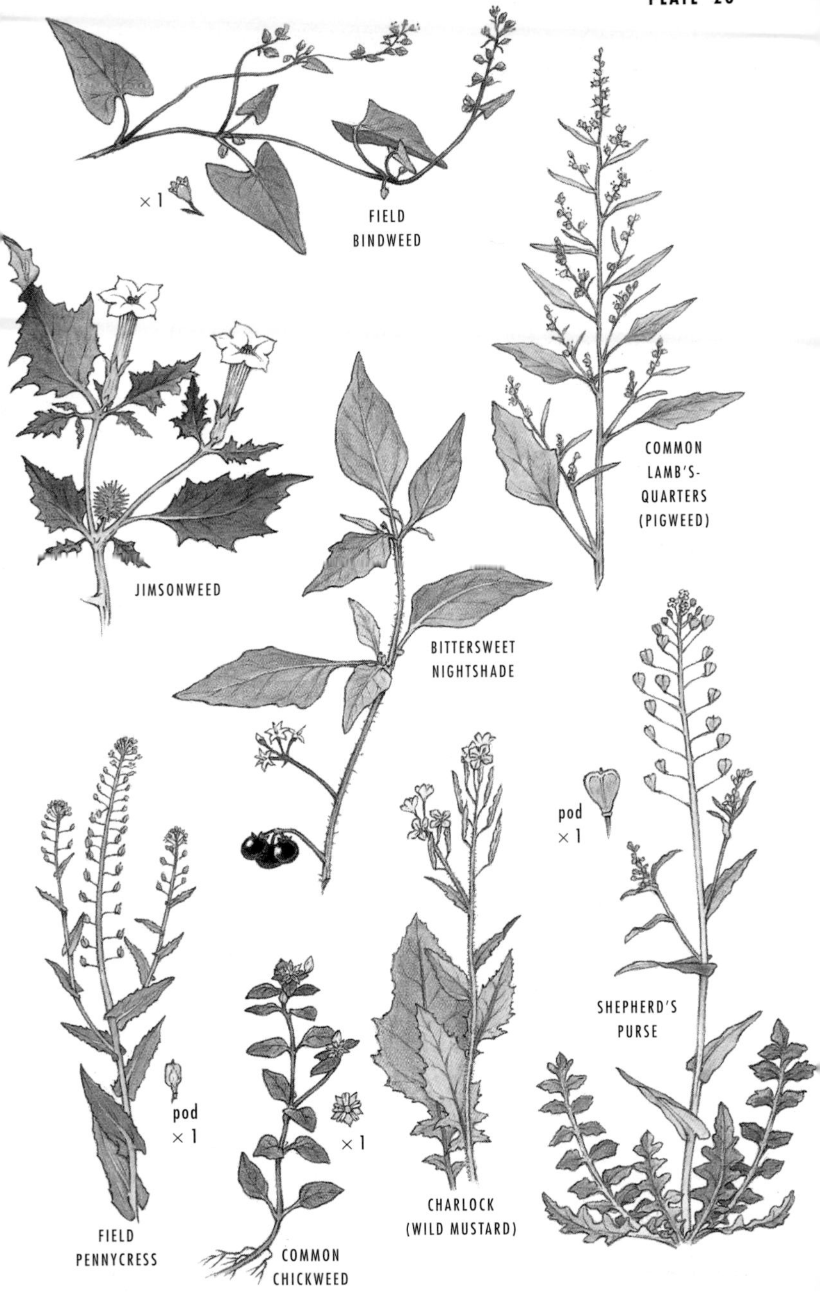

×1

FIELD
BINDWEED

COMMON
LAMB'S-
QUARTERS
(PIGWEED)

JIMSONWEED

BITTERSWEET
NIGHTSHADE

pod
×1

SHEPHERD'S
PURSE

pod
×1

×1

CHARLOCK
(WILD MUSTARD)

FIELD
PENNYCRESS

COMMON
CHICKWEED

HERBS OF OLD FIELDS — NATIVE COMPOSITES (P. 178)

The composite family has many native species that are abundant in old fields, including the diverse goldenrods and asters (see Pls. 23, 41), as well as others illustrated in this guide. The species shown here are common and widespread.

WILD LETTUCE

A tall (to 10 ft.) plant with a candelabra-type arrangement of yellow flowers. Seeds borne on feathery hairs, which aid in wind dispersal. Leaves are variable in shape but tend to be very deeply lobed. Blooms midsummer through fall.

BLACK-EYED SUSAN

Tall (to 3 ft.); sunflowerlike, single flowerhead with deep brown center and bright yellow rays. Leaves oval, slightly toothed. Leaves and stem hairy. Blooms early summer through fall.

GOLDEN RAGWORT

Yellow, daisylike flowers in flat-topped clusters. Basal leaves wide, heart-shaped; upper leaves oval, deeply and irregularly lobed. Stem to 3 ft. tall. Prefers wet meadows, swamps, woodlands. Blooms early spring through midsummer.

SWEET JOE-PYE-WEED

Wide, flat-topped spray of fuzzy, pinkish purple flowers. Leaves radiate from stem in fours. Leaf arrowhead-shaped, toothed. Leaves smell like vanilla extract when crushed. Stem green, to 6 ft. tall. Blooms midsummer through fall.

TALL IRONWEED

True to its name—up to 10 ft. tall. Deep violet, somewhat feathery flowers in a candelabra-like arrangement. Leaves lance-shaped, slightly toothed. Prefers moist sites, wet meadows. Blooms late summer through fall.

BONESET

Fuzzy white flowers in wide, flat clusters at tip of plant. Leaves opposite, united at base to surround stem, and slightly toothed. Stem hairy; to 4 ft. tall. Blooms midsummer through fall.

SNEEZEWEED

Yellow rays that bend backward on flowerhead. Center light yellowish brown, very rounded. Leaves oval, toothed. Blooms late summer through fall.

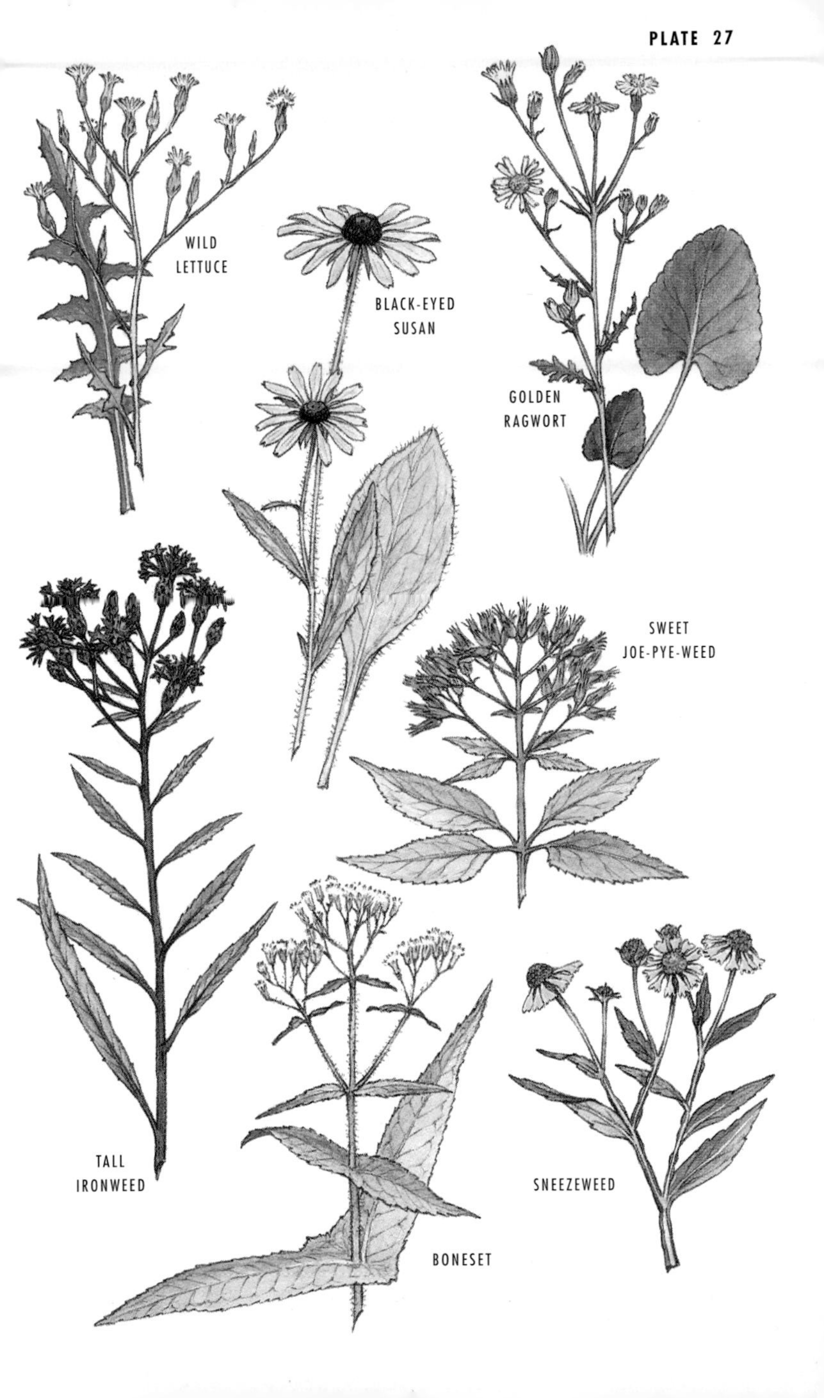
WILD LETTUCE
BLACK-EYED SUSAN
GOLDEN RAGWORT
SWEET JOE-PYE-WEED
TALL IRONWEED
SNEEZEWEED
BONESET

HERBS OF OLD FIELDS — ALIEN COMPOSITES (PP. 172, 175, 326)

Many alien composites are well established and abundantly represented in old fields. These are among the most common.

OX-EYE DAISY

Large white daisy with bright yellow center. Leaves dark green, deeply and irregularly lobed. Stem to 3 ft. tall. Blooms throughout summer.

COMMON TANSY

Wide cluster of flattened, dishlike, yellow flowerheads. Leaves fernlike, deeply and elaborately lobed. Aromatic. Stem to 3 ft. tall. Blooms mid- to late summer.

CHICORY

Purplish blue (sometimes pale blue or white), daisylike flowerheads with fringed tips on rays. Flowers stalkless. Stem nearly devoid of leaves. Basal leaves deeply lobed. Flowers open only in morning. Blooms summer through fall.

NODDING THISTLE

Large, rounded, drooping, purple-red flowerhead atop a very prickly stem up to 3 ft. tall. Purple bracts below flowerhead. Leaves deeply lobed, spiny. Blooms summer through fall.

COMMON SOW-THISTLE

Bright yellow, daisylike flower resembling that of hawkweeds. Leaves prickly, deeply and irregularly lobed. Can reach height of 8 ft. Two other alien sow-thistles closely resemble this species. Blooms summer through fall.

CANADA THISTLE

Numerous, small, light purple flowerheads. Leaves deeply lobed, prickly. Stem smooth, to 5 ft. tall. Very common. Blooms mid- through late summer.

KING DEVIL

One of many hawkweed species, some native, some alien. Native hawkweeds have leaves ascending stem. Alien species have leaves in basal rosettes. This species has bright yellow flowers on a tall (to 3 ft.) stem. Leaves elongate, smooth-margined. Both leaves and stem very hairy. Blooms late spring through summer.

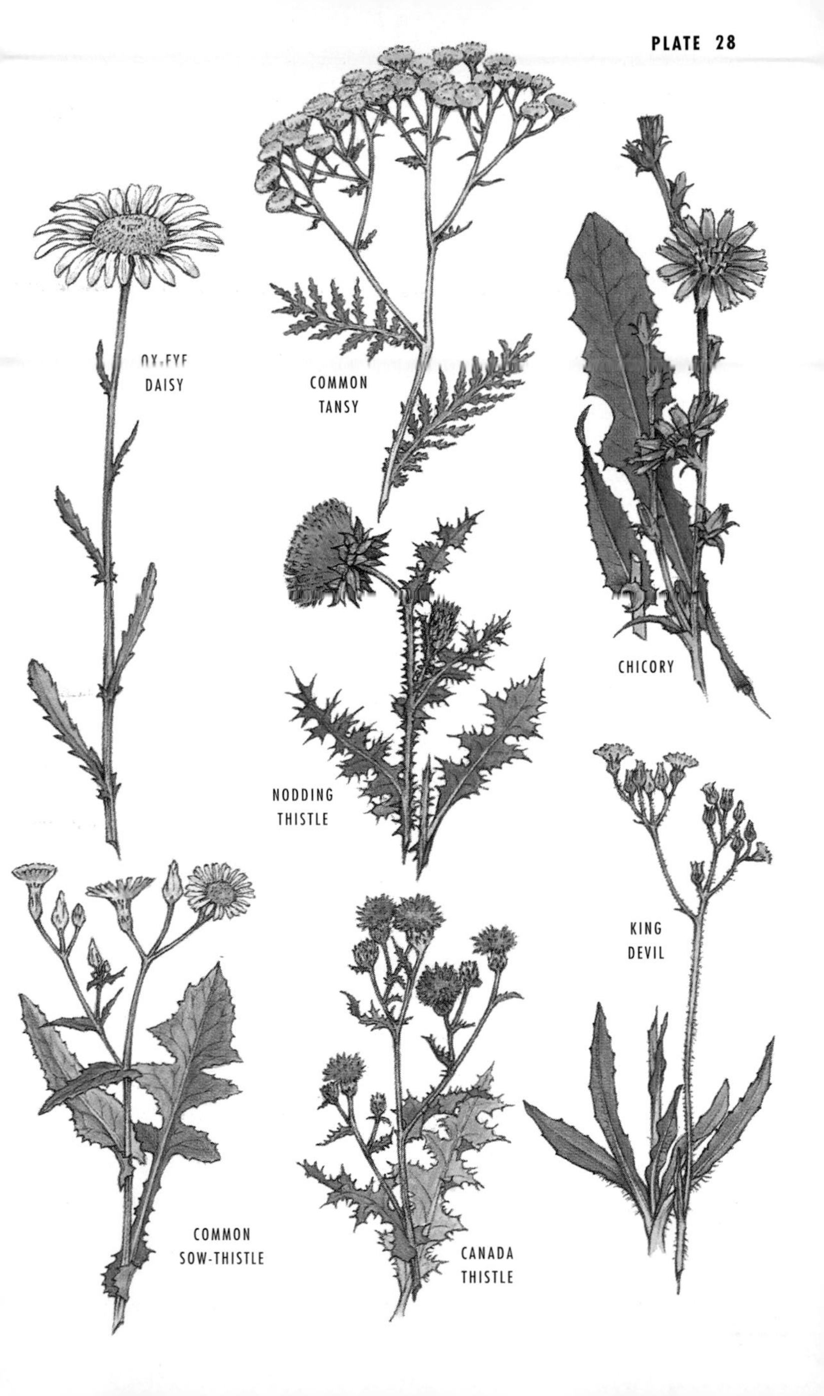
OX-EYE
DAISY
COMMON
TANSY
CHICORY
NODDING
THISTLE
KING
DEVIL
COMMON
SOW-THISTLE
CANADA
THISTLE

PLATE 29

HERBS OF OLD FIELDS — ALIEN PERENNIALS (P. 180)

BUTTER-AND-EGGS

To 3 ft. A weedy snapdragon with a spike of tubular, yellow, "spurred" flowers, each with an orange lip. Leaves lance-shaped, untoothed. Blooms spring through fall.

WINTER CRESS

A mustard. Blooms from early spring through late summer. Bright yellow flower clusters; pods below. Deeply lobed leaves.

VIPER'S BUGLOSS

To 2½ ft. Hairy stalk; blue trumpetlike flowers with long red stamens. Leaves, stem very hairy. Blooms early summer through fall.

COMMON ST. JOHNSWORT

To 2½ ft. Bright yellow 5-petaled flowers with tiny black dots. Dense (bushy) stamens. Leaves opposite, elliptical, untoothed. Blooms from early through late summer.

COMMON TALL BUTTERCUP

The tallest of the eastern buttercups; to 3 ft. Four overlapping, shiny, yellow petals. Leaves palmate and highly dissected. Stem hairy. Blooms late spring through summer.

SHEEP SORREL

To 12 in. Tiny, green, beadlike flowers, which turn reddish brown. Leaves arrow-shaped. Blooms summer through fall.

CURLED DOCK

To 4 ft. Brownish green flower clusters on spikes. Leaf margins wavy. Blooms throughout summer. Seeds tend to remain on dried plant in fall.

FIELD GARLIC

Pinkish white flowers and small bulblets. Pungent garlic odor and taste. Long, grasslike leaves. Blooms throughout summer.

STINGING NETTLE

Very irritating to skin. Entire plant covered with dense, stinging hairs. Touching plant causes itching, burning. Opposite, arrow-shaped, toothed leaves. Tiny greenish flower clusters grow from leaf axils. Blooms throughout summer.

ENGLISH PLANTAIN

Tiny, whitish, conelike flowerhead on a long stalk. Leaves elongate ovals, untoothed. Blooms summer through fall.

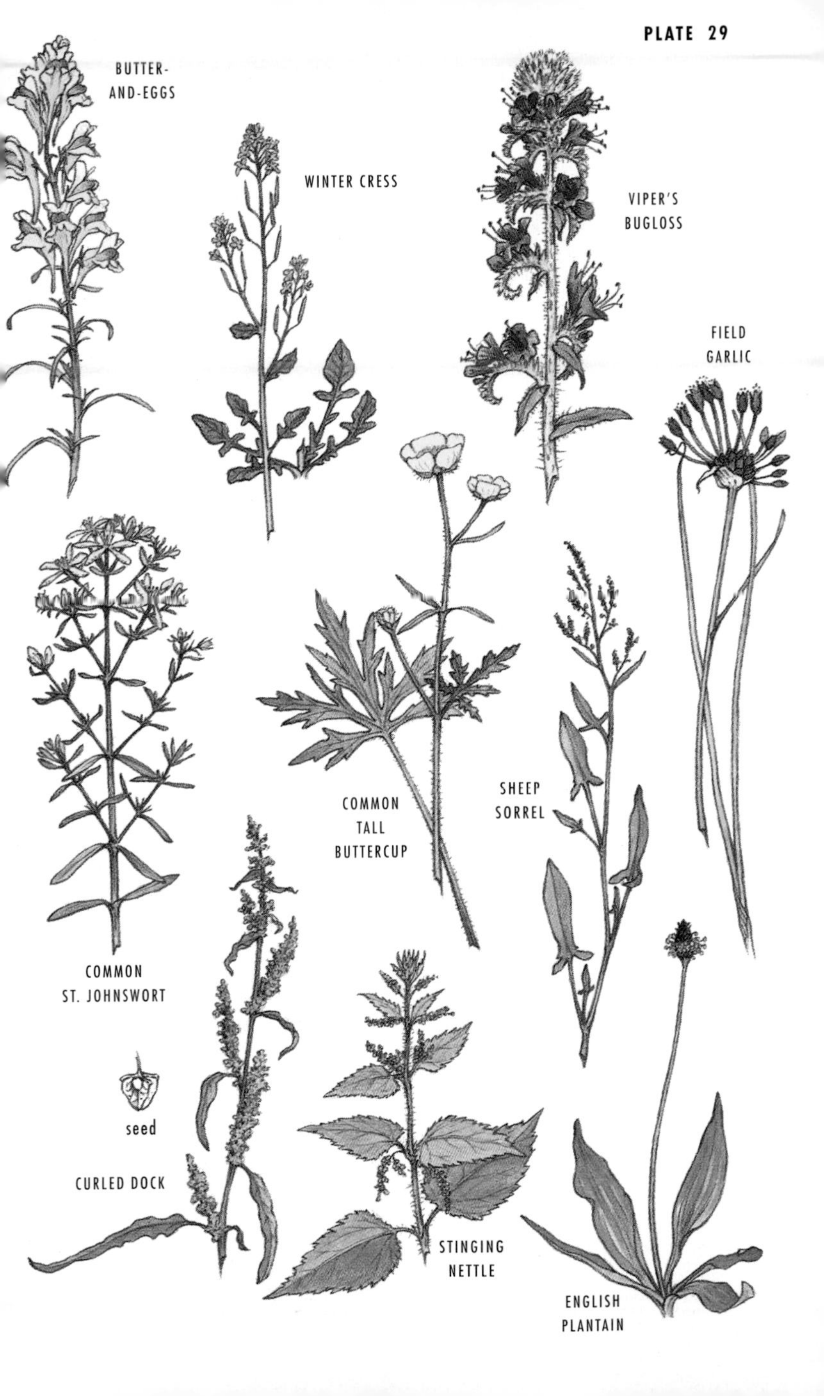
BUTTER-AND-EGGS
WINTER CRESS
VIPER'S BUGLOSS
FIELD GARLIC
COMMON TALL BUTTERCUP
SHEEP SORREL
COMMON ST. JOHNSWORT
seed
CURLED DOCK
STINGING NETTLE
ENGLISH PLANTAIN

HERBS OF OLD FIELDS — LEGUMES (P. 183)

The legumes are a large plant family (Leguminosae) that includes peas and beans. Many legumes, including some woody species, inhabit old fields. Most familiar are the many clover species, both native and alien, which have 3 leaflets per leaf, arranged triangularly. All legumes have compound leaves, and some grow as vines. Some have bacteria in their roots that fix nitrogen from the atmosphere.

BLACK MEDICK

Spreads as a vine. Small, bright yellow flowers, similar to those of **Hop Clover.** Pods black. Leaflets rounded. Blooms early spring through early winter. Alien.

BUSH-CLOVER

Small, purple flowers, few in number. Grows as a shrub (to 3 ft. tall) or a vine. Multibranched. Leaflets oval, on stalks. Blooms mid- to late summer. Native.

WHITE SWEET CLOVER

Common along roadsides. Tall (to 8 ft.), upright plant with clusters of small white flowers at branch tips. Leaflets oblong and notched. Very similar to **Yellow Sweet Clover,** which has yellow flower clusters. Both bloom throughout summer, though White Sweet Clover blooms into fall as well. Both alien.

RABBIT'S-FOOT CLOVER

Fuzzy, pale pink flowers. Leaflets slender, oblong, untoothed. Blooms spring through fall. Alien.

RED CLOVER

Bright red flowers that attract Bumble Bees. Leaflets with a pale V. Blooms spring through summer. Alien.

NARROW-LEAVED VETCH

Several species of vetch are common. All have compound leaves and tend to grow as vines. This species has reddish purple flowers that develop into green pods. Blooms spring through fall. Alien.

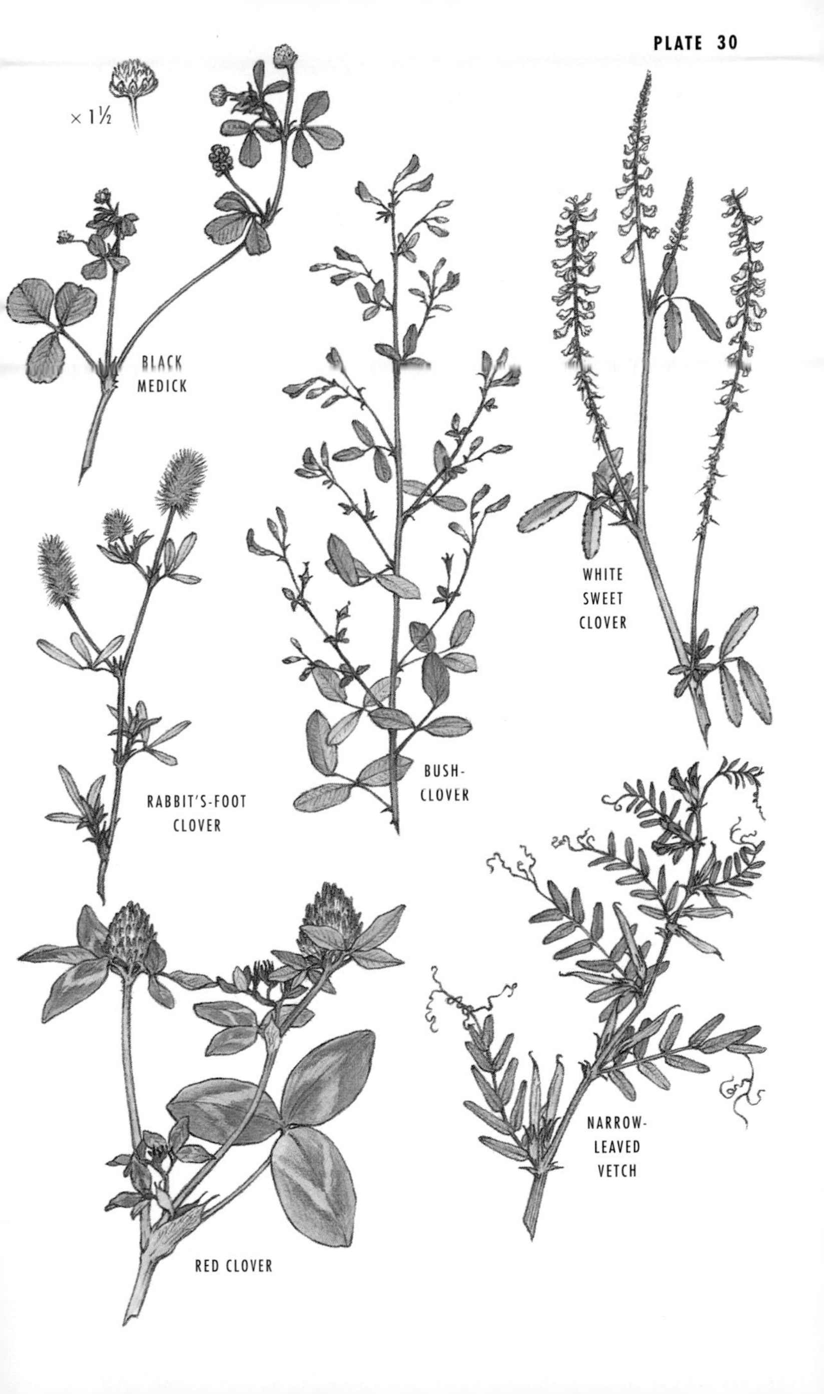
× 1½
BLACK MEDICK
WHITE SWEET CLOVER
RABBIT'S-FOOT CLOVER
BUSH-CLOVER
RED CLOVER
NARROW-LEAVED VETCH

VINES OF OLD FIELDS (P. 187)

Vines grow best in habitats with much sunlight. Old fields provide ideal circumstances for vine growth. The species shown here are among those most common and widely occurring.

FOX GRAPE

Large, heart-shaped, rounded leaves, many with 3 poorly defined lobes. Flowers greenish, very nondescript. Fruits violet (see also Pl. 50). Favors open woods, edges, thickets. **Summer Grape** is very similar, but its fruits are black, not violet. Native.

BITTERSWEET

Flowers are small and greenish; fruits are bright orange. Leaves oval, slightly toothed. Entwines around woody vegetation, fences. Native.

GREENBRIER (CATBRIER)

Very shiny, bright green, leathery leaves, with smooth margins. Flowers in tiny, white-green clusters. Berries deep blue. Stem green, with sharp green thorns. Favors open woods and thickets. Native.

POISON-IVY

Leaves shiny, in groups of 3, oval, pointed, slightly toothed. Leaves turn bright red in fall. Flowers small, green. Fruits gray or white. Grows densely in thickets, climbs tree trunks. Native. *All parts of the plant are* ***toxic*** *to skin at all seasons.*

HEDGE-BINDWEED

A wild morning glory. Flowers large, trumpetlike, pink or white. Leaves arrowhead-shaped. Grows densely in early old fields, entwining around other plants. Native. Also see Field Bindweed, Pl. 26.

TRUMPET-CREEPER

Bright orange-red, trumpetlike flowers. Compound leaves with oval, pointed, toothed leaflets. Native. See also Pl. 14.

VIRGINIA CREEPER

Compound leaves, with 5 oval, toothed leaflets arranged palmately. Leaves turn orange-red in fall. Flowers small, greenish clusters, fruits deep blue. Climbs tree trunks, fences, etc. Native. See also Pl. 50.

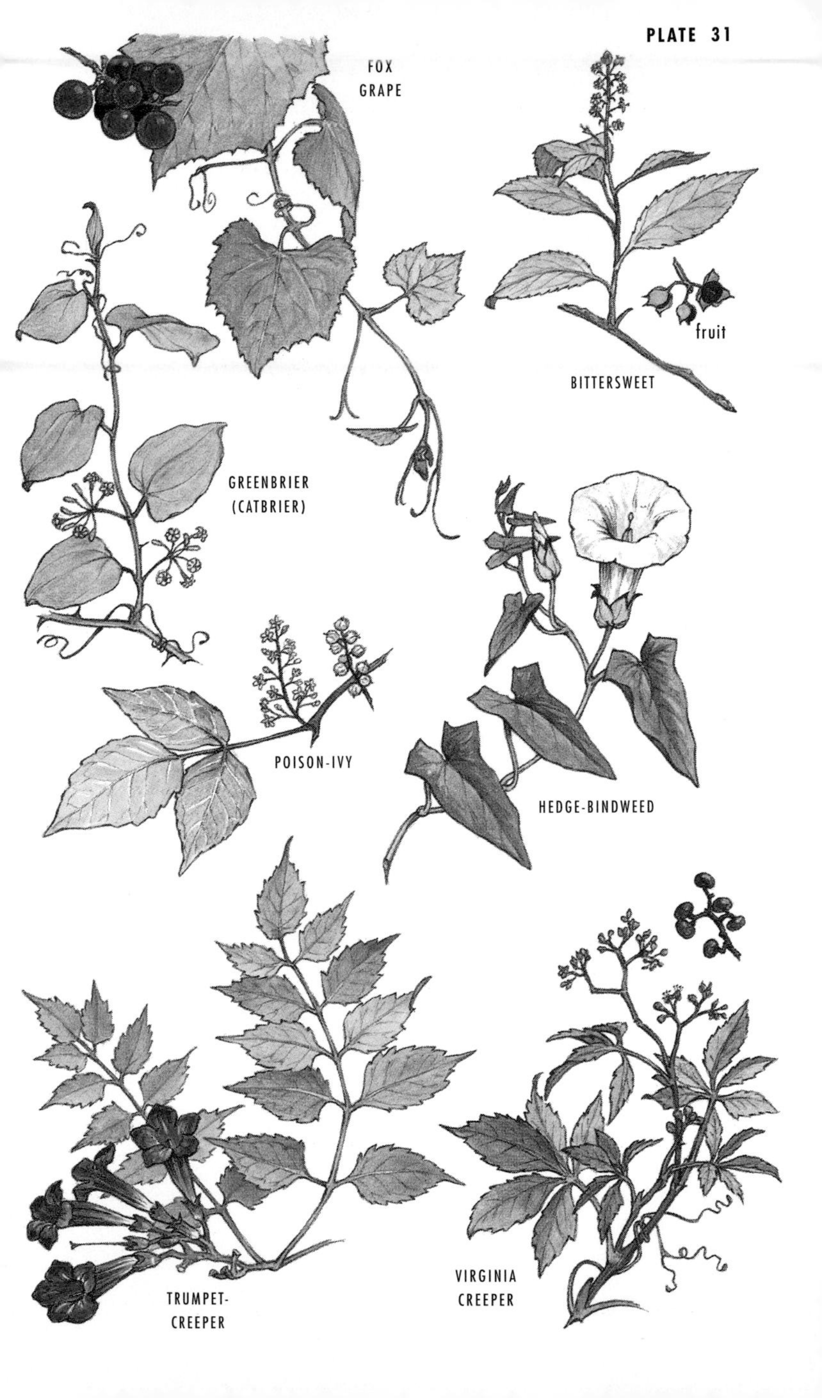
FOX
GRAPE
BITTERSWEET
fruit
GREENBRIER
(CATBRIER)
POISON-IVY
HEDGE-BINDWEED
VIRGINIA
CREEPER
TRUMPET-
CREEPER

SHRUBS AND A FERN OF OLD FIELDS (P. 190)

Shrubs are among the first woody invaders of old fields. They often form dense thickets interrupted by areas still predominantly covered by grasses and herbs.

STAGHORN SUMAC and SMOOTH SUMAC

Sumacs are spreading shrubs with compound leaves containing many toothed leaflets. These species are quite similar, but Staghorn has branches covered with velvetlike hairs, whereas Smooth Sumac is exactly that—smooth. Flowers in dense, greenish yellow clusters; fruits in dense, hairy, red clusters (see Pl. 50). Both species native.

MULTIFLORA ROSE

Several rose species are common invaders in old fields. This species, an alien, has escaped from cultivation. Stems prickly; leaves compound, with oval, toothed leaflets. Flowers small, white. Fruits (rose hips) red, not clustered.

DOWNY SERVICEBERRY or SHADBUSH

A shrub or small tree with oval, toothed leaves. Flowers with 5 white petals, red anthers. Fruits deep purple (see Pl. 48). Native.

HAWTHORNS

There are 35 hawthorn species. Many hybridize, making field identification difficult. Hawthorns are shrubby, small trees. All have large spines on the branches. Leaves lobeless or with poorly defined lobes, and with teeth. The species illustrated is Downy Hawthorn. It has white, 5-petaled, cup-shaped flowers, in clusters. Fruits fleshy, deep red.

MEADOWSWEET

Tall (to 5 ft.), with a spike of whitish pink flowers. Stems reddish brown. Leaves alternate, oval, toothed, with small leaflets at axils. Similar to Steeplebush, but the latter has a woody stem. Both bloom through summer. Native.

COMMON JUNIPER

Dense, spreading shrub with prickly needles in whorls of 3 and tiny, waxy, blue-gray cones. Old fields, especially with dry, rocky soil. Native.

BRACKEN FERN

To 3 ft. A fern of open, sunlit areas and dry uplands. Fronds broad, triangular, with 3 large leaflets.

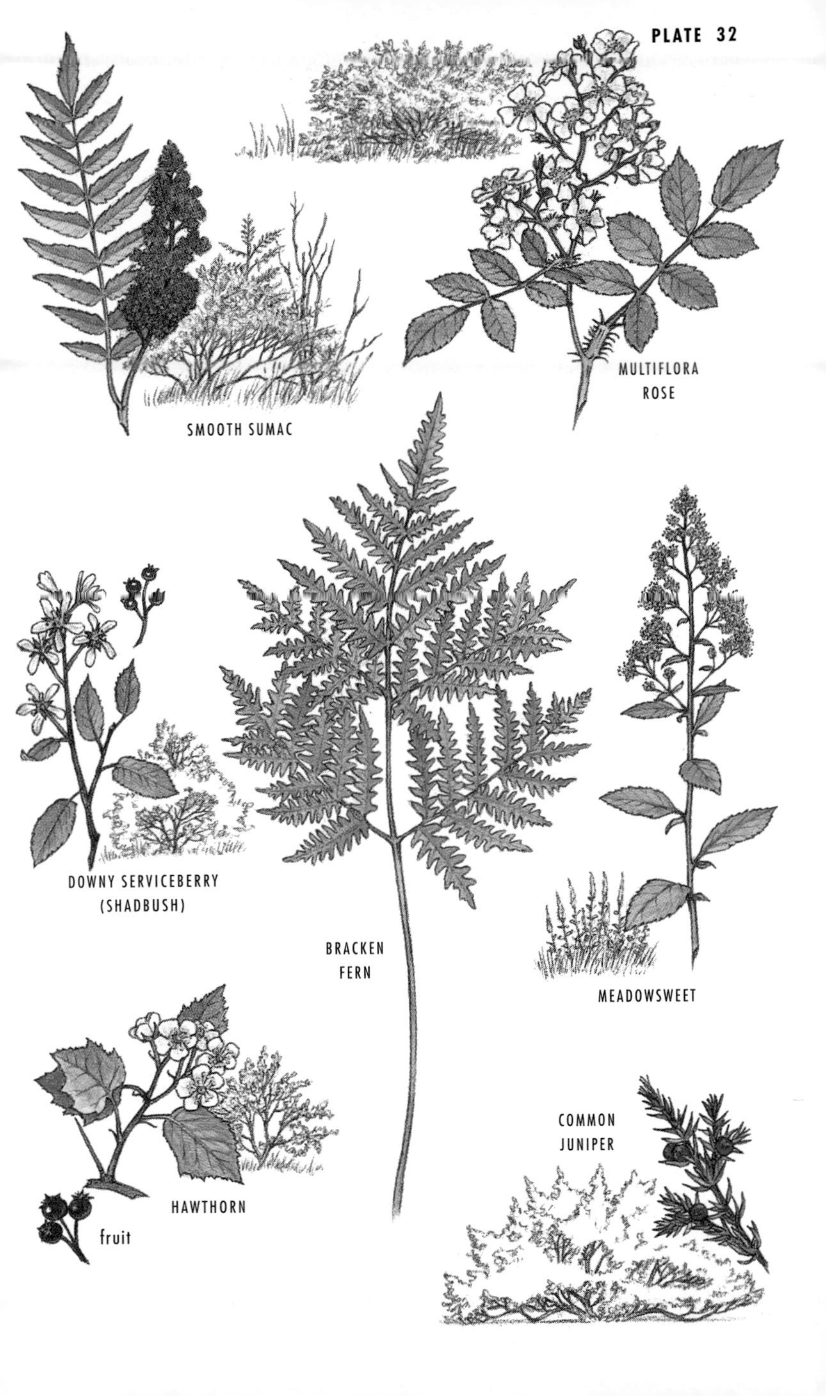
SMOOTH SUMAC
MULTIFLORA ROSE
DOWNY SERVICEBERRY (SHADBUSH)
BRACKEN FERN
MEADOWSWEET
HAWTHORN
fruit
COMMON JUNIPER

WAXMYRTLES AND HEATHS (PP. 190, 192)

These shrubs are generally found on acidic, sandy, dry soils (though many heaths are found in bogs). Some are quite aromatic. Some have leaves toxic to animals. Leaves generally feel leathery and waxy, and many species are evergreen.

NORTHERN BAYBERRY

Oblong, untoothed, aromatic leaves in dense clusters at branch tips. Berries pale blue and waxy. Grows as dense shrub in old fields, sandy areas, dunes. Yellow-rumped Warblers and Tree Swallows feed heavily on berries in fall. Native. Also see Waxmyrtle (Southern Bayberry), Pl. 12. Waxmyrtle family.

SWEET-FERN

Superficially resembles a fern frond. Leaves lance-shaped, with rounded lobes. Fragrant when crushed. Stems hairy. Grows as a shrub in old fields. Native. Waxmyrtle family.

COMMON HIGHBUSH BLUEBERRY

A tall (to 12 ft.), dense shrub. Leaves oval, untoothed. Flowers bell-like, pinkish white, in clusters. Fruits deep blue (see Pl. 48). This species has a wide tolerance range; it is found in swampy woodlands and dry, open old fields. Native. Several other Highbush Blueberry species occur, as do several Lowbush Blueberry species. Heath family.

BLACK HUCKLEBERRY

Small, bell-shaped flowers; reddish, in clusters. Leaves covered with tiny dots of resin (visible with hand lens). Fruits black or deep blue. A woodland shrub. Native. Heath family.

MOUNTAIN LAUREL

An evergreen, understory shrub in many forests, often forming dense thickets. Also on heath balds in Appalachians. Flowers in dense clusters, generally pink but with red at base of "bowl" and anther tips. Leaves very shiny. Native. Heath family.

SHEEP LAUREL

An evergreen shrub. Favors forest understory or old fields. Leaves droop. Flowers deep pink; not located at branch tips. Native. Heath family.

NORTHERN
BAYBERRY
SWEET-
FERN
BLACK
HUCKLEBERRY
COMMON
HIGHBUSH
BLUEBERRY
fruit
MOUNTAIN LAUREL
SHEEP
LAUREL

TREES OF OLD FIELDS (P. 193)

These species grow best in sunny environments and thus are common invaders of old fields. They grow fast but are usually shaded out by other tree species. A few, however, may remain in the canopy of closed forests for some time.

EASTERN RED CEDAR

Conical-shaped, evergreen, needle-leaved tree; to 50 ft. Needles scale-like, aromatic (with a faint odor of gin). Cones berry-like, bluish, waxy. Bark reddish, in strips. Common in old fields on calcareous (lime-rich) soil.

BLACK LOCUST

To 80 ft. Compound leaves, with dark green, oval, untoothed leaflets. Flowers white, in dangling clusters. Pods brown, flattened, about 3–4 in. long. Bark grayish and deeply furrowed and ridged. Branches have sharp paired spines.

QUAKING ASPEN

To 70 ft. Leaves heart-shaped, toothed; bright green above, pale below; on long stalks. Leaves turn golden yellow in fall. Bark smooth, greenish gray, furrowed in older trees. Often invades after fire or on sandy, dry soils.

PIN or FIRE CHERRY

Small (to 30 ft.), rapidly growing invader, especially in burned or clear-cut areas. Leaves alternate, oblong, pointed, toothed; turn yellow in fall. Bark reddish gray and smooth in young trees, darker and furrowed in older trees.

GRAY BIRCH

Small (to 30 ft.), often shrubby tree. Leaves alternate, arrowhead-shaped with sharp teeth; turn yellow in fall. Flowers illustrated on Pl. 35. Bark gray, smooth, furrowed at base of trunk. Invades on dry, sandy soils.

WHITE ASH

To 80 ft., with a very straight trunk. Leaves compound, with pale green, elliptical, pointed, toothed leaflets; turn yellow in fall. Flowers illustrated on Pl. 35. Seeds hanging downward in winged clusters. Bark blackish gray, with ridges and deep furrows. Invades old fields on mesic (moist) sites, persists in mature forest.

scale-like
needles
prickly
needles
EASTERN
RED
CEDAR
pods
BLACK LOCUST
QUAKING
ASPEN
PIN (FIRE)
CHERRY
GRAY BIRCH
staminate
flowers
pistillate
flowers
seeds
WHITE ASH

Adaptation

There are two kinds of questions that can be asked about the processes of natural history, "how-type" and "why-type" questions. When you observe a soaring hawk, a woodpecker hammering on a tree, or a brilliantly colored wildflower, you may wonder how the hawk manages to remain airborne without flapping its wings, how the woodpecker's head can withstand the severe blows of drilling into wood, and how the flower manages to produce colors in such a striking arrangement. All of these are how-type questions. They can be investigated and be subjected to relatively straightforward scientific analysis. The aerodynamics of flight can be applied to the hawk's anatomy to understand how the bird soars on warm thermal air currents. The physics of hammering can be examined relative to the anatomy of a woodpecker's head to understand how the bony structure of the skull cushions the impact of the hammer blow. The biochemistry of flower pigmentation can be unraveled in the laboratory to explain how the reds and violets are synthesized. But, *why* does the hawk soar, why does the woodpecker drum, and why are flowers brilliantly colored? These questions deal with adaptation. To answer them is to reveal the most significant aspects of natural history.

Both how- and why-type questions are important, but their focus is different. How-type questions focus on process. Answers to how-type questions deal with proximate or immediate mechanisms that operate the system under investigation. Why-type questions deal with reasons why something is as it is. What advantages are provided to the organism in terms of its survival? Why-type questions deal with ultimate reasons for the patterns of nature. In the chapters that follow I discuss many ultimate questions, asking why this and why that. This chapter is meant to serve as a primer on seeing nature in a way that sharpens your focus on why-type questions.

The Woodchuck — Hibernation and Adaptation

Consider the fact that the Woodchuck hibernates during the cold winter months. How does it manage this physiological feat? By observation and experimentation we learn that Woodchucks fatten up before they begin to hibernate, enlarging their stores of an energy-rich substance called brown fat that will carry them through their long fast. We learn that they are sensitive to changes in day length and that their hormonal systems change prior to the onset of hibernation. We measure the dramatic decrease in their respiration, heart rate, and body temperature as they drop deeply into their winter state of suspended animation. In other words, with time, patience, and some good experimental technique, we can learn how a Woodchuck manages to enter a state of hibernation. But, we still don't know why.

You might think that the answer to why a Woodchuck hibernates is obvious: winter is cold, and hibernation avoids cold. A small rodent is better off sleeping in a cozy den than being outside trying to find food in cold snowy weather. However, neither Gray Squirrels nor White-footed Mice, both of which occur in the same woodlands as Woodchucks, and both of which are smaller than a Woodchuck, hibernate. To further complicate matters, the Eastern Chipmunk nearly hibernates but does not quite enter a state of true hibernation. Rather, it sleeps deeply but awakens off and on throughout the winter. True hibernators like the Woodchuck do not awaken until spring. If hibernation is so advantageous, why don't all little mammals do it? In fact, relatively few do.

Why-type questions, such as "why hibernate?," are difficult puzzles to solve because the answers lie deep in the history of the

Fig. 25. Woodchuck, awake and in hibernation.

species. The most common answer to a why-type question is that the characteristic, be it the ability to hibernate, soar, or drum on a tree trunk, is an *adaptation*. An adaptation is any anatomical, physiological, or behavioral characteristic conferring survival value (termed "fitness") that contributes to eventual reproductive success. If hibernation is examined from this perspective, the answer to why woodchucks hibernate is this: In the past, probably millions of years ago, Woodchuck ancestors possessing the ability, perhaps slight at first, to sleep deeply for at least part of the winter, were the ones that survived the best. They reproduced and left the most offspring within the overall Woodchuck population. Whatever genetic influences accounted for the ability to enter hibernation became more and more abundant within the population as more and more hibernators survived and reproduced their genes. Thus, hibernation became a fixed trait in the species because it provides a very high *fitness* (survival value) for those animals that do it.

The science of genetics answers part of the riddle of why Gray Squirrels and White-footed Mice do not also hibernate. Although they are rodents, and thus are related to Woodchucks, they do not all share exactly the same genes. Squirrels only breed with squirrels and mice only with mice. The trait for hibernation simply never evolved in either squirrels or mice. An adaptation in one species is never guaranteed to evolve in all species, even if it would be advantageous.

The "story" of why the Woodchuck hibernates is hard to test in the laboratory. It is not possible to go back through time and see how hibernation evolved in Woodchucks. The difficulty with answering why-type questions is that very often we must rely on indirect evidence to support our explanation. Occasionally, we become fortunate enough to answer why-type questions thoroughly, but often an educated hypothesis is the only recourse.

Why ask why-type questions?

Our goal is to understand nature. Just as it is satisfying to be able to identify various species of plants and animals, so it is satisfying to understand something of the ways they function in nature. Why-type questions are *ultimate-type* questions. They identify the most interesting aspects of natural history, those of adaptation and survival. The answers to ultimate-type questions reveal the actual fabric that holds nature together. Being able to ask and answer ultimate questions about natural history adds a new and powerful dimension to your understanding of nature.

A biannual event of major significance to observers of natural history is the spring arrival and fall departure of migratory birds. The farther from the equator you go, the higher the percentage of migratory bird species. Migration is defined as a regular movement by a population or entire species from one part of its range to another part. Migratory birds move between their southern wintering ranges and their northern breeding grounds. In North America, of the 650 breeding bird species, 332 (51%) winter in the American tropics. Many others migrate to the southern states and still others migrate locally, such as to a lower elevation. Migration is a very broad phenomenon, not confined to birds. There are migrant fish, mammals, and insects. Migrations occur in virtually all habitats.

Bird banding has revealed that many birds return to nest in exactly the same locations every year. More recent studies have demonstrated that this same *site fidelity* is characteristic of many species on their wintering grounds as well. The typical migratory bird is loyal to both its local nesting and wintering sites, and endures a flight often of thousands of miles to move from one to the other. How and why do birds migrate?

How do birds migrate?

How do birds manage the feat of migration? What problems must a bird overcome in order to successfully migrate?

To migrate, a bird must be able to do all of the following:

1. It must be able to efficiently move from one part of its range to the next. Because they can fly, birds are adapted to migrate. Flight provides an ability to move relatively rapidly from one area to another. It is no mystery why proportionately more species of birds migrate and species of mammals hibernate. Most mammals cannot fly. To migrate, they must walk, a much slower endeavor. The only flying mammals are bats, and many bat species do migrate. Some also hibernate.

2. Birds must have fuel in order to migrate. They must be able to store sufficient energy for long overseas flights. Many bird species store large amounts of fat near their breast muscles, where the added weight is well balanced for maintaining stability in flight as well as being near both the heart and flight muscles. This fat is burned to supply calories of energy for migration. To make fat, birds feed intensively prior to beginning their migratory flights and the food is quickly metabolized into fat. Birds of some species must stop periodically during migration to rebuild their fat stores, feeding intensively for several days during a migration

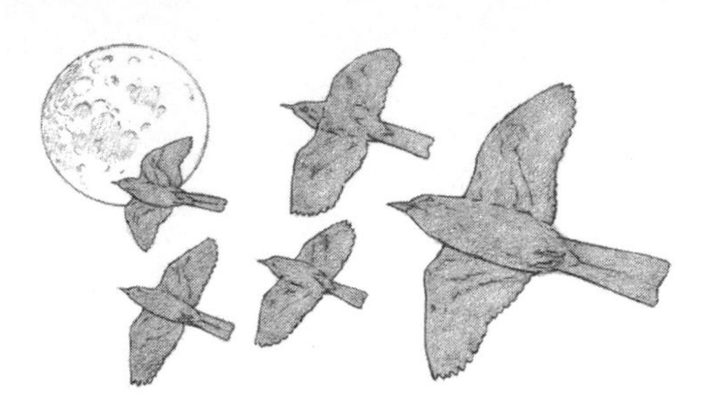

Fig. 26. Flock of passerines migrating at night.

stop-over. Although this explains how birds obtain energy to migrate, why do they utilize fat rather than sugar or protein? A gram of fat yields approximately twice the energy of a gram of protein or sugar, so fat, weight-wise, is the most economical fuel for a bird to carry.

3. Birds migrate either by day or by night. Many songbirds, such as the thrushes, warblers, orioles, tanagers, and finches, migrate at night. Nocturnal migration enables these normally diurnal birds to feed during the day, their normal foraging time. Night is also cooler and thus night flight is less energy-demanding. You can hear the chips of night migrants as they pass overhead, usually at low altitude. By focusing binoculars or a spotting telescope on the full moon, you can often see the silhouettes of night migrants as they pass across its face. Diurnal migrants include swifts and swallows, which feed on airborne insects, and hawks, which utilize daytime thermal currents to achieve energy-efficient soaring flight.

4. There must be a biological urge to migrate. Birds exhibit a clear restlessness prior to the initiation of a migratory flight. If they are caged, nocturnal migrants like thrushes and sparrows will not sleep at night but will instead fly against one section of the cage, the section that corresponds with the direction in which the bird would normally migrate. Migratory restlessness is apparently triggered by an interaction between the bird's hormonal system and brain. Day length appears to be an important cue in establishing the onset of migratory restlessness. Day length varies less in the tropics, and it is not totally clear what signals birds use

to stimulate onset of the flight north to their breeding grounds.

5. There must be some sort of sophisticated navigational system. A bird obviously must be able to find its way over thousands of miles of featureless ocean at night. Birds have the ability to navigate using both the sun and stars as reference points. Such an ability requires both a sense of direction and a sense of elapsed time, skills that are presumably encoded in birds' genes and localized within the avian brain. Some birds are sensitive to latitudinal changes in the earth's magnetic field, a skill that is also a navigational aid. Birds have very sharp vision and use landmarks such as sea coasts and mountain ranges during migration. Birds must possess well-developed memories for local details (including characteristic odors), since many species are faithful to certain sites on both their nesting and wintering grounds.

6. Migrants must be able to exploit varied habitats. Birds must not be overspecialized to a particular habitat type. The Blackburnian Warbler nests in boreal spruce-fir forests and winters in mountain cloud forests in South America. In between it experiences many different forest types, as well as barrier beach shrubs, fields, urban parks, and other habitats. The American Redstart is a familiar nesting warbler of most eastern deciduous forests. It winters primarily in lowland tropical rain forests and thus must be adapted behaviorally and anatomically to forage successfully in two quite different habitats. Many species, such as orioles and tanagers, that are primarily insectivorous on their breeding grounds switch to feeding on fruits during migration and then to a diet of nectar on their tropical wintering grounds. The Eastern Kingbird is territorial in summer (when it feeds on insects) but, on its South American wintering area, it lives in large flocks and feeds on fruits.

7. Migrants not only experience varied habitats, but they must interact with varied species. Migrants must be able to compete against different sets of residents on both their summer and winter grounds. Interactions among species may be particularly intensive in tropical areas that have a high diversity of resident species, including many that are predatory.

For example, the white tail band of the Eastern Kingbird may serve as a signal to maintain the cohesiveness of the winter flocks. The waxwings, also fruit-eaters that live in large flocks, all have yellow tail bands. The Kingbird's tail band, though not an adaptation on the bird's breeding ground, is probably adaptive on its wintering ground.

The questions above must be answered to explain *how* birds migrate. Note that nowhere in the answers to those questions have we dealt with *why* birds migrate.

Why do birds migrate?

Given the complex problems posed by a biannual migration, why do it? This question has no definitive answer, but we can make some educated guesses. Remember, why-type questions are more difficult to answer directly than how-type questions. Earlier, we examined the problems birds must solve in order to migrate. Now let's take a look at the problems birds would face if they didn't migrate. Put slightly differently, what advantages does migration provide?

1. Migration allows birds to exploit abundant food. Insects and other invertebrates abound in the temperate zone during the summer growing season. Immense numbers of caterpillars, grasshoppers, mosquitoes, and myriads of other insects, as well as spiders and worms, populate fields and forests. High in protein content, these hordes of six-, eight-, and no-legged creatures represent abundant potential food for birds. Moving north for the temperate summer provides access to this resource. Reproduction requires that a bird find an abundant source of protein for its growing young. This need is met more easily where animal food is plentiful.

2. Migration may allow birds to reduce competition among species as well as reduce the probability of predation. Birds leaving their tropical wintering areas also leave their tropical competitors and predators. Insect food is probably no more abundant in the tropics than in the temperate zone in summer and may well be less abundant. The high flush of arthropods in the temperate zone during breeding season, coupled with the lower overall diversity of birds in the temperate zone, may minimize competition among the bird species. In addition, there are fewer potential nest predators in the temperate zone than in the tropics. No monkeys, toucans, or army ants occur in the Eastern Deciduous Forest.

3. The temperate summer provides longer days, enabling birds to find more food per day, which allows nestlings to grow more quickly. In the tropics, day length is approximately 12 hours year around. In the temperate zone there are up to 5 or more additional hours of daylight each day in the summer. This allows nestlings to fledge sooner and therefore reduces the probability of nest predation. It also permits birds to produce a second brood and sometimes a third. Clutch (brood) sizes are greater in temperate areas than in the tropics, a possible reflection of both the greater amount of food and the smaller risk of predation in the temperate zone.

Despite the risks birds encounter during long-distance migration, the combination of arthropod (food) abundance, fewer competing species, fewer predators, and longer days seems to make the benefits outweigh the risks. Still, like any adaptation, migration need not be universal to be adaptive. Some species in the temperate zone and most in the tropics never migrate.

Insect-eating Plants

Why do insectivorous plants occur in bogs?

The various species of pitcher-plants, sundews, and the Venus' Fly-trap all have two things in common: they capture and digest insects and spiders and they inhabit bogs. Is there any reason why this highly unusual plant characteristic of carnivory should occur only among some bog-dwelling species?

All carnivorous plants are well adapted to capture insects. Special insect-capturing structures occur on each type of insectivorous plant. Understanding *how* these structures function does not explain *why* the plants that bear them occur where they do.

Insects and spiders are rich in protein. When protein is digested it breaks down into compounds containing much nitrogen, which is essential for plant growth. Insectivorous plants thus obtain nitrogen, a vital nutrient, from their arthropod prey.

Bogs are highly acidic environments (see p. 68). Acidity greatly retards decomposition because it discourages growth of fungi and bacteria. Bogs characteristically fill with dead plant and animal material that decomposes extremely slowly, if at all.

Since acidity retards decomposition, sources of simple nitrogen are not available in bogs. The nitrogen remains "locked up" in the undecomposed and partially decomposed litter that fills the bog. It is advantageous to insectivorous plants to be carnivores, because carnivory enables the plants to tap a needed source of ni-

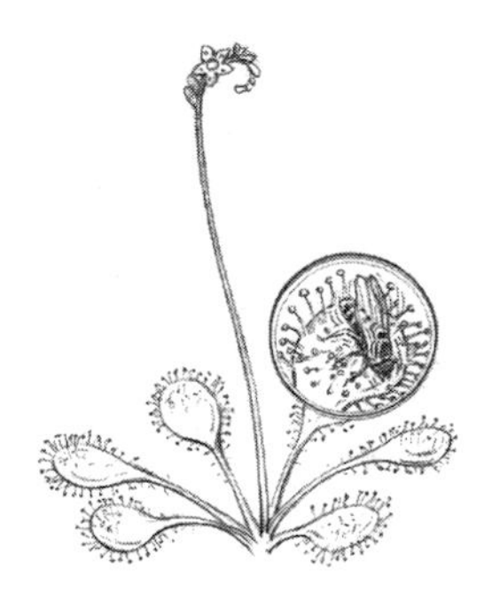

Fig. 27. Sundew with insect.

trogen. The energy an insectivorous plant "invests" in the specialized anatomy, biochemistry, and "behavior" is repaid in ready access to nitrogen. Only in a severely nitrogen-poor environment would such an elaborate specialization as carnivory be adaptive to the plant. It would be of no advantage for a plant to be carnivorous in environments rich in nitrogen. Only by understanding the circumstances under which insectivorous plants live can we understand why carnivory in plants evolved.

The relationship between how- and why-type questions is very well illustrated by the Venus' Fly-trap. This plant rapidly closes its vise-like spiked leaves to entrap flies and other insects. How does the plant manage to close its leaves so quickly? Sensitive hairs, triggered by the insect, cause the insect trap to close. These hairs, in order to work, require acid, a commodity that abounds in the plant's bog habitat. Why is the plant insectivorous? Because of the effects of its acidic environment on reducing the availability of nitrogen, and the need to procure adequate amounts of this essential nutrient.

Plant Defenses

Why don't herbivores eat all the leaves?

In summer forests and fields are green, as plant growth abounds. Leaves and stems, rich in energy, would seem both obvious and easy targets for plant-eating animals, the herbivores. Close inspection of leaves usually reveals at least some damage by herbivores and sometimes this damage can be extensive. But total defoliation is very unusual. Defoliation is only associated with dramatic outbreaks of species such as the Spruce Budworm and the Gypsy Moth. How do plants manage to fend off animals? Another way of posing this question is to ask what adaptations plants have evolved to aid in protecting them against herbivores.

Mechanical Defenses

Some plants are armored with thorns and sharp-edged leaves. Some have leaves covered by tiny spines called trichomes. These mechanical defenses discourage some of the larger herbivores, like deer and rabbits, as well as certain insects such as caterpillars. Fields in which cattle graze are often characterized by very short stubby vegetation, with the exception of thistle, which grows tall and abundantly. Cattle avoid the spine-covered thistle leaves and thus the thistles prosper while the other non-spiny plants are kept short by the cattle.

Mechanical defenses are not uncommon but they nonetheless occur only on a minority of plant species. You can examine 10 or

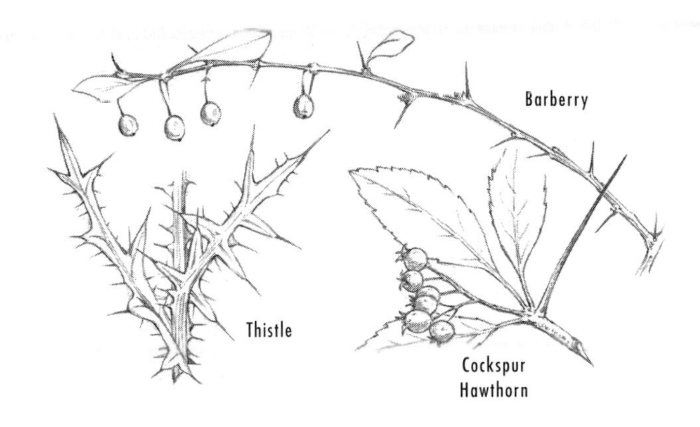

Fig. 28. Plant armor.

20 plant species around your home and find thorns, spines, and sharply pointed leaves on only a few. Most plant-feeders are insects, small enough to simply walk around a thorn. Even large herbivores like White-tailed Deer display little reluctance to chew thorn-laden branches, especially in times of food shortage. Mechanical defenses alone offer limited protection against herbivory.

Chemical Defenses

Although you won't be able to see them, the most effective plant defenses against herbivores are chemicals. Leaves, stems, and roots often contain varied types of chemicals, called defense compounds, that aid in repelling herbivores.

Poison-ivy and Poison-sumac, for example, both contain highly volatile oils capable of inducing severe skin irritation. Stinging Nettle leaves are covered by tiny hairs that exude an irritating acid when the plant is touched.

Few plants have leaves as large as Skunk Cabbage. You will notice that these massive leaves are rarely chewed. They contain calcium oxalate, a chemical producing a strong burning sensation on the lips. Skunk Cabbage also has a strong fetid odor, a by-product of its chemical defense system (see also p. 89).

Ginseng (see Pl. 13) has a thick tuberous root, a tempting target for subterranean worms and insects. However, the root contains a powerful chemical that acts as a heart stimulant in humans and serves as an effective defense against worms and insects.

Fig. 29. Foxglove.

A similar type of chemical, called cardiac glycoside, is found in Foxglove and plants in the milkweed family (see p. 345). Digitalis, a potent heart stimulant, is obtained from Foxglove. In low concentrations, cardiac glycosides induce vomiting in vertebrates. In milkweeds the cardiac glycoside is contained in a white, syrupy latex. Certain beetles and other insects cut leaf veins, preventing the poisonous latex from reaching the leaf extremities (edges), and thus rendering these parts of the leaf edible. In this manner these insects circumvent the chemical defenses of the milkweed. This is an example of a *counteradaptation.*

Spicebush leaves are highly aromatic, especially when broken. A member of the laurel family, Spicebush (see Pl. 10) and its relatives are protected by hard-to-digest phenolic compounds, many of which are the key chemicals of well-known spices.

Oaks are among the many kinds of plants containing bitter-tasting chemicals called tannins, which can interfere with the digestion of proteins. Tannins are produced as the leaves mature and help protect them from herbivores. The tea-like color of some streams is caused by the slow release of tannin from decomposing leaves.

Balsam Fir produces a chemical that mimics the growth hormones of several insects that prey upon the plant. Insects, upon ingesting this chemical, often fail to complete their development; thus the chemical acts as an insecticide.

Maple leaves have tough sheaths around their veins and deposit glasslike crystals in the leaf and on the surface. In addition, they store defense compounds such as tannin. Small caterpillars, photographed with an electron microscope, have shown patterns of extensive tooth wear after feeding on maple leaves.

Animal Responses to Defense Compounds

Although plants are generally protected by defense compounds, many insects and other herbivores have evolved the abilities to tolerate and/or detoxify plant defense compounds. The caterpillar of the Spicebush Swallowtail butterfly feeds with no difficulty on the phenol-laced leaves of the Spicebush. The important thing to note is that the Spicebush Swallowtail caterpillar feeds on relatively little else. Insects have tended to cope with defense compounds by becoming specialized to feed on only one family of plants. You will learn more about the Spicebush Swallowtail in Chapter 7.

The ability to tolerate defense compounds varies not only among species, but even among subspecies. The Eastern Swallowtail butterfly has both a northern race and a southern race. The northern race can and does eat aspen leaves, which contain phenols as defense compounds. The southern race cannot tolerate eating aspen leaves—doing so reduces caterpillar survival.

Plant defense compounds represent an obstacle to herbivorous insects. Within a particular species, any individual genetically capable of eating a kind of plant that others of its species find difficult or impossible to digest will gain energy unavailable to others. This individual will stand a comparatively greater chance of reproducing and thus its genes for dealing with the toxic plant will spread. This process results in various species becoming specialists on particular plants or plant families. For the plant, its defense compounds protect it against many, but not all insects. The entire process can be envisioned as an evolutionary cat-and-mouse game. A plant evolves a defense compound, affording it protection and thus making it successful competing against other plants. Eventually, one or more types of insect develop the ability to tolerate the defense compound and specialize on the plant.

Unlike insects, mammalian herbivores, if they eat leaves at all, tend to eat a diversity of plant species but avoid eating too much of any particular toxic plant. By ingesting only small amounts of specific defense compounds, these large animals, such as White-tailed Deer, can detoxify or eliminate the defense compounds. Beavers feed heavily on aspen bark and juvenile shoots, which have high concentrations of salicin, a defense compound.

Both insects and mammals prefer to eat young leaves and shoots, as these have the least concentrations of defense compounds and are most tender.

The Decomposer Food Web

Because plants are generally successful at keeping herbivores from overindulging, the world remains green. Most of the energy contained in bark, wood, leaves, and roots is not transferred to animals while the plant is alive. Only when the leaves fall in autumn or the tree dies does the bulk of the energy begin to be tapped by animals called decomposers (see Chapter 8). The result of plant defenses is that most energy in forests moves from plants to decomposers, not from plants to herbivores.

Other Adaptations

Why are leaves green?

Leaves are green because they contain a pigment called chlorophyll. Chlorophyll is responsible for photosynthesis, the process by which the plant combines water and carbon dioxide with energy from the sun to make high-energy sugar compounds. Without photosynthesis, life as we know it could not exist. All animals (with the exception of some organisms in deep ocean communities supported by bacteria that gain energy from a process called chemosynthesis, which is not light dependent) are ultimately dependent upon plants to capture a part of the sun's immense energy output. Only the plants, using photosynthesis, are able to convert solar energy into chemical energy. You cannot stand out under the rays of the sun and get fat, but a plant can and does.

But, is the color green an adaptation? The answer is no. Green is but one color in the full rainbowlike spectrum of wavelengths of light emitted by the sun. Each color is associated with a particular wavelength. Red wavelengths are longer than green ones, but green wavelengths are longer than blue ones. The green color of a plant occurs because chlorophyll reflects visible light at green wavelengths. Chlorophyll absorbs light and uses it in photosynthesis only at red and blue wavelengths. Although chlorophyll is a very significant plant adaptation, the green color is a mere by-product of chlorophyll's physical chemistry. For plants, green wavelengths are essentially useless and are simply reflected, making the plant look green.

This example is important because it focuses on the fact that not all characteristics of plants and animals need be directly adaptive. You could create any number of "adaptive scenarios" to explain why plants "should" be green. However, green coloration, perhaps the most distinctive feature of plants, is not in itself an adaptation.

Why isn't all bark alike?

Bark characteristics differ among tree species. Beech bark is very smooth and gray. Red Maple bark is smooth and grayish when the tree is young but becomes ridged and dark as the tree grows old. White Oak bark is pale and flaky. Black Oak bark is deeply ridged and very dark. Why do the differences exist?

Are these differences in bark adaptive? If so, what adaptations are provided to the tree by different bark characteristics? Bark is clearly adaptive. It is important as a support tissue and as a protection against herbivores. Bark color and texture are, however, quite questionable as adaptations. Perhaps light-colored bark reflects more light than dark-colored bark and thus aids in keeping the tree cool. The opposite argument could also apply. Perhaps darker bark aids in warming the tree as it absorbs sunlight. Tropical trees live in very warm environments and tend to have light-colored bark. Trees in higher latitudes tend to have bark that is proportionately darker in color, but many exceptions exist. Bark texture is even more dubious as an adaptation. Perhaps smooth bark aids in preventing lichens and mosses from colonizing the bark. Perhaps smooth bark requires less energy to manufacture. Perhaps deeply ridged bark is more difficult for boring insects to penetrate.

The possibility is strong, however, that bark characteristics *per se* are not adaptive. Trees, like all other life-forms, are made by constellations of DNA molecules called genes. Genes interact in extremely complex ways. Bark itself is important for the tree to survive, but the specific appearance of the bark may be a by-

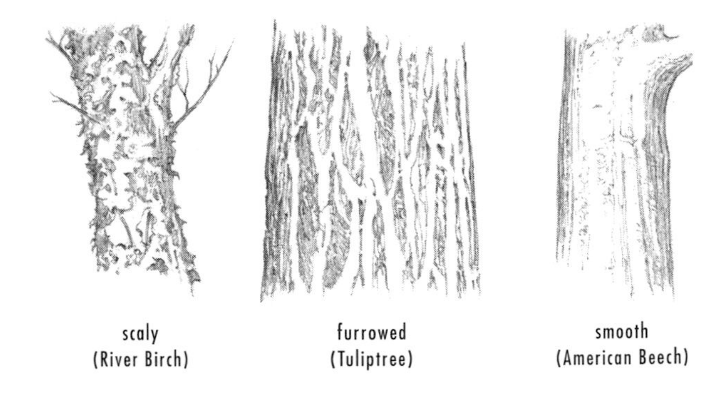

Fig. 30. Varying bark patterns.

product of the multitudes of interactions among the thousands of genes that make the tree. Thus the answer to the question, "Why isn't all bark alike?" could simply be that all genes are not alike and different combinations of genes make different types of bark. As in the previous example, it is important to realize that not every characteristic of a living thing need be adaptive. The possibility also exists that some bark textures are directly adaptive while others are not.

Why do some spider webs have thick zigzag strands?

Orb weavers are common spiders that spin elaborate radiating webs. The webs of some species, like the common Garden Spider, contain a central thick zigzag strand that makes this part of the web highly visible. Spiders use much energy in spinning webs and the thick zigzag strand, because it is so clearly visible, is not an aid in catching flying insects. Why does the spider "waste" energy by making the central strand zigzag and thick? Could the thick central zigzag strand really be an adaptation? If so, how does it improve the spider's ability to survive?

The answer is that the spider probably is not wasting energy and the zigzag strand is indeed adaptive. The thick zigzag strand serves to warn flying birds and running mammals of the presence of the web. A bird, if it were to fly through the web, would destroy it and cause the spider to spend many hours of labor plus many calories of energy repairing the web. There is also an "opportunity cost": no food would be captured while the

Garden Spider. The thick, zigzag strand in this web helps protect the web from damage by birds and other animals.

spider repaired or replaced the damaged web. It is thus more prudent for the spider to "warn" the birds, and the zigzag strand is a useful adaptation because it ultimately saves the spider more energy than it costs.

This example demonstrates how subtle an adaptation may be. Although many characteristics may not be adaptive, it could be very misleading to reject adaptation if the adaptive function is not immediately obvious.

How Adaptation Occurs — Natural Selection

Every observer of nature sooner or later comments upon how precisely plants and animals are adapted to their particular environments. The process responsible for adaptation is called *natural selection.*

Natural selection was first described during the last century by two English naturalists, Charles Darwin and Alfred Russel Wallace. Both men independently arrived at identical conclusions about how adaptation occurs, but it was Darwin who, throughout most of his lifetime, assembled a vast array of diverse evidence to support his contention that natural selection led to adaptation and evolution.

Every population, whether plant or animal, consists of individuals that are mostly genetically distinct. You can see this easily among people. With the exception of identical twins, we do not all look alike. We look different in part because of our environments. Some of us eat more than others and so are heavier. But we also look distinct because of our genes. You resemble your parents more than you resemble others because you share more genes with your parents than with strangers. Although you share many genes with each parent, you do not have exactly the same combination of genes as either parent, so you look like yourself, not like an exact copy of either your mother or father. You are even less genetically similar to a cousin and still less similar to a non-family member. Only if you are an identical twin is there another person with exactly your combination of genes. Genes produce various traits, and different genes in different combinations produce different traits among individuals. The cumulative effects of both genetic and environmental influences produce the individual organism, which is called a *phenotype.*

All Red Maples do not look alike nor, upon close inspection, do all Eastern Chipmunks. Just as there is genetic variability among humans, so there is among all species. Some Eastern Chipmunks are slightly larger in body size than others. Some may be more resistant to infectious bacteria. Some may be more able to tolerate starvation or cold. Genes are largely responsible for these differences.

Both Darwin and Wallace realized that whether or not a trait is "good or bad," i.e., adaptive or not, depends upon the environment in which the trait exists. A few examples will illustrate this point.

A few bacteria were genetically resistant to antibiotics such as penicillin long before these drugs were ever discovered and put to medicinal use. Many antibiotics are naturally occurring substances, produced by fungi. However, only when antibiotics were widely used against disease-causing bacteria did the trait of antibiotic resistance become highly adaptive, and those bacteria possessing the trait survived, while those lacking it perished in tremendous numbers. Now, unfortunately, antibiotic resistance is only too common among species of pathogenic bacteria. These bacteria prospered, because genes for antibiotic resistance were adaptive and survived disproportionately, as only their carriers survived to reproduce.

Another example of natural selection is perhaps the best known of any. In England during the Industrial Revolution, soot from the factories blackened tree trunks in areas around industrial centers. A few individuals of the Peppered Moth possessed a genetic mutation that made them darker than the usual light gray phenotype. These melanistic individuals were, by chance, well camouflaged against the blackened bark, while the normal lighter individuals stood out and were much more easily spotted by birds, their principal predators. Melanistic individuals survived better, reproduced, and the gene producing melanism spread in the popula-

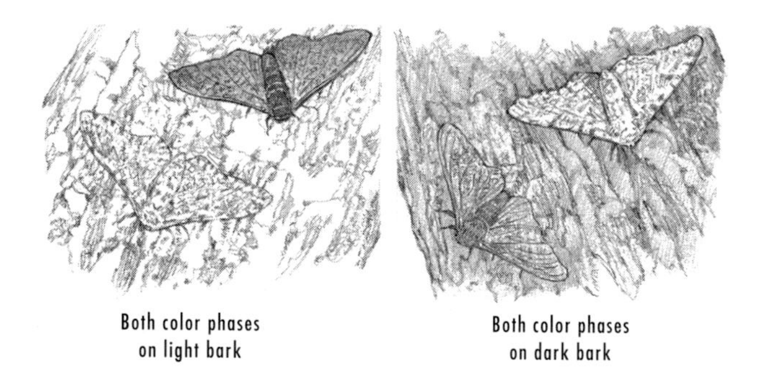

Fig. 31. Peppered Moth.

tion. Within a few moth generations, most members of the species were melanistic. The phenotype had changed in response to an environmental selection pressure, namely the changed appearance of tree trunks. Dark moths were "better" than light moths only in polluted environments. As coal was burned less and environments became less polluted during the present century, light-colored moths have increased relative to melanistic moths.

As a final example, consider the case of the Medium Ground Finch, one of 13 species of Darwin's Finch that are found only on the Galápagos Islands in the Pacific Ocean, 700 miles (1000 km) off the coast of Ecuador. Darwin observed these odd finches when he visited the Galápagos and, years later, he argued that each finch species evolved from a common ancestor that had colonized the islands many years previously. Recent research on the Medium Ground Finch has demonstrated natural selection. On Daphne Major Island, large finches with large beaks survived a severe drought much better than smaller finches with smaller beaks. Males, which average larger in size than females, survived better than females during the drought period. The birds with the largest bills were able to crack seeds that were too large and hard for the smaller-billed birds to crack. These were virtually the only seeds available during the drought, so the smaller-billed birds starved. Large bills were not adaptive until the drought, at which time they conferred high "fitness" (survival value).

Darwin and Wallace argued that all populations of living things, whether Common Ragweed or Moose, have the ability to increase in numbers beyond what the environment can support. The high populations, coupled with the limited resources of the environment, cause a struggle for existence. The struggle may be extremely subtle. A grass plant may be better able to tolerate lead in the soil than others within its population. A Blue Jay may be more resistant to an avian virus than others in its population. Both the grass and the Blue Jay have an advantage in a particular kind of environment.

Because of the struggle for existence, not all individuals in a population can survive and reproduce. Darwin and Wallace coupled this observation with the observation that individuals differ because of genetic variability. Both naturalists reasoned that those genetic variants most suited to the particular environment would have the highest probability of survival to reproductive age. This process of differential survival and thus differential reproduction is *natural selection*. If the environment stays the same, it selects out extreme variants and maintains the status quo. If the environment changes, those individuals most suited to the change survive best, and the species evolves a different phenotype (e.g., bill size may increase). Natural selection is the process by which species "track" their environments, in the genetic sense.

Adaptation is no guarantee of survival. Today's adaptation may be tomorrow's liability. The Giant Panda is precariously close to extinction because it eats bamboo almost exclusively, and bamboo has a curious reproductive pattern: whole bamboo forests produce flowers only every hundred years or so, in synchrony, and then the plants die off. Bamboo forests have recently "crashed" in China and will take time to recover, so the Giant Panda has far less to eat and populations are threatened. In addition, the habitat of the Giant Panda has been significantly reduced. It now occupies a much more limited range than in the past. Because of its specialized diet, this unique animal is probably doomed if its bamboo forests disappear.

Natural selection is really a two-step process. First, through the occasional mutation of individual genes and through the formation of new combinations of genes during reproduction, new variants form. This process is generally random. It is never certain which new genetic combinations will appear in a population, and a particular environment does not cause an "appropriate" mutation to occur. Second, some variants survive better than others, and which individuals survive depends in large part on the characteristics of the environment. Because the particular stresses imposed by the environment (climate, predators, parasites, competitors) determine which variants survive best, this process is not random but is, instead, selective. A cold environment will select for a different series of traits than a warm environment. Natural selection is continuous.

Adaptation is often highly opportunistic. The Seed Bug feeds exclusively on the flowerheads of Ragwort in southern old fields. Each male Seed Bug is aggressive and defends one Ragwort flowerhead, which resembles a small yellow daisy. Males both feed and mate on the Ragwort heads. Whenever another insect lands on a Ragwort flowerhead, the resident male Seed Bug attempts to copulate with it. This behavior, which obviously evolved as part of the mating process, acts to drive away intruding insects, unless, of course, the intruder is a female Seed Bug! Experimental removal of resident Seed Bugs allowed 4 or 5 additional insect species—species ordinarily repelled by Seed Bugs—to visit each flowerhead. The courtship behavior of the Seed Bug, itself adaptive, has evolved yet another function, that of territorial defense.

Patterns of Spring

To the naturalist, winter often seems prolonged and spring all too short. Spring is a season of awakened activity. Dormant plants and animals reappear and reproduce. Buds suddenly open and flowers blossom, heralding the spring. Some flowers on trees are quite small and inconspicuous, and are pollinated by the gusty spring winds. Other flowers, some on trees, most on shrubs and herbaceous plants, are brightly colored, attracting varied insect pollinators. Birds, having migrated south, return to establish and defend territories with vigorous songs, court their mates, and begin nesting.

As warmth replaces cold and days lengthen, forests and fields turn from brown to yellow-green to deep green. Another growing season begins. Insects, reactivated after winter's end, seem to suddenly reappear in profusion, keeping pace with the rapidly emerging vegetation. The forest floor is brightened as scores of wildflowers, the spring ephemerals, suddenly bloom and are visited by newly active Bumble Bees and Honey Bees. Stream and pond life is especially vibrant. Crayfish, snails, freshwater clams, and flatworms, as well as many larval and adult insects, can be observed by looking in vernal ponds and woodland streams. Frogs and salamanders are actively courting and laying eggs at night, presenting a challenge for the naturalist who would observe their life cycles.

The most remarkable characteristic of spring is how different one day can be from the next. Warblers and other migrating birds can abound today and be gone tomorrow. Flowers open and leaves unfurl over a period of just a few days. Animals of all kinds are more easily seen as many are dispersing and attempting to establish new territories. A diary tracking the changes from one day to the next will reveal how much can happen within a short time in the spring.

Spring is a time when you can easily regret missing a single day afield. Even spring nights can be particularly productive for the naturalist. Woodcock courtship occurs at dusk, when Whip-poor-wills, Chuck-will's-widows, and owls are vocalizing. As warm fronts from the south blow north, migrating birds can be heard calling as they pass overhead. In wetlands and vernal ponds the frog and toad chorus fills warm spring nights with sound, and spring rains bring frogs and salamanders out on the roads as they seek breeding pools.

FLOWER STRUCTURE AND POLLINATION (PLS. 35, 36)

The blooming of flowers signifies the arrival of spring. Cherries, dogwoods, magnolias, maples, laurels, rhododendrons, and redbuds, as well as numerous herbaceous species of forests and fields, all have colorful, conspicuous flowers. But others, such as oaks, birches, hickories, elms, ashes, walnut, beech, grasses, and sedges have greenish yellow, inconspicuous flowers. As you observe a selection of flowers, note whether or not the flower is "colorful," "obvious," or "inconspicuous," and see if flowers emerge before the leaves unfurl. Furry Pussy Willow catkins, tiny red clusters of Red Maple flowers, and broad blossoms of Flowering Dogwood adorn branches before the leaves open. Contrast this pattern with that of species such as American Beech and oaks, where flowering is less conspicuous and occurs essentially simultaneously with leaf opening, the flowers preceding the leaves barely or not at all. (Binoculars are helpful in observing flowers of canopy tree species.) Try watching individual Honey Bees and Bumble Bees as they feed on nectar and pick up pollen. The behavior of bees is closely related to the flowers they feed on and pollinate (see "Bee Behavior," p. 277).

FLOWER STRUCTURE

Take a close look at several species of flowers, avoiding, for the moment, the composites (see p. 175). An excellent choice for seeing flower anatomy is a lily or magnolia blossom. Flowers are reproductive organs. The objective of pollination is to unite the sperm-making pollen of one plant with the egg-containing ovary of a second plant of the same species. In many species, such as Tuliptree, magnolias, maples, and most herbaceous plants, both male and female organs are in the same flower (see Fig. 32). The *stamens* produce sperm-containing pollen, and the *pistil* houses the egg-containing ovary. In other species, such as birches, oaks,

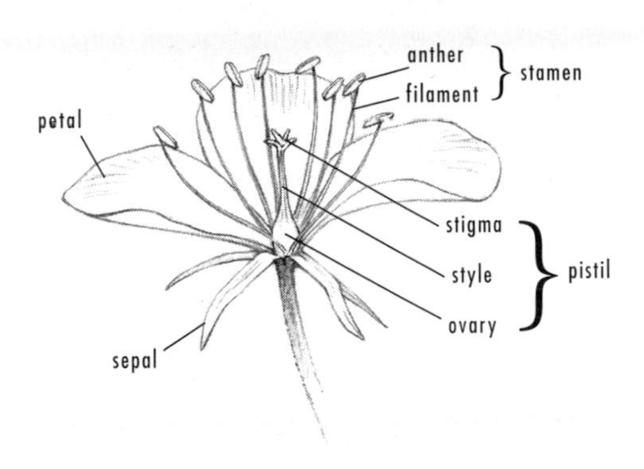

Fig. 32. Parts of a typical flower.

and hickories, male and female organs are in different flowers but both male and female flower types occur on the same plant. Still others, like the willows and poplars (including aspens and cottonwoods) have male and female flowers on separate trees.

Flower anatomy, like animal anatomy, varies among species, but the following is a basic pattern, readily visible in many species of trees, shrubs, and wildflowers. Though the reproductive organs are the essential parts of any flower, they are not usually the most showy parts. It is the *petals,* which are actually modified leaves (in an evolutionary sense), that make the flower stand out. Petals form a *corolla,* which surrounds the reproductive parts (stamens and pistil). Beneath the petals and often alternating with them are other modified leaves called *sepals.* The sepals together make up a structure called the *calyx.* The *pistil,* which houses the ovary, is divided into an elongate tubular *style* and a pollen-receiving tip called a *stigma.* Each *stamen* produces pollen and is composed of an elongate *filament* tipped by the pollen-containing *anther.*

Monocots and Dicots

There are two major groups of flowering plants. One, called the *dicots,* includes most of the world's flowers, and a second group, called the *monocots,* includes the palms, grasses, and such familiar flowers as the lilies, irises, orchids, and arums, such as Skunk Cabbage and Jack-in-the-pulpit. The key to separating a dicot from a monocot is to note the number of petals present on the corolla. Dicots have petals in multiples of fours and fives, and monocots have petals in multiples of threes. Dicots also differ from monocots by the pattern of veins in the leaf. Dicots have netlike venation, with the veins radiating out from a main central vein, and monocots have distinctly parallel leaf veins.

Wind Pollination

Early spring winds are ideal for blowing pollen from plant to plant. Consequently, many trees (especially canopy species) are pollinated by wind, and have flowers located high in the canopy. They open before the leaves emerge, helping assure maximum pollen dispersal.

Many wind-pollinated tree species (see Pl. 35) have male (staminate) flowers in the form of elongate, drooping catkins. Poplars, birches, oaks, hickories, walnuts, hazels, and alders all have staminate male catkins that hang downward, where the wind can easily blow pollen from them.

In birches, male and female flowers are on the same plant. Male catkins typically occur as elongate, hanging clusters, usually greenish. Female birch flowers are on shorter, thicker, hairier catkins, and they tend to be solitary rather than clustered. Poplars, such as aspens and cottonwoods, have male and female flowers on separate plants. As in birches, male catkins are elongate and drooping, and occur in clusters, but female catkins also are elongate and hang downward. Female catkins are both hairier and shorter.

In oaks and beeches, male catkins are grouped in conspicuous drooping clusters. Female flowers are not catkins, but instead are small, inconspicuous two- to four-petaled structures at the junction of branchlets near the outermost leaves.

Hickories and walnuts have elongate, drooping male catkins occuring singly or in small clusters. Female flowers are small and inconspicuous at the branch tips.

Ash flowers are either bisexual (both sexes on the same flower) or unisexual (sexes on separate flowers). In White Ash, male flowers are in dense clusters, lack petals, and have 2–3 stamens per flower. Female flowers occur in separate clusters on the same tree and are elongate and very exposed, each with a single pistil.

Elms have clusters of small and inconspicuous bisexual flowers.

In addition to the species just described, all conifers, such as the pines, spruces, and firs, also are pollinated by wind and most species have both male and female cones present on the same plant. Male staminate cones and female pistillate cones both occur in clusters at tips of separate branches. The fertilized female cone containing the seeds matures and remains on the tree until seeds are dispersed. Grasses and sedges are the other major wind-pollinated groups of plants.

Why do some plants rely on wind pollination?

Many groups of wind-pollinated tree species have separate male and female flowers. Separate flowers aid in insuring cross-pollination in plants that cannot depend on the systematic movements of pollen-laden insects. Having male and female flowers on separate trees also enhances cross-pollination.

In contrast, insect-pollinated flowers have a distinct advantage by having male and female organs on the same flower (see p. 274). Insects pick up pollen from every flower they visit and thus maximize cross-pollination.

Wind pollination is most adaptive in open habitats. Wildflowers on the forest floor are not exposed to wind of sufficient force to disperse pollen effectively. However, canopy trees with flowers located high above the ground, very exposed to spring winds, can efficiently utilize wind for pollen dispersal. Clusters of male catkins, located near branch tips, gain maximum exposure to wind. It is advantageous for flowers to open prior to leaf opening because such a pattern assures wind exposure. Other wind-pollinated plants, such as grasses and some herbs of old fields (i.e., Common Ragweed), live in very open, windswept areas. Wind-pollinated flowers are inconspicuous, non-odorous, and not usually colorful. None of those traits would serve to enhance wind pollination.

Insect Pollination

In the temperate zone, conspicuous, showy flowers are almost always pollinated by insects. Flowers provide nectar for insects that in turn carry pollen from flower to flower as they make their daily rounds. Insect pollination is one of nature's most widespread forms of mutual interdependence. Nectar is of no direct use to a plant. Making it is costly (because it is very rich in sugar) but that is the price that evolution has levied on many plants to help ensure reproductive success. Insect pollination is by far the most important form of pollination in wildflowers and it is also very common among shrubs and trees (see Pl. 36). In most cases, petals are the most brightly colored parts of flowers, but in some species, such as Flowering Dogwood, the colorful petal-like structures are actually modified sepals, called bracts. The bright colors make flowers obvious to cruising insects; some flowers also emit strong odors that serve as beacons for insect pollinators. Willows and Tuliptree are among the many species pollinated by bees. Beetles are the principal pollinators of large magnolia flowers, and various fly species pollinate Sassafras. Flowering Dogwood is pollinated by a variety of insects, as are many species of spring wildflowers.

Most insect-pollinated species have male and female sex organs in the same flower. A ring of colorful petals or sepals surrounds the reproductive parts and when an insect visits the flower it comes in contact with both stamens and pistil. Such an arrangement promotes cross-pollination.

Though you will not be able to observe pollen closely without a microscope, pollen grains from insect-pollinated flowers usually are sticky and often have protuberances, all the better for attaching to the leg hairs of a passing insect.

How Flowers Attract Insects

The arrangement of flowers helps attract pollinators. Depending on species, flowers may be clustered in flat, plate-like structures (called umbels and corymbs), elongate stalks (spikes, racemes, or panicles), or on widely spreading stalks (cymes). These arrangements concentrate flowers, making the total image more vivid and increasing the likelihood that an insect will visit several flowers on the same plant. Many plants locate flowers at the tips of long stalks that wave high over the plant. The composite flowers are unique in that they concentrate hundreds of tiny individual disk flowers tightly into a conspicuous flowerhead (see p. 175).

Different flower shapes and positions attract different pollinators. For instance, within the lily family, the orange Wood Lily produces open, upright flowers that are conspicuous in dry open fields and along woodland edges, well positioned for pollination by butterflies. The Canada or Wild Yellow Lily inhabits moist meadows, streamsides, and bogs and its flowers droop downward, making it difficult for butterflies to reach. It is pollinated by bees.

Many flowers, like those of Southern Magnolia, are open and easily accessible, but some flowers require considerable effort for an insect to gain entry. Bees must push their way into Turtlehead, and Bumble Bees, which are husky insects, must use their strength to literally pry apart Closed Gentian blossoms to get to the nectar. Bees vibrate their wing muscles vigorously to shake pollen loose from blueberry and Bittersweet Nightshade blossoms. Long, tubular flowers are not easily accessible to bees with short tongues, but long-tongued species have no difficulty. However, short-tongued bees may "cheat" by cutting a hole in the flower base to reach the nectar. Carpenter Bees "rob" blueberry plants in this manner, securing nectar but not carrying away any pollen.

Many flowers may have colors arranged in converging stripes or other patterns that act as "nectar guides," leading the insect to the center of the flower. Often, as in the case of Moth Mullein and Viper's Bugloss, anthers are colored differently from petals. Yellow Loosestrife has a circle of red dots around the center of its otherwise yellow blossom. Irises have boldly veined sepals, the centers of which are a different color from the periphery. Flowering Raspberry has deep pink petals surrounding a yellow center and Painted Trillium has white petals that are red at the base, leading the insect to the yellowish interior. Many composites, such as Black-eyed Susan, have flower clusters with distinctly colored centers.

Some bee-pollinated flowers look deceiving to human eyes. They appear uniformly white or yellow, and indeed, to our eyes, they are. However, bees see in ultraviolet wave lengths as well as visible light. These flowers also have distinct nectar guides leading bees into them, but these patterns are only visible in ultraviolet light, thus we cannot see them. The bees, however, do. For example, Marsh-marigold, which looks merely bright yellow to human eyes, absorbs UV light very strongly in the center and much less so on the periphery. To a bee, the flower looks like it has a dark center spot, somewhat like a Black-eyed Susan. Many composites, such as some of the sunflower species, look uniformly yellow to us, but have ultraviolet patterns that, to bees, make them look dark in the center.

Many plants vary the amount of nectar contained in their flowers. On a single plant some flowers may contain much nectar, others little or none. Further, an individual flower may be nectar-rich on a given day and nectar-poor on another. This pattern has been described as a "nectar lottery," which forces insects to keep trying flowers until they "hit a winner." Such conditioning greatly promotes cross-pollination by forcing the insect to contact many flowers. You can observe the nectar lottery by watching a bee move from plant to plant. It will sample plant after plant, touching down very briefly and often finding the flowers nectar-poor, but eventually stopping to feed when it "hits the jackpot."

Major Groups of Insect Pollinators

Bees and wasps are among the most important pollinators. Some, like Honey Bees, are colonial and communicate the distance and direction of a nectar source to their co-workers in the hive by performing an intricate "waggle" dance. Others, like Carpenter Bees, are solitary. Bees' tongues vary in length among species but all are efficient at extracting nectar, and their legs (and often their bodies) are hairy, able to transport large sticky masses of pollen. Moths and butterflies also rank high as pollinators. These insects have particularly long tongues, well adapted to probe elongate, tubular flowers. Among the flies, the syrphid flies, which closely resemble bees, are important pollinators, as are the carrion flies, which are attracted to offensive-smelling plants. Only a few beetles are pollinators.

Bee-pollinated Flowers

Flowers that are pollinated principally by bees tend to be open and accessible, but some are partly closed or irregular in shape, insuring that only certain bee species can pollinate. Many legumes are pollinated by bees and have flowers arranged in such a way that the bee must depress the lower keel petal to reach the nectar. In doing so, it acquires pollen on its underside. Clovers are pollinated by various bees. Red Clover is mostly pollinated by Bumble Bees, but White Clover is pollinated by the smaller Honey Bees. The tongues of Honey Bees are too short to probe deeply enough into Red Clover to reach nectar. Other flowers, such as Common Milkweed (see p. 342), Squirrel-corn, Dutchman's-breeches, and Mertensia (Virginia Cowslip), are structured to favor large Bumble Bees.

Jewelweeds, snapdragons, violets, and columbines are other examples of bee-pollinated wildflowers, while willows and Tuliptree are examples of bee-pollinated trees. Many bee-pollinated flowers are blue, violet, or yellow, and, as mentioned above, some reflect ultraviolet wave lengths.

Bumble Bee. Bumble Bees pollinate many kinds of flowers, such as asters, shown here, and milkweeds. Individual bees, however, are highly selective about which species of flowers they pollinate.

Bee Behavior

Select a meadow or woodland with wildflowers in bloom and watch individual bees as they forage for nectar. Though the number of flower species may be numerous, a single bee will tend strongly to select only a single species of flower. A Bumble Bee feeding on Red Clover will move methodically among the Red Clover plants, ignoring all others. Each bee becomes a specialist, "majoring" in one species of flower, having selected a major flower species during its initial foraging excursions. In the case of intricate flowers with complex shapes, studies have shown that bees learn how to enter a flower by trial and error, and you can watch the bees learn as you observe them. A novice bee will move almost randomly around a flower, searching for a point of entry. It will take up to 20 seconds to figure out how to get inside the flower. Often it will give up, having failed to find or force entry, and will move to another flower without having fed. An experienced bee will move directly to the nectar source, usually within 2 seconds. Novice bees also sample more than one flower species before making a selection. You may see a bee fly to several species, wandering around on the surface of each. Chances are good that this individual is a novice, in the process of selecting its major. It requires about 7 foraging trips for a bee to become proficient and choose its major. Each bee will also select a "minor" species, to which it can go if its major is not in bloom. Within a given bee species, individuals vary in their selection of majors. Not all Bumble Bees are Red Clover majors. Some major in Jewelweed, others in Common Milkweed. Individual bees (of the same species) also vary in the way they deal with the same flower species. In plants such as Butter-and-eggs or Jewelweed, the flowers have long nectar spurs, and some Bumble Bees will enter the flower directly, while others will pierce the base of the spur to get at the nectar. Long-tongued bees select deeper flowers, while short-tongued species select shallow or open flowers.

Butterfly- and Moth-pollinated Flowers

Most butterfly-pollinated flowers are upright, fragrant, tubular, and red, yellow, or orange. Butterfly flowers include species of phlox, orchids, pinks, asters, goldenrods, milkweeds (including Butterflyweed), and thistles. Moths also feed on tubular flowers but are attracted to white, night-blooming species such as Jimsonweed. The yellow Evening-primrose is also pollinated by moths. In our area, few trees are pollinated by butterflies or moths.

Jack-in-the-pulpit. This forest understory wildflower is pollinated by flies.

Fly-pollinated Flowers

Many fruit-producing trees as well as many wildflowers are pollinated by the bee-like syrphid flies, which seek nectar, as bees do. Sassafras, a forest understory tree, is also pollinated by flies. Fly-pollinated flowers are often whitish and shallow. Other flies pollinate such rancid-smelling plants as Carrion-flower, Red Trillium, Skunk Cabbage, and Jack-in-the-pulpit. These flowers are not all brightly colored but probably rely mostly on odor to attract the insect. Flies are presumably "fooled" by these flowers, which emit the odor of rotting carcasses. When the flies lay eggs, they accidentally pick up pollen in the process.

Flowers Pollinated by Several Kinds of Insects

Early spring wildflowers such as Hepatica, Trout-lily, Bloodroot, Spring-beauty, Cut-leaved Toothwort, Wild Oats, Mayapple, and trilliums tend to be pollinated by more than one kind of insect, though Bumble Bees will usually predominate, especially on cold days (see below). If you observe closely you will note several species of bees (especially solitary bees rather than Honey Bees) and flies landing on these wildflowers, most of which are white or pink in color. You will also notice that both large and small bees, as well as flies of various sizes, may visit the same flower. Watch and record insect visitors on a warm spring day and you will soon appreciate that early-blooming forest wildflowers are generalists, pollinated by many insect species.

The blooming period is short in spring, and many species flower simultaneously. These "spring ephemerals" bloom when weather is often cold or inclement, keeping insect activity low. It is to the advantage of these species to be generalists, able to be pollinated by more than just a single kind of insect, because this characteristic enhances the probability of cross-pollination during the short spring, when risk of failure (no pollination) is highest. In addition, most spring ephemerals are self-compatible, so that a single insect, by wandering about on a single flower, succeeds in pollinating it. Seeds are produced even if cross-pollination by insects does not occur.

The Importance of Bumble Bees

Virtually all of the early spring flowers are pollinated by Bumble Bees, which are active earlier in the spring than most other insect pollinators. Bumble Bees, highly tolerant of cold, can begin foraging early in the morning, even in predawn hours, and keep foraging well into late afternoon. Flying Bumble Bees can maintain a high body temperature even on very cold days because they are covered by dense "fur" that effectively insulates them. The muscle activity of flight generates body heat. On a typical spring day you will see Bumble Bees in the early morning and late afternoon and Honey Bees in the warm noonday sun. During very warm days, Bumble Bees remain inactive during the hottest hours.

Pollination in Old Fields

Annuals and perennials of early-stage old fields tend to be pollinated by a diverse assemblage of insects. Some goldenrod and aster species can be pollinated by as many as 20 insect species, and some herbs by even more. In one old field in southeastern Michigan 57 species of herbs and shrubs were visited by 134 species of bees over a 35-year period. Some abundant flower species were visited by 28–49 bee species, including 3 families of long-tongued bees and 3 families of short-tongued bees. It is very difficult to separate bee species in the field, a task further complicated by the number of flies that mimic bees (see p. 350)! As mentioned above, individual bees do not have catholic diets; instead they become "majors" on one particular herb species. Many early pioneer plants (colonizers) produce little nectar but they flower for long periods, often up to 50 days. Many of these plants are also capable of self-pollination, thus a single plant can produce fertile seed even in the absence of another of its species. This is of advantage, since in early-stage old fields, a given plant may be the only member of its species present. Later-stage successional fields harbor herb species that are more specific in the type of insects that can pollinate them. A good example of such a species is Common Milkweed (see p. 341).

Why are so many plants pollinated by insects?

The tremendous diversity of flowering plants, especially dicots, is probably due, at least in part, to the efficiency of insect pollination. When flowering plants first appeared during the Mesozoic Era, or Age of Dinosaurs, they represented a food source for insects. During the major part of the Age of Dinosaurs, from 120 to 60 million years ago, flowering plants diversified and flourished. Their high reproductive success was likely due to the specialized feeding habits of various insects that visited specific flowers, and, in doing so, effectively cross-pollinated the plants. Both flower and insect have, in example after example, become locked together in an ecological interdependence. Without the flowers, the insects would lack nectar, but without the insects, flowers would be sterile. Such a relationship, where both parties benefit and each depends on the other, is yet another example of evolutionary mutualism.

Insects are more precise carriers of pollen than wind, and it is hardly surprising that in environments where wind is not very prevalent (such as the interior of a forest), insect pollination is the rule.

Cross-pollination is most efficient when particular insects consistently are attracted to particular flowers. Therefore, flowers pollinated by insects tend to have certain distinct shapes, many of which provide efficient access or produce a fragrance attractive only to certain insects. Flower specialization is a way of evolutionarily "cornering the market" on a particular insect group and using it to move pollen about. Plant evolution has been affected by competition among plants for pollinators, a competition that has resulted in specialized flower anatomies, varied nectar production, and seasonal spacing of flowering. Flowers that open early get first crack at pollinators but risk pollinator inactivity due to poor weather or cold. Flowers that open in mid- or late summer do so when insect activity is at its peak, and thus when competition for pollinators is at its highest. These flower species tend to be most specialized, attracting a specific pollinator. This is why there are so many different shapes and colors of flowers in a summer meadow.

Sometimes, however, it is disadvantageous to restrict nectar access (and thus pollination success) to a specific pollinator. Early spring wildflowers and annuals and perennials of early-stage old fields are better served by being pollination generalists. By accepting any of a range of insect species, including both small and large bees and various fly species, these plants, which either bloom for short periods or have very brief life cycles, increase the probability of cross-pollination. The fact that most of these species can pollinate themselves certainly hedges their bets and helps assure that even marginal insect activity will still result in pollination.

AMPHIBIANS IN SPRING

Amphibians include toads and frogs, which together are termed the anurans, and salamanders. One other group, the caecilians, are burrowing amphibians that do not occur in our area. Amphibians are not normally thought of as terrestrial animals. Many, however, live in forests and fields and are very commonly heard and seen in spring.

An amphibian is a vertebrate animal with scaleless, soft, moist skin and toes lacking claws. Frogs and toads, with their large heads, chunky bodies, and long jumping hind legs, are very distinct, but salamanders are sometimes confused with lizards, which are reptiles. Reptiles have scaly dry skin and toes with claws.

The most important characteristic of an amphibian is that it leads a double life, undergoing a complete metamorphosis from larva to adult. The larval animal is aquatic but the adult of many species is essentially terrestrial. Amphibians generally must reproduce in water. In most species, eggs are laid in gelatinous masses or sheaths after fertilization.

In frogs and toads (with two exceptions) fertilization is external, with the male clasping the female and depositing sperm as the female sheds eggs. In most salamanders, fertilization is internal. Males deposit sperm-containing gelatinous capsules called spermatophores, which females pick up with special muscles in the cloaca, the opening through which all waste products as well as eggs pass.

In spring, watching amphibians will teach you about metamorphosis and complex life cycles, courtship and mating, habitat specialization, and color phases in animals.

For accounts of individual species, refer to *A Field Guide to Reptiles and Amphibians.*

Finding Amphibians

Most amphibians are nocturnal. The heat of the day dries their skins, which must remain moist, so amphibians are active at night. Often, many frogs, toads, and salamanders can be observed on wet roads at night during spring showers and on rainy nights throughout the summer.

Most frogs and toads occur in wet habitats such as swamps, marshes, ponds, and lakes, especially near woodlands. Many breed in forests, using temporary vernal ponds that fill only after heavy spring rains and snowmelt. Some, however, like adult toads and the Wood Frog, are common far from water. In spring frogs are often found in dew-covered fields and wet meadows.

Salamanders are found in streams, ponds, and woodlands. Woodland species are rarely found in the open, usually being concealed beneath rocks and logs. Some species even enter damp cellars. Certain species of the mole salamanders (such as the Spotted Salamander) can be seen in large numbers at their vernal breeding ponds on warm, rainy spring nights (see below).

To find amphibians you should venture forth at night with a good-quality flashlight or headlamp. Locate a pond where frogs and toads are breeding. Since these animals are highly vocal, locating such a pond should not be difficult. Salamanders can be observed during the day by carefully turning over rocks and decaying logs, especially in moist areas such as streamsides.

Frogs and Toads (Anurans, Pl. 37)

Anuran Calling and Mating

OBSERVATION: Begin to know frogs and toads by learning to recognize their voices. Just as spring sunrises are accompanied by a dawn chorus of birdsong, spring sunsets, and the dark hours following, are filled with anuran calls. Recordings are available of frog and toad calls to help you learn them.

Frogs and toads become active as the temperature warms and waters open. Southern animals begin calling weeks in advance of animals of the same species in northern states. For the most common anuran species there is a rough seasonal sequence as follows: The earliest vocalizers are Spring Peepers, followed quickly

Fig. 33. Calling toad.

by Wood Frogs. Soon American Toads begin their nightly trills, followed by the Pickerel and Leopard frogs, Cricket Frogs, and Fowler's Toad. After these species come the Gray Treefrog, Green Frog, and Bullfrog. The last to sing in spring is the Mink Frog, found only in the northern states and Canada. Southern states follow basically the same sequence but with a few more species (see p. 287).

Calling dates overlap extensively and a single pond may harbor six or more species, all calling simultaneously. Anuran calls are highly distinct among species and are rather easy to learn. Keep a record of the earliest calling dates of each species in your region.

Frogs often exhibit *countercalling,* when two or three frogs alternate, each calling in turn. This behavior prevents calls from overlapping, making the location of each caller easier for females to discover.

EXPLANATION: The difference in timing and especially the distinctiveness of each species' calls helps ensure that frogs and toads will only mate with others of their species. Hybrid animals cannot produce fertile offspring (because parental genes are not sufficiently complementary), and so represent a waste of parental reproductive effort. However, hybridization would tend to be a frequent event among anurans because species must find a pond in which to mate, forcing several species together. With some temporal separation of calling dates and clear differences in voice among species, hybridization is avoided. The adaptive value of anuran vocalization is to attract a mate *of the same species.* This function is particularly clear in the case of the American and Fowler's toads. Both species are similar in size and appearance but differ both in calling dates and voice. The American Toad, an early breeder, emits a high trill. The Fowler's Toad, a late breeder, utters a deep nasal bleat.

Although some species, like the Bullfrog and Fowler's Toad, breed throughout the spring and summer, some, like the Spring Peeper, breed in large numbers in temporary spring ponds. These species must synchronize their breeding early in order to ensure adequate time for egg and tadpole development prior to the drying up of the temporary pond. Very cold temperatures can freeze over the ponds, resulting in large-scale killing of eggs and tadpoles. Countercalling (see above) is common among peepers.

Anuran vocalization functions similarly to bird song. Some frogs, like the Bullfrog and Green Frog, use their calls both as a strong territorial signal and to attract potential mates. In other species, like the Spring Peeper and Pickerel Frog, calls function mostly as advertisements for mates, although Spring Peepers do react to the presence of other males by emitting a distinct call.

How Calling Is Done

Try to observe the throat of frogs and toads as they vocalize. Treefrogs and toads all call by greatly expanding a balloon-like air sac, which is part of their throat skin. True frogs (not treefrogs or toads) have a pair of inflatable air sacs beneath and between each eye and ear. Many frogs, however, like the Bullfrog and Cricket Frog, call without greatly expanded throat sacs. Throat sacs add volume and resonance to the call. Pickerel frogs call under water.

Anuran Mating Behavior

Although males and females are alike in size, males have proportionately larger tympanums (eardrums) than females, and so with some practice it is possible to separate the sexes. During mating, which always occurs in shallow water, males tightly grasp females around the upper or sometimes the lower body, a position termed *amplexus.* Females begin laying eggs and males release sperm, which is attracted chemically to the eggs. Since egg-laying can require several hours, amplexus is often maintained for a considerable time. Eggs, which are protected by a gelatinous cover, may be laid singly, in strings, or in masses. Egg-mass characteristics differ and can often be used to identify the species.

Anuran Life Cycles

OBSERVATIONS: Anurans undergo a dramatic metamorphosis during the course of their life cycles. Eggs hatch into tadpoles, fully aquatic fishlike animals that breathe oxygen in water through gills. Tadpole gills are covered by a flap of skin and are not readily visible. Tadpoles swim using a muscular tail. They are vegetarians and scavengers. Their digestive systems are proportionately longer than those of adult frogs and toads because plant tissue is more difficult to digest. As a tadpole matures into a frog or toad, its body changes form. Most noticeable is the development of legs, occurring well before the tail is reabsorbed. Gills disappear. Unlike insects, amphibians undergo metamorphosis while the animal remains active. There is no equivalent in amphibians to the pupal stage in insects. As the spring and summer progress, you can observe tadpole development by periodically netting tadpoles in shallow ponds.

EXPLANATION: What benefits are derived by anurans as a result of their complete metamorphosis? Why do they have such complex life cycles? Vertebrates in general have very simple life cycles. Complex metamorphosis is much less common in vertebrates than in insects. Among terrestrial vertebrates, only the anurans and salamanders have complex life cycles.

Amphibian metamorphosis may enable the tadpoles to thrive on a rich but temporary food source, namely the vegetation of vernal ponds. Tadpoles are vegetarians and scavengers, whereas all adult frogs and toads are strict carnivores. Temporary ponds can only be utilized briefly before they dry up. The tadpoles are confined to these ponds but benefit from the plant food in them. The adults, which have far greater mobility, disperse the species.

Metamorphosis is an ancient trait, however, dating back to the first amphibians that evolved more than 300 million years ago. Tadpoles probably evolved in large part because their eggs were laid in water. It is possible that modern amphibian metamorphosis is not in and of itself adaptive, but is "evolutionary baggage," left over in genes from eons ago.

Anuran Adaptations and Diversity

OBSERVATIONS: Examine a frog or toad and discover how its body form reflects the process of adaptation. Even a casual inspection reveals the animal's unique adaptations for jumping. The long, muscular hind legs used for propulsion are dramatically larger than the forelegs. The short front legs function for balance. All adult frogs and toads are predators. The head resembles a compressed crocodilian face: the eyes are raised up, allowing the animal to see while it remains just under the water surface. Nostrils are located on the upper surface of the snout. The mouth opens to a very wide gape. Sense organs are well developed, with prominent eyes and ear drums. Unlike crocodilians, however, the majority of anurans lack sharp teeth (though at least one South American species has teeth sharp enough to draw blood—and it eats other frogs!). Smaller species of frogs and toads gulp their prey, many species snare insects with their long sticky tongues, and larger species capture animals with their jaws.

Several distinct families of anurans exhibit the same basic body form. In our area, the most common families are the spadefoot toads, the true toads, the treefrogs, and the true frogs. Try to locate at least one representative of each family and compare them.

Spadefoot toads are terrestrial and fossorial (burrowing). They are hard to see because they burrow in sandy soil during the day and feed only at night. Heavy spring rains trigger the onset of breeding and this is the best time to observe spadefoots, since they are above ground. The key characteristics of spadefoot toads are the black, fingernail-like spade on each of the hind feet that is used in burrowing, and the eye, which has a vertical pupil, in contrast to that of all other anurans.

True toads are known for their warty skin, which helps adapt them to terrestrial habitats. The "warts" are glandular and poisonous if eaten. In addition, there are 2 prominent parotoid glands located behind each eye. Toads are poor jumpers but their noxious warts probably provide ample protection from predators.

Treefrogs are recognized by their enlarged toe pads, which are suction discs that enable the animal to cling on a vertical surface. Most treefrogs are smaller than true frogs and toads.

True frogs have moist skins and lack any trace of toe pads.

EXPLANATION: A comparison of a frog with a salamander demonstrates how radical the anuran body form is in comparison with the more typical and more conservative vertebrate body form. How the unique anuran body form evolved is not well recorded in the fossil record, though some very ancient froglike fossils are known. Once it appeared, however, the animals became very successful.

The four families described represent an *adaptive radiation* within the anuran group. From the basic anuran body form, itself a marvel of adaptation, four distinct families evolved, each adapted to exploit a different part of the environment. Spadefoot toads and true toads are adapted to dryness and spadefoots are specialized burrowers. Treefrogs are adapted to cling on vegetation and true frogs are adapted to life in ponds, marshes, and other wetland environments. Throughout the history of life on earth, whenever a successful new body form evolved, it soon seems to have radiated into many species.

Anuran Diversity and Latitude

OBSERVATIONS: Make a survey of the number of species of anurans in your region. Numbers of species vary with changes in latitude. Twenty-one species of anurans occur north of southern New Jersey and most of these species also are found in the South. However, 39 species, almost double the number, occur south of southern New Jersey, particularly along the Coastal Plain, and 21 of these do not occur at all in the North. As you travel from north to south you will find many more anuran species. For instance, only 2 species of true toads, the American and Fowler's, occur in the Northeast, but 5 species are found in the Southeast, including the endangered Houston Toad, found very locally in northeast Texas. Anuran diversity decreases further in eastern Canada, where only 10 species occur. This latitudinal decrease in species diversity from north to south is termed a *diversity gradient*.

ANURANS IN EASTERN CANADA — 10 SPECIES

American Toad	Bullfrog
Northern Chorus Frog	Green Frog

Lesser Gray Treefrog
Spring Peeper
Wood Frog
Northern Leopard Frog
Pickerel Frog
Mink Frog

ANURANS IN FLORIDA — 22 SPECIES

Eastern Narrow-mouthed Toad
Greenhouse Toad
Eastern Spadefoot Toad
Oak Toad
Southern Toad
Least Grassfrog
Southern Cricket Frog
Southern Chorus Frog
Ornate Chorus Frog
Greater Gray Treefrog
Spring Peeper
Green Treefrog
Pinewoods Treefrog
Squirrel Treefrog
Cuban Treefrog
Barking Treefrog
River Frog
Pig Frog
Bullfrog
Green Frog
Gopher Frog
Southern Leopard Frog

Among the treefrogs alone, there are 2 species in New England, 15 in the Southeast, and 34 in Costa Rica.

Latitudinal diversity gradients are found for most kinds of plants and animals. Ferns, conifers, flowering plants, insects, reptiles, amphibians, birds, and mammals all become more diverse toward the equator. A very pronounced increase in plant and bird species diversity occurs in Central American tropical rain forests, an area not included in this book.

EXPLANATION: Why are there more species in the southern latitudes? No definitive answer to this question as yet exists, but several explanations seem possible. Each taxonomic group may be responding

Pickerel Frog.

to a different set of factors and no general answer may apply to all.

All groups except birds and mammals, however, do have one characteristic in common. They are *ectothermic,* meaning that they cannot generate constant high levels of body heat. Although birds and mammals are *endothermic* and do maintain constant high body temperatures, most must obtain insects and/or plants for food and thus trends in bird and mammal species diversity may be directly affected by insect and plant diversity. Mammals, which depend upon insect food less than birds do, do not increase as markedly in diversity in southern latitudes (with the exception of bats, which have undergone a dramatic adaptive radiation in the tropics, evolving into nectar-, fruit-, fish-, and blood-feeders, as well as insect-feeders). Also, endothermy requires high metabolic rates, meaning that the animals must obtain large amounts of food. Any limit in food sources would severely restrict the number of endothermic animals that could be supported by the habitat.

Cold northern winters present a formidable obstacle for ectothermic animals. Cold temperature slows chemical reactions and ice can destroy cells. Winter hardening and bud dormancy in plants, hibernation and migration in animals, are all elaborate adaptations to cope with the effects of the cold stress of winter.

Therefore, one possible reason for decreased diversity in the North is that relatively few plant and animal species have succeeded in adapting to the severe climatic stresses. The farther north one goes, the more formidable is the winter climate. The South has a more equable climate, permitting larger numbers of species to find food and requiring less elaborate physiological adaptations.

An additional reason for the lower diversity in the North may be historical. The North was significantly glaciated as recently as 20,000 years before the present, an event that very significantly changed ecological communities and resulted in a large number of extinctions. Perhaps there has not been sufficient time to recolonize northern latitudes following the retreat of the glaciers.

Salamanders (Pl. 38)

Adaptations of Salamanders

Spring is an ideal time to learn about salamanders, many of which are common in forests. Usually, but not always, they remain near streams and vernal ponds. Mostly nocturnal, salamanders can be found during the day by uncovering rocks and decaying logs. (Be very careful not to disturb snakes when you do this!) The woodland salamanders all illustrate how a group of animals is adapting from an aquatic to a more terrestrial existence.

Salamander Life Cycles

Adult salamanders are easy to find beneath rocks and logs but, because they are essentially nocturnal and do not vocalize, their life cycle activities are more difficult to observe. Nevertheless, a knowledge of how these animals live their lives will add to your understanding of adaptation and help you observe life cycles directly in the field.

Many salamanders, because they are amphibians, have a life cycle similar to that of anurans (frogs and toads). After mating, females lay eggs that hatch into aquatic larvae. Following the larval stage the animals undergo a metamorphosis to the adult stage. Other salamanders, however, have life cycles that occur entirely on land (see pp. 292, 293).

Salamander Mating Habits

Salamanders are utterly silent; vocalization plays no part in their mating process. There is no such thing as learning the voices of salamanders. Because salamanders seek out the undersides of rocks and decaying logs, coincidental encounters between the sexes occur frequently. In addition, some species, like the Spotted Salamander, move *en masse* to breeding ponds during an evening of early spring rain. This synchronous movement assures that the sexes have access to each other.

Salamanders do not mate randomly. Males court females, an event termed a courtship bout, and females must accept a male before mating ensues. Males then deposit sperm-containing bodies called spermatophores that are picked up by the females, using unique cloacal muscles.

Why do salamanders—and other animals—engage in courtship behavior?

Why do newts and other salamanders expend so much effort at mating prior to the actual transferral of sperm to females? It would seem more efficient and less risky in terms of predators to merely "get it over with" in the briefest possible time. However, the advantage of courtship behavior in salamanders, as in other animals, is that it prevents mating with the wrong species and so avoids hybridization and waste of sex cells. Courtship results both in species recognition and synchronous receptivity to mating between male and female. Males initiate and perform most of the courtship, probably because they must make the initial commitment. If a male deposits spermatophores and the female refuses to accept them, that male is genetically a failure. If the female is unreceptive to spermatophores, she too is a genetic failure. Both animals must be simultaneously prepared or sex will not succeed. Such is the function of mating behavior.

The courtship and mating behavior of some common salamanders is described below.

RED-SPOTTED NEWT **PL. 38**

The Red-spotted or Eastern Newt is found in forest ponds and streams. The male begins courtship in the water by swimming past a female, making brief contact with her as he passes, repeating the process until he succeeds in grasping the female with his hind limbs. Male newts develop substantially thicker hind limbs than females and, during the breeding season, male hind limbs have a rough, almost sandpaperlike skin, which helps the male firmly clasp the female. This union is termed *amplexus*. The male may remain clasped to the female for nearly an hour before the next stage. (Sometimes the female newt will be highly receptive and prolonged amplexus does not occur.) Eventually, the male releases the female and moves ahead of her with his tail elevated in a posture called the *tail walk* before he deposits the spermatophores. If the male has been successful in courtship, the female follows and accepts the spermatophores. Once inside the female, sperm may be stored or released. A female may store sperm of several males. The animals go their separate ways following mating.

LUNGLESS SALAMANDERS **PL. 38**

Look for lungless salamanders in forests under decaying logs, in leaf litter, and along streams. A common and widespread species is the Red-backed Salamander. Lungless salamanders have a different form of courtship from that of the newts. A male rubs the underside of his mouth against the female, especially near the female's cloaca. Glands in the chin area secrete a substance that induces the female to accept spermatophores.

SPOTTED SALAMANDER **PL. 38**

The Spotted Salamander breeds in vernal ponds throughout eastern woodlands. Males deposit spermatophores on twigs, leaves, or other objects, where females pick them up. Females lay eggs in vernal ponds. The jelly-like egg masses, each containing up to 100 large eggs, are easy to spot attached to twigs. The egg cases of these salamanders can be differentiated from those of Wood Frogs, which lay eggs at the same time, because the salamander eggs are surrounded by an outer envelope of jelly. This species is among the earliest salamanders to breed, perhaps due to the temporary nature of some of the spring ponds.

The Spotted Salamander is unique in its habit of mass reproduction. Following a warm spring rain, Spotted Salamanders converge on breeding ponds and a single pond may host more than 100 animals. The animals aggregate and engage in a frenzied behavior of rubbing against one another. This activity is accompanied by deposit of spermatophores by males.

Salamander Eggs

OBSERVATIONS: Just as salamander courtship is a challenge to observe, salamander eggs, with few exceptions, are a challenge to find.

Among the forest-inhabiting salamanders, three patterns of egg laying occur:

1. Most species of **MOLE SALAMANDERS** lay sizeable gelatinous egg masses in vernal ponds (see Spotted Salamander, above). These are perhaps the easiest salamander eggs to find. Eggs are small and transparent, and float on the surface or are attached to vegetation. Following mating by large numbers of these salamanders, many egg masses may be seen in a single vernal pond. One exception, the Marbled Salamander, lays its eggs on land, but on flooded sites. The female remains with the eggs until hatching. All mole salamander eggs hatch into fully aquatic larvae.

2. **DUSKY SALAMANDERS** lay egg clusters containing 10–20 eggs beneath objects on land or in water. These salamanders provide parental care—the female remains with the eggs until hatching. The eggs hatch into fully aquatic larvae.

3. **LUNGLESS SALAMANDERS** lay eggs in damp places on land. Females usually remain with the eggs until hatching. The larval stage is completed *in the egg*, so when the eggs hatch, miniature adults emerge. The newly hatched animals do not remain with the mother.

EXPLANATION: The three patterns of egg laying described above illustrate adaptations to increasingly terrestrial reproduction. The mole salamanders are most aquatic, and lay large numbers of eggs. Because the eggs are unprotected and generally visible, many perish. Most larvae that do hatch are eaten by aquatic predators. However, of all the eggs laid by a female during her lifetime, *only two* need to survive and reproduce for the population to remain stable (one to replace the female and one to replace her mate). Considering that adult mole salamanders usually survive for several breeding seasons, the high egg and larval mortality caused by predation, cold, and so on does not threaten the species with extinction. However, in recent years acid rain, which changes the chemistry of the water and interferes with egg and larval development, could decimate mole salamander populations.

Dusky salamanders also lay eggs in water, although some lay eggs alongside streams and not in them. Fewer eggs are laid per female than is the case with mole salamanders, but the eggs are concealed under rocks and other objects where they are less subject to discovery, and females actively defend the eggs. Larvae are subject to high predation rates but eggs are less at risk than in the mole salamanders. Notice that as terrestrial egg-laying develops, so does some egg protection.

Many species of lungless salamanders lay small numbers of eggs on land and females remain with the eggs. It is interesting to note that the Marbled Salamander, a member of the mole salamander family, also lays terrestrial eggs that females remain with and defend. Eggs are at risk from predators and dryness. Attention from the female helps ensure survival to hatching. Also, among some of the lungless salamanders, the larval stage occurs in the egg, an obvious adaptation to terrestrial reproduction. Because the larval stage is within the egg, hatching time is longer, another reason natural selection has made it adaptive for the female to care for the eggs. Terrestrial salamanders lay relatively few eggs, but put much energy into parental care.

Salamander Larvae

Larval salamanders resemble adults but are smaller and totally aquatic (with the exception of newt efts—see below), with prominent finned tails and external gill tufts. As they undergo metamorphosis to the adult stage, their legs increase in size and their tail fins and gills are reabsorbed. Two unusual patterns of larval development occur, one confined to the Red-spotted Newt and one shared by several species.

1. The **RED-SPOTTED NEWT** (Pl. 38) has an aquatic larval stage followed by a terrestrial juvenile **RED EFT** stage, which is eventually followed by an aquatic adulthood. In most populations the eggs hatch into aquatic larvae whose greenish yellow color pattern resembles that of the adults. However, these larvae metamorphose into bright orange-red animals, called red efts, which are fully terrestrial, even to the point of having dry thickened skin, somewhat like toad skin. Efts emerge in late summer but live on the forest floor for up to 6–9 years before becoming aquatic adults. The transformation from eft to adult involves maturation of sex organs as well as changes in skin chemistry and characteristics and color changes. Efts are easiest to find by searching forest trails following a heavy rain.

Why do red efts exist? Both the early larval phase and adult phase of the Red-spotted Newt life cycle are aquatic. The terrestrial eft stage is an adaptation for dispersal. Efts do not move rapidly but a single eft can cover a lot of ground, for a salamander, in the course of up to 9 years. Adult newts are pond-dwellers and the eft stage enables members of the species to colonize new ponds, while helping to prevent overcrowding in ponds with established populations.

Why are efts red? Red efts are an example of *warning coloration* (see pp. 359 and 379). Their skin glands contain both noxious and poisonous substances that act to discourage predators. If a predator picks up an eft in its mouth it has a highly unpleasant experience, an amphibian equivalent to tasting a hot red pepper. The bright red color is thought to benefit the eft by providing an easy signal for predators to remember. When a red eft is confronted by a predator it adopts a defense posture with its hind legs erect and its tail held over its head. The **RED SALAMANDER**, a different species that is not noxious to predators, is believed to be a mimic of the red eft (see p. 345).

2. Among some salamanders, occasionally a larval form fails to undergo a complete metamorphosis to adult. The animal grows and matures sexually and thus is able to reproduce but it remains aquatic and retains its external gills and tail fins throughout its life. This odd developmental pattern is called *neoteny.* Neoteny occurs among the **MOLE SALAMANDERS**, especially the Tiger Salamander, a few of the **BROOK SALAMANDERS**, all blind **CAVE SALAMANDERS**, some of the newts, and all **MUDPUPPIES** (see Pl. 38). It is normal only among the mudpuppies and blind cave salamanders and is otherwise rare and local.

Neoteny probably developed by a genetic accident, but it allows the sexually mature adult to continue living in an aquatic habitat. Provided the salamander can find food and survive against predators, the neoteny trait can be advantageous. If the neotenous animal succeeds well, the trait not only persists but spreads.

Lungless Salamanders

OBSERVATIONS: Most species of woodland salamanders belong to the same family, the lungless salamanders. A group of air-breathing vertebrates exists without lungs! Although you cannot observe this characteristic without dissecting the animal, be assured that it is true. The animals remain by day under moist rocks, leaves, or decaying wood. If you place a lungless salamander in the open, it will hastily move under cover using a combination running gait and tail thrashing, suggesting an animal attempting to "swim" on land. If you touch the salamander you can feel how moist the skin is.

EXPLANATION: Lungless salamanders breathe through their skins. The moist skin provides a membrane for gas exchange, the taking in of oxygen and excretion of carbon dioxide. As long as the animal remains moist, it can breathe. Drying will kill it because it prevents air passage through the skin; thus, a dried lungless salamander dies of suffocation. Why are the lungless salamanders lungless? How might this condition be an adaptation to life in forests? The answer is that the odd condition may have little to do with forest life. Rather, lunglessness is most likely related to living in fast-moving streams, the ancestral habitat of these salamanders. Loss of lungs resulted in loss of buoyancy and thus less likelihood of being washed away by fast-moving current. Lungless salamanders could remain protected among the rocks on the stream bottom. When the group became more terrestrial, they did not "re-evolve" lungs. They carried their lunglessness with them, as a form of "evolutionary baggage." They adapted to their need to remain moist by hiding from dryness.

BIRDS IN SPRING

Spring brings the return of migrant birds, flooding north to establish territories and breed. Spring is an ideal time to observe bird behavior. You will be able to observe flocking, territorial, courtship, and breeding behavior among various species.

Migrant Waves and Weather

OBSERVATIONS: Abruptly on a spring morning following a warm front from the south, trees in a forest, park, or backyard may be filled with warblers, thrushes, orioles, tanagers, and grosbeaks. Up to a dozen or more species of warblers may flit about in the same tree while several thrush species hop on the ground. The next day, after the migratory wave has passed, the trees seem empty by comparison. In several days another wave appears, again following a warm front. Waves contain different assemblages of species. Some birds migrate north earlier than others and thus migrant waves change in species composition from early to late spring. Many waves occur before trees are fully leafed out, making it much easier to observe small birds such as warblers. During a migrant wave, birds do not necessarily occupy their typical habitats. A Blackburnian Warbler that nests in coniferous trees may, while migrating, feed among the leaves of newly emerging oaks and maples. Bobolinks, birds of open grassy meadows, may be perched in park treetops.

EXPLANATION: Migrants must be able to survive the journey north. To do

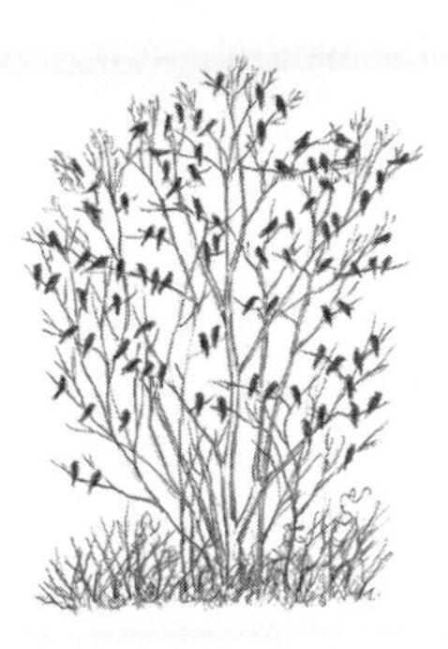

Fig. 34. A roosting flock of blackbirds.

so means that they must find food readily after an all-night flight. Migrant waves correlate closely with warm fronts from the south because such a weather pattern assures a higher probability of insect activity. Also, birds are helped by tail winds from the frontal system, making their flight more energy-efficient. Stalled cold fronts result in temporary migration bottlenecks. Several days of unseasonably cold weather followed by a warm front will precipitate a major migrant wave. In fall, the opposite conditions, namely a northern cold front, portend a strong southerly flight of migrants. See p. 253 for more on migration.

Blackbird Flocks

OBSERVATIONS: One of the earliest signs of spring is the return of blackbird flocks. Red-winged Blackbirds, Common Grackles, and Brown-headed Cowbirds, having wintered together in flocks sometimes numbering in the millions, leave their wintering areas in the southeastern states and fly north, often in flocks of many thousands. Bobolink flocks return after wintering on the pampas of southern South America. Flocks typically contain several species, with the exception of Bobolinks, which usually flock only with other Bobolinks. Blackbird flocks migrate north mostly at night, feeding during the day. A flock may remain for several days in one area while the birds refuel before continuing northward. Massive noisy aggregations fill still leafless trees as the birds roost in woodlands, around swamps, and along wet meadows. The early flocks contain almost all male birds, a fact that can be easily confirmed since Bobolinks, Red-winged Blackbirds, and Brown-headed Cowbirds are sexually dimorphic (see p. 300). Males that breed in southern and midlatitudes arrive back before males that nest in northern areas, producing a "leapfrog pattern" of migration. Flocks containing predominantly females arrive after the males have established their territories (see p. 298).

Why do birds flock?

Flocking in birds, like schooling in fish, helps individuals avoid predation. Migrating birds visit areas unfamiliar to many, if not most, individuals of the flock, especially the many first-year birds hatched during the previous summer and making their first migration northward. Flocks are visually obvious and often very noisy, seemingly attractive targets for predators. However, flocking makes it difficult for a predator both to go undetected and to be able to focus its concentration on any individual bird. You can test this yourself by attempting to keep your eyes on one specific bird as the flock swirls in the air before you.

Flocks may be *intraspecific,* composed of a single species (such as the Bobolink) or *interspecific,* composed of several species. Flocking is very common among many species outside of breeding season.

Flocking provides several advantages besides predator protection. First-year birds can learn where favorable habitats are and can remain floaters in the population if they are unable to procure territories (see p. 300). Flocks provide a means of discovering food sources. However, competition among individuals could be heightened by flocking, so there is a potential cost as well as a benefit. Flocking entails very few additional costs to individual birds, other than the stresses imposed by being close to others at a time when territorial behavior is becoming increasingly strong.

Fig. 35. Male Red-winged Blackbird singing.

Flocking makes competition for territories very intensive because individuals arrive on the breeding grounds together.

Bird Vocalization

Observations: Bird sounds fill woodlands and fields in spring. One of the major challenges facing the naturalist is to be able to separate and identify the many different vocalizing species. The best time to hear birds is at daybreak. As migrants arrive, each day's *dawn chorus* becomes more and more exciting.

Most bird species make basically two kinds of sounds. One is the call note. This is usually brief and nonmelodious. The dry *dee-dee-dee* of the Black-capped Chickadee is an example. The other is the *song*, which is usually loud, melodious, and often repeated many times. The sweetly whistled *fee-bee, fee-bee* song of the Black-capped Chickadee is quite different from its call notes. Just as human accents vary regionally, call notes and songs may vary for a single species throughout its range. Common Yellowthroats, Rufous-sided Towhees, and Song Sparrows sound a bit different from north to south. Although call notes may sound to you to be much the same for a particular species, they actually vary with the situation. Chickadees emit a high *zee* note when they detect a predator. This call signals the members of the flock to become quiet and remain motionless until the predator has moved away. Many species exhibit a series of different call notes, each of which confers its own meaning. Listen closely to detect these different notes. Both sexes use call notes.

Bird song is most frequent in the early morning and late afternoon until dusk. Midday is the time when the least singing occurs, although some birds, like the vireos, sing even then. Some species, like the Ovenbird and Northern Mockingbird, sing both day and night. Nocturnal species, such as Whip-poor-wills and Screech Owls, vigorously vocalize during hours of darkness. In most species only males do the singing, often from high exposed perches.

Explanation: Call notes and songs have different functions. Call notes are given throughout the year as communication among mated pairs, family groups, or birds that forage together. Call notes serve as warning and alarm signals when birds are threatened (see "Predator Mobbing and Harassment," p. 307), they often form a part of distraction displays (see p. 386), and they are used to relocate chicks and fledglings if they become separated from the parents. Call notes carry well for short distances.

In most species songs are almost entirely restricted to the breeding season and thus are heard in spring and summer only.

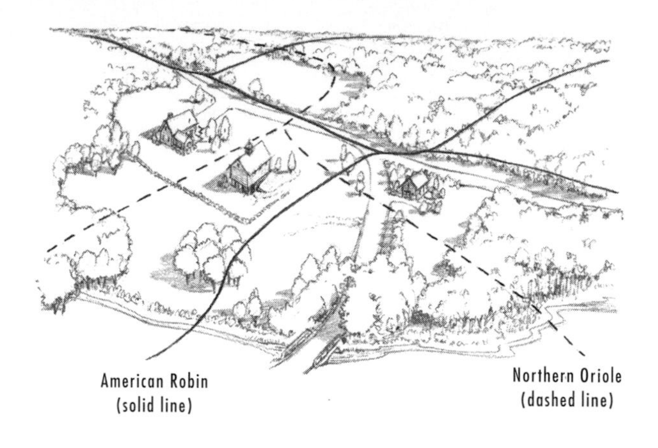

Fig. 36. Territory maps.

Songs function as territorial signals (see "Territoriality," below). Males sit atop often exposed perches and sing over and over with utmost vigor. The songs alert other males of the same species that the real estate is occupied, and warns them to stay away. You can often witness attacks on intruding males as a singing, territorial male will suddenly leave its song perch and drive away another bird. Songs also signal females that there is a male on territory who is prepared to mate. Singing does occur as birds are migrating, so not all singing males are necessarily on a territory. Some may be just passing through. Singing usually persists well after the territory is established and during the nesting period.

Territoriality

OBSERVATIONS: In most species, male birds return before females. These males attempt to establish territories for their exclusive use, often for both feeding and breeding, but sometimes for breeding only. A territory is a defended area in which males and often both sexes actively drive away others of the same species and, sometimes, of other species. You can identify a territory by observing a male bird return to sing atop the same perches day after day, often chasing intruder males. Territories can even be mapped by patiently noting where a singing male spends its time. Males occasionally do battle over territories, although these bouts of physical combat are usually very brief in duration. Male birds are generally conspicuous when on territory.

Species such as the American Robin or Gray Catbird often use the same nest location in succeeding years. Many species show *nest-site fidelity*, nesting repeatedly in the same location.

EXPLANATION: Defense of a territory benefits a mated pair by providing them with exclusive use and intimate knowledge of an area at a time when they will have to provide not only for themselves but for helpless nestlings. Foraging is both more efficient and safer when birds know the terrain in detail. Nest-site fidelity, which was first demonstrated when John James Audubon banded Eastern Phoebes at Mill Grove, Pennsylvania, aids birds in that they have less to "relearn" when they return to breed.

The costs of territoriality seem to be born mostly by males. Much energy is required for the territorial defense and there are risks of predation from being obvious in song and plumage. In some species females choose specific territories, not specific males, while in others, females choose specific males as mates. Males compete against one another for territories and to assure access to females. This is why males return before females and establish and defend the territory. In many species males are the more brightly colored of the sexes, a probable evolutionary result of territoriality and male-male competition (see "Sexual Dimorphism," p. 300). Song helps reduce physical combat between males. Fights do not last long and it is more common to see a male chasing an invader than actually fighting with a rival. Once a male establishes a territory he is virtually invincible and cannot be driven away.

Females bear the major responsibilities of nesting, although males of many species aid in both nest-building and feeding of young. Once nesting has begun, females can be equally aggressive in territorial and nest defense. In species as different as the House Wren and the Northern Goshawk, both sexes are extremely belligerent when their territories are invaded.

Territory is usually *intraspecific*, meaning that a bird defends a territory against others of its species, but not against other species. This is because different species do not share the same needs and so do not compete extensively for resources. Only within a species do all individuals share almost identical needs. A Wood Thrush and an Ovenbird are both forest birds, but the former builds a nest in the understory while the latter builds a ground nest. They do not eat the same things or use the same nest sites and nesting materials. Wood Thrushes are no threat to Ovenbirds and vice versa. Both species, however, are vigorous defenders of territory, each against its own species.

Territory sizes vary among species. Large birds like Great Horned Owls and Red-tailed Hawks require territories measured in square miles. Small passerines (songbirds) require much less space. During years of abundant insect food, many species have smaller territories, suggesting strongly that food availability is a major determinant of territory size.

Many species defend a territory in which they both feed and nest. Others defend nesting territories but share feeding areas. The Chimney Swift and various swallow species nest colonially and disperse widely to catch food on the wing. In these birds, the only defended territory is the immediate vicinity of the nest.

You may occasionally notice a bird skulking in the undergrowth within an already established territory of another individual. This bird is probably a non-territorial individual called a *floater.* Should any calamity befall the territorial bird, the floater will quickly replace it. Floaters are typically inexperienced first-year birds that have yet to establish territories. They do not challenge established males and remain quiet and secretive. Virtually all bird species studied so far have floaters in their populations.

Sexual Dimorphism (Pl. 39)

OBSERVATIONS: A survey of breeding birds of a forest or field will quickly reveal that in some species the sexes look identical, in others they look slightly different, and in still others, they look very different. When the sexes look alike, they are *monomorphic,* and when they look distinct, they are *dimorphic.* In dimorphic species, the males are almost always the most brightly colored of the sexes. Females are colored in more subdued tones.

MONOMORPHIC SPECIES

Red-tailed Hawk
Broad-winged Hawk
Red-shouldered Hawk
Owls
Killdeer
Mourning Dove
Cuckoos
Eastern Kingbird
Great Crested Flycatcher
Eastern Pewee
Least, Acadian, Willow, Alder, Yellow-bellied flycatchers
Horned Lark
Swallows (except Purple Martin)
Blue Jay
American Crow
Chickadees
Tufted Titmouse
Brown Creeper
Gray Catbird
Northern Mockingbird
Brown Thrasher
most Vireos
Cedar Waxwing
Common Grackle
Eastern Meadowlark
Song Sparrow
Field Sparrow

Chimney Swift
Wrens
Chipping Sparrow
European Starling

SLIGHTLY DIMORPHIC SPECIES

American Kestrel
Accipiters (female hawks larger than males, though alike in plumage)
Woodpeckers
Purple Martin
Nuthatches
Kinglets
Blue-gray Gnatcatcher
Ruby-throated Hummingbird
Wild Turkey
Ruffed Grouse
Northern Bobwhite
American Robin
Eastern Bluebird
Northern Junco
White-throated Sparrow

HIGHLY DIMORPHIC SPECIES

Orioles
Tanagers
Red-winged Blackbird
Brown-headed Cowbird
Bobolink
Grosbeaks
Northern Cardinal
American Goldfinch
Indigo Bunting
Rufous-sided Towhee

Sexual monomorphism and dimorphism are not consistent within families or habitats. For instance, among the blackbirds, warblers, and sparrows, each has both monomorphic and dimorphic species.

MONOMORPHIC WARBLERS

Worm-eating Warbler (woodlands)
Ovenbird (woodlands)
Northern and Louisiana waterthrushes (woodlands)

SLIGHTLY DIMORPHIC WARBLERS

Northern Parula (woodlands)
Black-and-white Warbler (woodlands)
Black-throated Green Warbler (coniferous woodlands)
Yellow-throated Warbler (southern woodlands)
Yellow Warbler (brushy fields, swamps)
Prothonotary Warbler (southern swamps)
Kentucky Warbler (moist woodlands)
Blue-winged and Golden-winged warblers (pastures, brushy fields)
Yellow-breasted Chat (brushy fields)

Red-winged Blackbird. This male is feeding, not displaying, thus his red epaulets are partially hidden, making him appear far less aggressive to other male Redwings.

HIGHLY DIMORPHIC WARBLERS

Black-throated Blue Warbler (woodlands)
Blackburnian Warbler (coniferous woodlands)
Common Yellowthroat (brushy fields, marshes)
Hooded Warbler (moist woodlands)
American Redstart (woodlands)

EXPLANATION: Sexual dimorphism, when it occurs, is usually most extreme during the breeding season. Although some species, like the Northern Cardinal, maintain sexually dimorphic plumage year around, males of the majority of the extremely dimorphic species molt from their breeding colors to a plumage much more similar to that of the female after the breeding season. Sexual dimorphism is strongly related to courtship and breeding behavior.

Males of dimorphic species are very colorful and often display their plumage when singing. Red-winged Blackbird males will spread their wings, showing the red epaulets to their maximum size. Even in partially dimorphic species, the breeding season is a time when plumage distinctions between the sexes seem emphasized, having a role in courtship behavior. Golden and Ruby-crowned Kinglet males display their bright gold and red head feathers during courtship. The Ruby-throated Hummingbird displays its iridescent scarlet throat to the female.

Dimorphic differences between the sexes serve as signals between courting birds. Male Common Flickers have a black "mustache" of feathers on each cheek. This mustache tells the female that the male is, indeed, a male. If a male's mustache is painted over, the female will ignore its courtship advances, treating it as another female. If a female has a mustache painted on her face, the male will drive her away aggressively, treating her as another male invading his territory.

In highly dimorphic species males are very vigorous in territorial defense. Aggression between male Northern Orioles, male Red-winged Blackbirds, and others is very common and easy to observe. Bright colors in males may result from a process called *sexual selection*. Males must compete against other males to secure territories and attract females. Gaudy colors could intimidate other males as well as gain the attention of females. Sexual selection implies that the gaudiest males have been the winners in procuring both territories and females. Experiments with male Red-winged Blackbirds indicate that the red wing epaulets impress other males much more than females. When the red shoulder patches are experimentally painted black, males tend to lose their territories to other "normal" males. The larger the red epaulets, the more successful is the male. Female Red-wings choose territories, not males. In other species, however, females seem to be choosing males based on plumage characteristics (see p. 304).

Why are only males gaudy? The risk of predation is probably higher in such obvious birds. Females of dimorphic species spend more time brooding eggs and hatchlings than males do. Reproductive success is likely to be higher when females are more camouflaged than males. The threat to male survival brought about by gaudy plumage may be one reason why extreme plumage dimorphism is not more widespread.

Birds with monomorphic or very slightly dimorphic plumage seem to substitute complex behavioral display for gaudy plumage (see p. 306). Male woodpeckers drum loudly in spring. Male Ruffed Grouse strut around and display with fanned tail feathers and erected neck ruffs. Ruffed Grouse also drum on logs to attract females. Northern Mockingbirds are extraordinarily aggressive, and seemingly tireless, often singing throughout the night. Male Mourning Doves are persistent in their pursuit of females, following them as they feed on the ground. The male dove will court the female with neck feathers fluffed and tail spread.

Delayed Plumage Maturation in Songbirds

OBSERVATIONS: Young males of some songbird species do not acquire their full male plumage during their first breeding season (the spring following their hatching year). Instead, they closely resemble females. Though these males look different from older males, they are fully capable of breeding. However, they do not acquire the full male plumage until their second year as breeders. This phenomenon is called *delayed plumage maturation.* Those species in our area that show delayed plumage maturation are:

American Redstart
Scarlet Tanager
Summer Tanager
Rose-breasted Grosbeak
Blue Grosbeak
Indigo Bunting
Painted Bunting
Red-winged Blackbird
Brown-headed Cowbird
Northern Oriole
Orchard Oriole
Pine Grosbeak
Purple Finch
House Finch
Red Crossbill
White-winged Crossbill
American Goldfinch

Delayed plumage maturation accounts for the reason why you may see a nesting pair that appears to be two females, though in most of the above species, males are sufficiently distinct in plumage to be separable from females. For instance, the young Orchard Oriole male has an all-black throat, which the female lacks. Young male Rose-breasted Grosbeaks show some red on the breast and black on the head. Male Pine Grosbeaks are decidedly more orange on the rump and head than females. Male American Redstarts have an orange tinge to their yellow tail and wing patches. Male Red-winged Blackbirds have the beginnings of the red shoulder patches.

EXPLANATION: Why should inexperienced males of some species in their first breeding season resemble females? Does delayed plumage maturation have a function? Several ideas have been suggested as to why such a trait might have evolved. One is that the delay in acquiring the full adult plumage functions to protect young, inexperienced males from predators. By closely resembling the more camouflaged females, these males have an additional safety margin. Another suggestion is that these young males mimic females, which allows them to explore territories without as much threat of attack from established, full-plumaged territorial males. A related idea is that delayed plumage maturation signals adult males that the younger males are subordinate, and of no direct threat. Most species that have delayed plumage maturation are only territorial around their immediate nest sites. They share feeding areas with others of their species. Females of these species are thought to choose their male mates based on bright plumage characteristics, not territory quality, since the only territory defended is the nest site. Young, delayed-plumage males are no threat to fully plumaged males because they look like females, and are therefore not likely to be selected as mates! Delayed plumage maturation permits these young males to settle in an area and be unmolested by other resident males because males lacking the full adult plumage are not perceived as competitors. However, should there be a sufficient number of females present, delayed-plumage males, too, can succeed in breeding.

Courtship Feeding

OBSERVATIONS: When male and female birds establish a *pair bond* and initiate the process of nesting, the female will occasionally act as though she is begging food from the male. She will crouch, spread and quiver her wings, and often call, with her mouth wide open. This behavior is highly suggestive of a nestling gaping for food from its parent. Males usually feed the soliciting female, and the food is often a large, protein-rich insect. *Courtship feeding* is common among jays, chickadees and titmice, nuthatches, woodpeckers, thrushes, mimic thrushes, warblers, orioles, tanagers, finches, and sparrows.

EXPLANATION: Courtship feeding benefits *both* the male and female of the pair. The female gains energy from the food provided by her mate while the male contributes to the well-being of the female, who requires much energy and protein to make eggs. The female represents not only her own potential reproductive success but also that of the male. The energy he invests in giving her high protein food will pay dividends when she helps reproduce his genes. In addition, courtship feeding probably aids in cementing the pair bond between the two birds.

Why does the female act like a juvenile bird? Probably because this behavior automatically elicits the feeding response in males, just as when juveniles beg for food. The female is not being "coy" or submissive to her mate by acting like a juvenile. She is merely employing an economic way to signal for food. In fact, because it

Fig. 37. Titmice—courtship feeding.

is the female who receives the valued morsel, and the male who gives it up, she may dominate the male during this phase of courtship. Of course, it is to the male's ultimate genetic advantage to be dominated in this manner.

Woodcock Courtship Behavior

OBSERVATIONS: Except in the extreme Southeast, the Woodcock is a common nesting bird throughout eastern North America. It is one of the earliest nesters, and its courtship behavior is both elaborate and relatively easy to observe. Woodcocks feed and nest in moist woodlands but males establish courtship territories in open grassy fields or forest clearings. At dusk and throughout moonlit nights, males perform an intricate courtship flight over their territories. The ceremony begins on the ground as males call a very soft *cook-oo,* often inaudible unless you are very close. Following this call, males vocalize a repetitive nasal *preent.* Soon the male Woodcock is airborne and its nasal call is replaced by a melodious high-pitched twitter, a prolonged rolling sound produced by the unique structure of the flight feathers. The bird spirals upward until it abruptly begins rapid descent to earth. In the waning daylight you will often lose sight of it as it flies upward. As the male suddenly begins its slip-slide descent, it emits a high-pitched, chirping *zleep zleep,* its actual song. After the bird returns to earth, there is a moment or two of silence before the bird begins its *preent* and the courtship flight is repeated.

Fig. 38. Woodcock and its courtship flight.

EXPLANATION: Why do male Woodcocks perform such an elaborate ritual? The courtship is adaptive because without it, males would not secure mates. Following the courtship flight, male and female Woodcocks mate and the female then leaves and attends to all the nesting. The male continues to display on its courtship territory, attempting to attract other females. Courtship in Woodcocks is a form of sexual selection in which behavior, the elaborate courtship flight, replaces gaudy plumage.

Predator Mobbing and Harassment

OBSERVATIONS: American Crows, Red-winged Blackbirds, Common Grackles, Blue Jays, Eastern Kingbirds, and many small passerine (songbird) species commonly mob predators such as owls, hawks, weasels, and foxes. Some of these species, namely crows, grackles, and Blue Jays, are themselves the victims of mobbing by small birds such as chickadees. Mobbing is usually heard before it is seen because mobbing birds are extremely vocal, emitting vigorous call notes, which sound (to the human listener) like excited alarm notes. If you hear a loud, concentrated cacophony of bird call, approach carefully and you may see an owl or weasel surrounded by various passerines.

Crows and blackbirds usually mob in single-species flocks, though you may see a single grackle, Red-winged Blackbird, or Eastern Kingbird harassing a hawk. If you are in a pine forest and hear many crows calling at about the same spot, approach cautiously and search for an owl. Chances are that an owl is in the

Fig. 39. Eastern Kingbird harassing a hawk.

midst of the crow mob. Smaller passerines mob in interspecific (mixed-species) flocks. Chickadees, wrens, nuthatches, and other passerines surround and harass a daytime perched owl. For this reason, it is often possible to attract these passerine species by imitating a Screech Owl call. These small birds also mob potential nest-robbers such as crows, jays, and grackles. Eastern Kingbirds are particularly pugnacious and harass airborne hawks, owls, and crows.

Mobbing birds rarely attack the predator; they merely harass it. However, the harassment is often sufficient to cause the hawk or owl to fly. When it flies, the mobbing birds pursue it with enhanced vocal efforts. The volume of their calls increases dramatically.

EXPLANATION: Mobbing isolates and identifies a predator at relatively little risk to the individual mobbing birds. It is important to remember that mobbing birds do not actually engage in combat with large predators, though a Kingbird may, in fact, strike a hawk. Predator detection is beneficial to each bird and thus each bird is serving itself, as well as other birds which may include its kin, by participating in the mobbing. Mobbing and harassment can help drive a predator from a particular area. The actual motivation for mobbing and harassment behavior seems centered on a combination of fear and aggression by the mobbers. They practice the strategy that a good defense is a vocal offense.

MAMMALS — DISPERSAL AND ROAD KILLS

OBSERVATIONS: Many species of mammals are more frequently seen in spring than at other seasons. Squirrels, opossums, armadillos, skunks, Muskrats, Woodchucks, Raccoons, and foxes are commonly seen as road kills.

EXPLANATION: Young mammals are forced to disperse from their family home range to seek a home range of their own. These animals, most of which are inexperienced, are more easily seen because

Fig. 40. Road-killed Muskrat.

they are temporary vagabonds attempting to become established in unfamiliar terrain. Because of their inexperience, many fall prey to predators and to automobiles.

Why do young mammals expose themselves to the risks of dispersal? Probably because they have little choice. Many experienced mammals use the same den year after year. If juveniles were permitted to remain after they mature, they would pose a potential competitive threat to the parents. They would strain the food-producing capacity of the home range and they might attract more predators. It is to the parents' advantage to drive them out.

Oddly, it may be to their own advantage to be driven out. If they are not permitted to breed while on their parents' home range, they are better off taking the risk of establishing their own real estate.

WIND-POLLINATED TREES (P. 272)

WHITE PINE

Pollen generated from small, clustered, staminate (male) cones at branch tips. Pollen is yellow and produced in such large amounts as to form a dense dust. Pollen fertilizes pistillate (female) cones, in which seeds mature. All conifers are pollinated by wind. Also see Pl. 6.

BLACK WALNUT

Male flowers are greenish and hang as slender catkins. Female flowers are greenish, at branch tips. Both male and female flowers on same tree.

NORTHERN RED OAK

Male and female flowers on same tree. Male flowers hang as greenish catkins, in clusters near branch tips. Female flowers are very small, at leaf axils. Also see Pl. 8.

GRAY BIRCH

Male and female flowers on same tree. Male catkins yellowish, elongate, hanging from branch tips. Female flowers cone-like, upright, at leaf axils in from branch tips. See also Pl. 34.

AMERICAN BEECH

Male and female flowers on same tree. Male flowers hang in clusters below branches, in yellowish balls. Female flowers are very small, surrounded by reddish scales, at leaf axils near branch tips. Also see Pl. 7.

AMERICAN ELM

Male and female parts on same flower. Flowers tiny, white with brownish bases and brownish stamens; in clusters near branch tips. Also see Pl. 1.

WHITE ASH

Male and female flowers are on separate plants—each tree is either male or female. Male and female flowers look similar; both are reddish brown, without petals. Female flowers occur in paired clusters near branch tips. Male flowers are in paired clusters, farther inward. Also see Pl. 34.

SHAGBARK HICKORY

Male and female flowers on same tree. Male flowers hang as greenish catkins, in clusters of 3. Female flowers are small, singular, greenish, at branch tips. Also see Pl. 8.

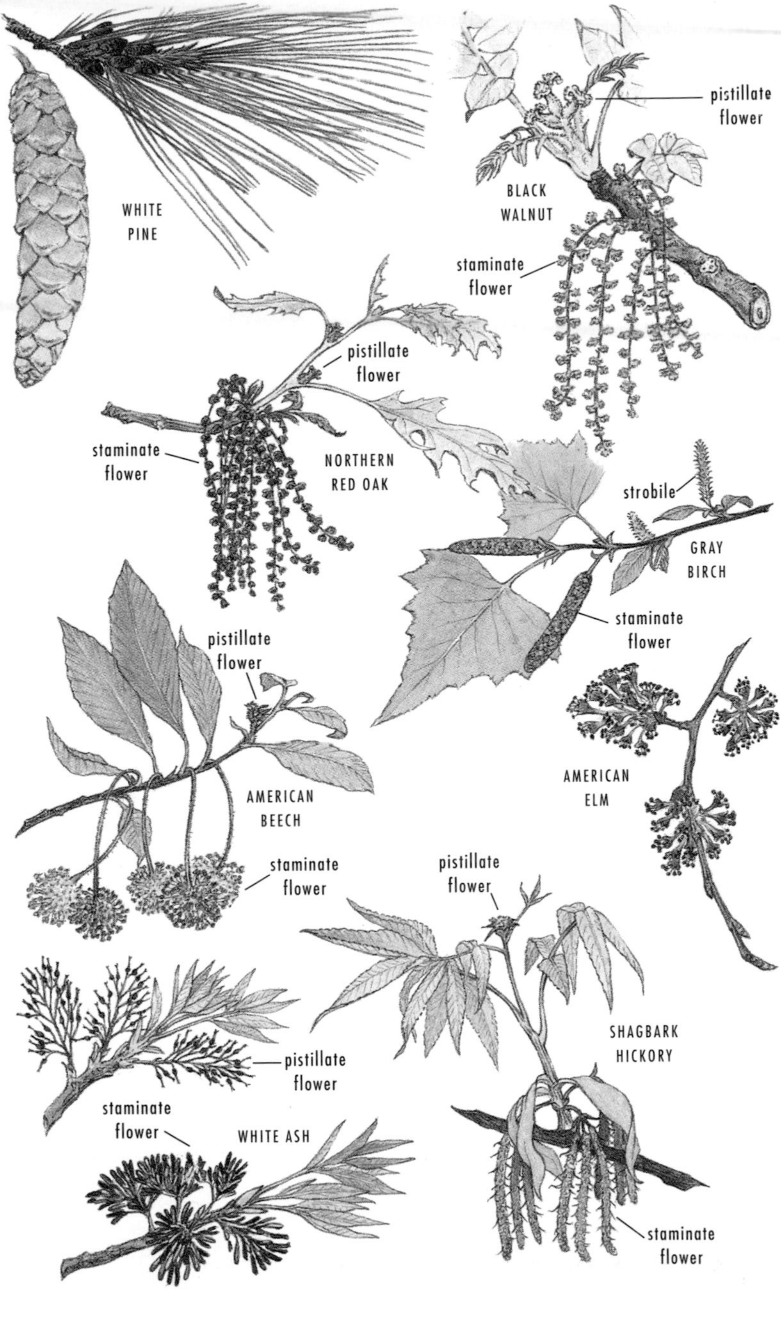

WHITE PINE

BLACK WALNUT
pistillate flower
staminate flower

NORTHERN RED OAK
pistillate flower
staminate flower

GRAY BIRCH
strobile
staminate flower

AMERICAN BEECH
pistillate flower
staminate flower

AMERICAN ELM

SHAGBARK HICKORY
pistillate flower
staminate flower

WHITE ASH
pistillate flower
staminate flower

INSECT-POLLINATED TREES (P. 273)

These species are pollinated by insects. Most have very obvious, brightly colored flowers. In some cases, the flowers are fragrant. Most insect-pollinated species flower when leaves are fully open. Most flowers contain both male and female sex organs, and most provide nectar for insects. When an insect visits a flower to drink nectar, it picks up pollen from that flower, and deposits pollen from flowers recently visited. See text for details.

SOUTHERN MAGNOLIA

Very large, white, single flowers at tip of branch. Petals (usually 6) form a cup. Three sepals. Considered to be a very primitive flower. Pollinated primarily by beetles. See Pl. 14 for description of tree, Pl. 49 for fruit.

PUSSY WILLOW

Male and female flowers on separate joints. Male flowers are catkins that open very early, well before leaves appear, but persist after leaves open. Flowers feel velvety and are covered with soft white hairs. Bracts produce nectar. Female flowers similar but with greenish pistils. Pollinated primarily by bees.

TULIPTREE

Large, single, upright, tuliplike flowers with 6 light green petals, each with an orange base. Similar to Magnolia in shape; both trees belong to the same family (Magnoliaceae). Pollinated primarily by bees. See Pl. 13 for description of tree.

SASSAFRAS

Small, yellowish green flowers clustered at branch tips, open before leaves. Male and female parts are on separate flowers, and usually on separate trees. Pollinated primarily by flies. See Pl. 1 for description of tree, Pl. 49 for fruit.

FLOWERING DOGWOOD

Actual flower is small, with 4 yellowish green petals, clustered and surrounded by 4 large white bracts. Flowers open before foliage. Pollinated by a variety of insects, including flies and bees. See Pl. 1 for description of tree, Pl. 49 for fruit.

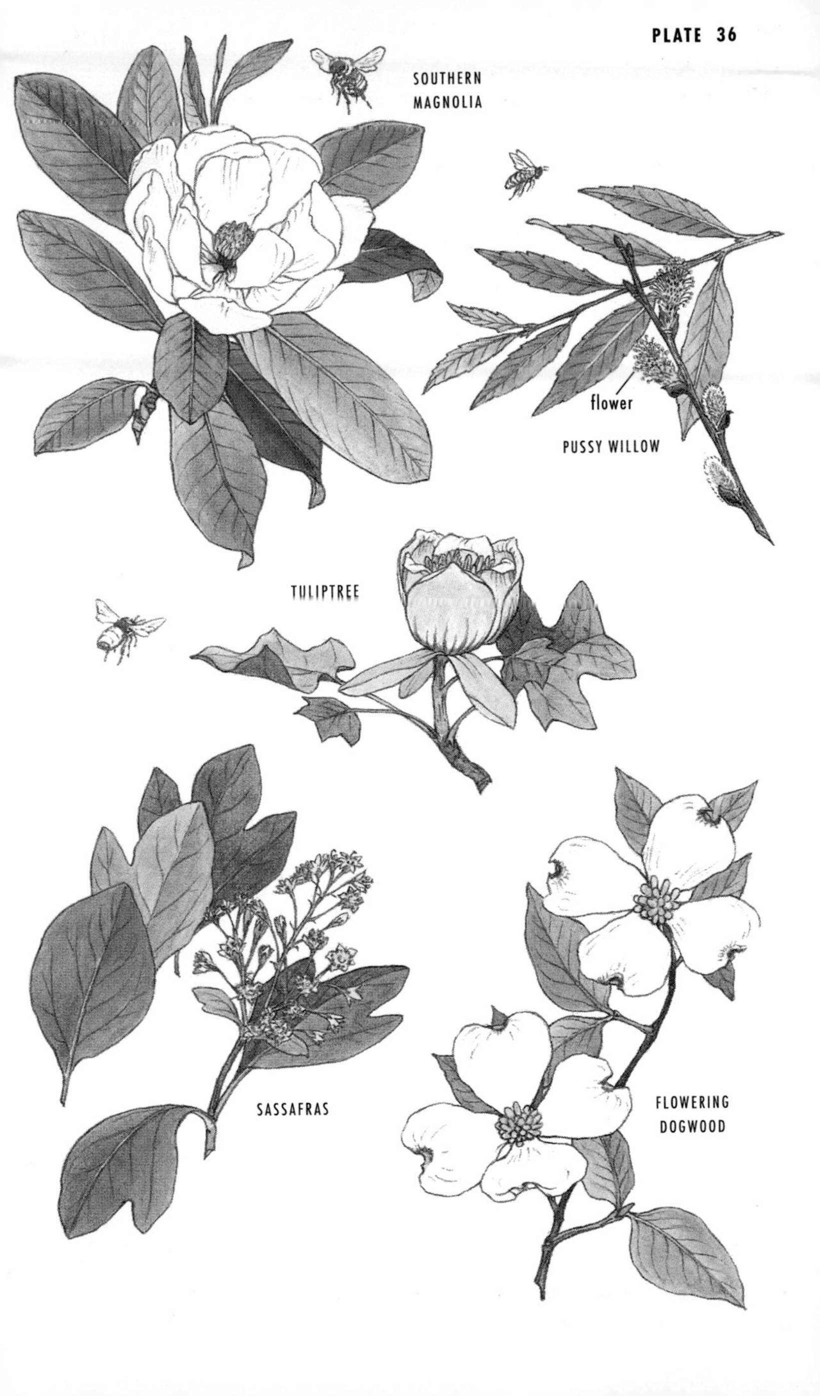
SOUTHERN MAGNOLIA
flower
PUSSY WILLOW
TULIPTREE
SASSAFRAS
FLOWERING DOGWOOD

SELECTED FROGS AND TOADS (P. 282)

Each of the 5 animals on this plate represents a major family of anurans (frogs and toads).

AMERICAN (COMMON) TOAD

Toad family. Toads are rounded and squat. They do not perform long leaps but move by short hops. Adults are terrestrial. Bodies brownish, with "warts" on skin. Skin secretions are irritating and poisonous. Large parotoid gland behind each eye; large ear drum (tympanum). Eggs in beadlike strings. American Toad ranges throughout the East except the Deep South, where the similar **Southern Toad** occurs abundantly. Call a long musical trill.

EASTERN SPADEFOOT TOAD

Spadefoot family. Spadefoots resemble true toads (above) but have vertical pupils and a minute, hardened "spade" on the inner part of the hind foot that serves as a digging organ. They live in burrows in dry, sandy soils during the day and are active at night. Spadefoot toads breed in temporary ponds. Voice a low, harsh *caw* that carries well.

SPRING PEEPER

Treefrog family. Treefrogs, as the name implies, are arboreal. Each toe has a small suction cup, an adaptation for clinging to trees, branches, and grass blades. Spring Peepers breed as early as November in Deep South, in early spring in North. Identified by brown cross on back. Call a high, short, whistled trill. Eggs in short strings.

GREEN FROG

True Frog family. Body usually green (may be brown or bronze), with scattered black spots. Ridge runs from eye to thigh. Voice a single note, suggesting a banjo twang.

WOOD FROG

Another **true frog** (see above). Mostly terrestrial, on forest floor in mesic (moist) forests. Black behind eye. Call a ducklike series of quacks. Not found in Deep South, but occurs in northern Canada and Alaska—the most northern anuran. Breeds in early spring. Eggs in rounded masses attached to vegetation.

SPRING
PEEPER
× 1
× ½
AMERICAN
(COMMON) TOAD
EASTERN
SPADEFOOT TOAD
× ½
spade
WOOD FROG
× ½
GREEN FROG
× ½

SELECTED SALAMANDERS (P. 288)

Each of the four species on this plate represents a different major family of salamanders.

RED-SPOTTED or EASTERN NEWT

Newt family. Skin rough, not smooth as in all other salamander families. Some newts are terrestrial, but the eastern species has both an aquatic and a terrestrial form. Aquatic form greenish yellow, with red spots on sides. Terrestrial form (eft) is salmon-red. Aquatic form is found in ponds, lakes, swamps. Eft inhabits moist forest floor. Throughout eastern North America.

SPOTTED SALAMANDER

Mole Salamander family. These salamanders tend to have thick bodies and blunt snouts. They are among the larger, more robust salamanders. Many, like the Spotted Salamander, have dark bodies with colorful patterning. The sides have a series of vertical indentations called costal grooves. Spotted Salamanders are common in hardwood forests throughout the East. They breed *en masse* in vernal woodland ponds (see text). Eggs in rounded masses, attached to submerged twigs, grasses.

MUDPUPPY

Mudpuppy family. Large salamanders (Mudpuppy reaches 18 in.) with external, deep red gill tufts and a vertically flattened tail. Fully aquatic. Found in streams, rivers, lakes, and permanent ponds. Mudpuppy does not occur in Deep South, but other, similar species do. Mostly nocturnal.

RED-BACKED SALAMANDER

Lungless Salamander family. A very large family with about 80 North American species, many found only in certain areas in the Appalachian Mountain region. Lungless Salamanders breathe through moist skin, lack lungs. Well-defined costal grooves. Found in streams, rivers, and on forest floor under decaying logs. Red-backed Salamander, very common in middle Atlantic states and Northeast, occurs in two color phases, red-backed and lead-backed (not shown). Eggs in small clusters, placed beneath logs. Female guards eggs. The most widespread southern *Plethodon* is the **Slimy Salamander** (not shown), which is black with sides spotted yellow, white, or gray.

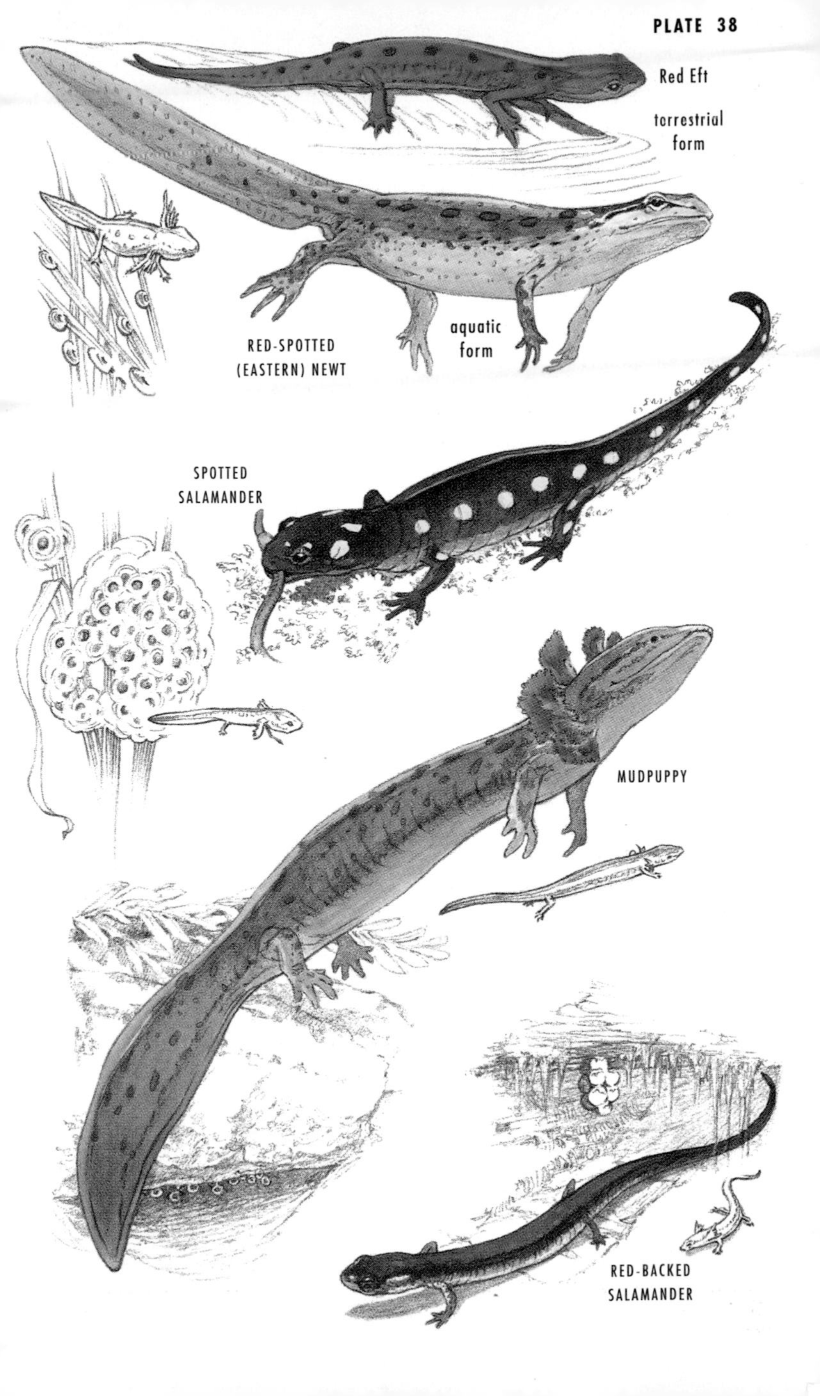
Red Eft
terrestrial form
aquatic form
RED-SPOTTED (EASTERN) NEWT
SPOTTED SALAMANDER
MUDPUPPY
RED-BACKED SALAMANDER

SEXUAL DIMORPHISM IN BIRD PLUMAGE (P. 300)

Plumage sexual dimorphism, in which male and female look quite different, is common to some but not all bird species. Reasons for sexual dimorphism relate to competition among males for territories and females, to selection of males by females, and to the different roles played by males and females in courtship, mating, and nesting. See text for details.

Highly Dimorphic Species

SCARLET TANAGER

Male brilliant scarlet with glossy black wings and tail; female olive above, yellow on breast. Tanagers nest in forests, usually in oaks. Male's song is similar to Robin's but more slurred, deeper.

NORTHERN (BALTIMORE) ORIOLE

Male bright orange with a black head; female greenish yellow. Immature male (not shown) resembles female but is darker on head and back. Orioles build a deep, baglike nest in shade trees. Male's song a rich, whistled warble. Call note a dry chatter.

BOBOLINK

Breeding male black below with an ocher-yellow neck and white rump and wing patch. Female sparrowlike — tan, with streaking on head and body. Males in fall closely resemble females (see text). Male sings a bubbly, vigorous song in flight. Call note a metallic *clink.*

Slightly Dimorphic Species

HAIRY WOODPECKER

A robin-sized, black and white woodpecker, with a white back and a long, straight bill (see Downy Woodpecker, Pl. 51). Male has red on neck. Call a strident, loud *peek!*

Monomorphic Species

GRAY CATBIRD

Uniformly gray, with a black cap and rufous underparts. Call a catlike mewing. Song a variable warble.

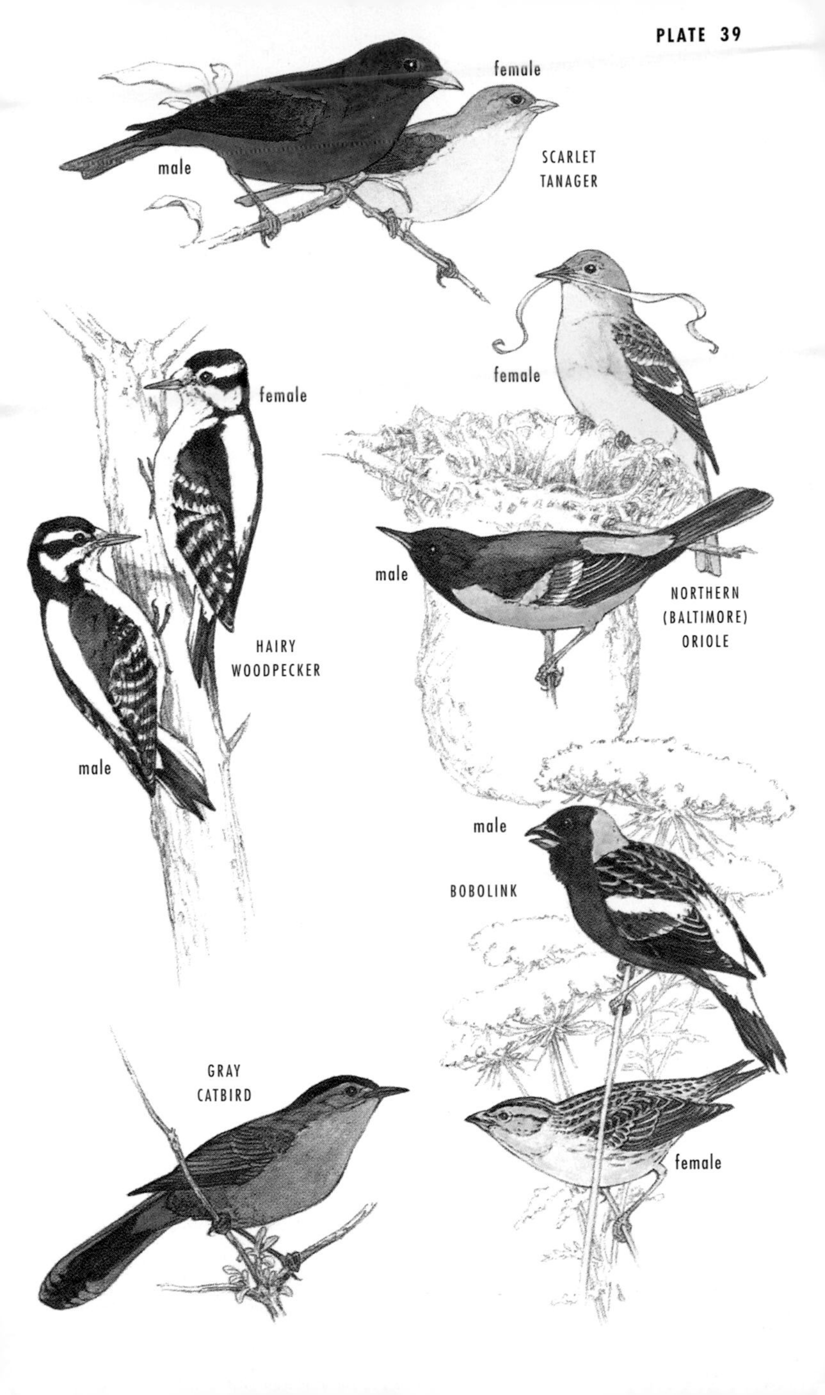
female
male
SCARLET
TANAGER
female
female
HAIRY
WOODPECKER
male
male
NORTHERN
(BALTIMORE)
ORIOLE
male
BOBOLINK
GRAY
CATBIRD
female

Nature in Summer

Summer is the time when the growing season is at its fullest. Among plants, the productive efforts that began with the longer days and warming temperatures of spring now reach their peak. Plants have grown as much as they are going to grow for a given year. The cycle of growing new leaves, flowering, attracting pollinators, and fruiting is completed in summer. Summer is a time when relations between the plants and the myriads of insects that fill the warm summer air are most readily visible. Examine a milkweed or a goldenrod and see for yourself what Charles Darwin meant when he wrote in his journal in 1839, "It is difficult to believe in the dreadful / but quiet / war of organic beings, going on in the peaceful woods, & smiling fields. — "

How much energy should a plant devote to reproducing in a given year? Why are woodland wildflowers smaller than species that grow in old fields? How are seeds dispersed? What function do fruits have? Why are some fruits produced in summer, some in fall? How do fruits attract migrating birds? These questions fill the mind of the naturalist who ventures afield in summer.

Among insects, self-preservation is a constant issue. As insect abundance peaks, so too does insect predation. But, the would-be prey fight back. Numerous patterns of camouflage and mimicry among insects and spiders, be they prey or predators, challenge the most observant naturalist. Snakes too are animals that are both predators and prey. The patterns displayed both in terms of color and behavior by serpents are lessons in natural selection awaiting the less timid among natural history buffs (see cautions on pp. 373 and 375).

Summer is the time of bird nesting. Territories are stabilized and the business of raising this year's broods is at its peak. Locating nesting birds is challenging, but hours spent watching the nesting process pay many rewards in terms of understanding and information. Nesting patterns vary. Some birds are cavity-nesters, requiring hollow trees, while others make open nests. Searching for birds' nests and observing the cycle of avian reproduction can occupy many pleasant hours. Another major pattern, most evident toward late summer, is the initiation of bird migration. Birds that mere weeks or even days earlier were staunchly defending territories and feeding nestlings now group in flocks, often containing several species. The birds gorge on newly formed fruit or forage actively on the still-bountiful insect crop. Bird song stops. Woodlands are quiet except for the droning of cicadas and katydids. Plumages of birds change. Males lessen in brilliance, or even come to closely resemble females. These are the avian patterns of summer.

A walk on a sultry summer day, with hand lens to view flowers and insects and binoculars for birding, will reveal interactions ranging from the subtle to the obvious. None is uninteresting.

TREES: GEOMETRY, LEAF SIZE, AND LEAF ANGLE

Summer is the time of full leaf. Trees are collecting sunlight through microscopic structures called chloroplasts, which are packed inside leaf cells. Each chloroplast abounds with the green pigment chlorophyll, making the leaf green (see p. 262), and, more importantly, trapping the sunlight's energy to make sugar molecules. For photosynthesis plants require water, carbon dioxide, and sunlight. Minerals are also essential, though less directly involved in the basic process. Biochemically, photosynthesis is complicated and cannot be directly observed in the field. However, the process begins in leaves. The patterns by which leaves are arranged on trees, as well as the size, notching, and lobing of leaves contribute to the photosynthetic success of the tree. Even the angles by which leaves hang from the tree are significant. Take a close look at trees, both in open fields and in forests.

Multilayer Trees

Trees that grow in open areas such as meadows and old fields have a leaf geometry termed *multilayered* (see Fig. 41). Leaves are individually small and are arranged randomly from branch to branch. There is considerable space for light to penetrate between leaves on a branch. A single branch from a multilayered tree casts less of a shadow than that of a monolayered tree (see below). Leaves not only are small but in many species are deeply notched or lobed. Examples of multilayered trees are pines, Gray Birch, aspens, ashes, Pin Oak, Sassafras, and Red Maple. In general, trees with little shade tolerance are multilayered.

Monolayer Trees

Trees that grow in deep shade have a leaf geometry called *monolayering* (see Fig. 41). Leaves are arranged close together and little light penetrates between them. A branch from a monolayered tree casts a more substantial shadow than does a branch from a multilayered tree (see above). Monolayer trees occur in the interior of mature forests. Typically, the largest leaves on the tree are at the base, where the shade is deepest. Large leaf size is especially apparent in seedlings and young saplings. As you look at seedlings and saplings in a forest, you will notice that their leaves are much larger and much less notched or lobed than those of the same species growing in the canopy. These characteristics are particularly easy to see in oak and maple saplings. In general, leaves from monolayered trees are less lobed or notched than those from multilayered trees. Examples of monolayered species are Sugar Maple, Eastern Hemlock, American Beech, and Flowering Dogwood.

Why are some trees multilayered and others monolayered?

To understand the significance of layering patterns, consider the costs and benefits of leaf size, shape, and location on the tree relative to light availability, carbon dioxide uptake, and wind stresses. The function of a leaf is to receive solar energy—the leaf is nature's solar panel. When we set up solar panels on houses, we make them large and flat, to expose as much surface area as possible to the sun. Leaves are flat, but hardly large. Why don't trees make just a few very large leaves? The advantage of a huge leaf is offset by the cost of replacing it should it be damaged by wind or other accident. It is probably best in the long run to

Fig. 41. Leaf-layering of trees.

make many small leaves, each representing less of an investment of energy. Nature can be cruel and replacement costs are best kept low, if possible. Another important advantage to having small leaves is that they can be layered. One layer can top another, and another, and so on. Light becomes attenuated as it passes through a tree, but it is still sufficiently bright to be utilized by the lower, more shaded leaves. Such a pattern of multilayering makes good use of bright sunlight. Since carbon dioxide is evenly distributed throughout the air, even moderate light can be adequate for full photosynthesis. Multilayering is a definite advantage to tree species in open areas.

Leaves arranged in a multilayered pattern are most efficient if they are small, permitting sunlight to move through the boughs of the tree to leaves on lower branches. Deeply notched or lobed leaves also aid in allowing sunlight to penetrate. In multilayered trees, anywhere from 15% to 45% of the sunlight manages to penetrate beneath the average branch. This gives many leaves a chance to photosynthesize. On average, leaf area is 2–3 times ground area in multilayered trees. In other words, light is carefully filtered by the tree, attenuated gradually to shade, such that for every square foot of ground surface, 2–3 square ft. of leaves have utilized the light. Trees grow quickly in open areas, not merely because sunlight is abundant but because the trees use it very efficiently.

In shady forests, multilayering is much less advantageous because of the deep shade that characterizes the lower forest strata. Shade-tolerant species tend to be monolayered, with leaves arranged in a rather precise (non-random), non-overlapping pattern. Monolayering makes maximum use of the low amount of sunlight and avoids problems imposed by self-shading. Monolayered trees have a leaf area only about 1–2 times that of the ground area. Their lower leaves are conspicuously larger and less lobed than their upper leaves. Especially characteristic are the large leaves of seedlings of trees like Sugar Maple and various interior forest oaks. These leaves show less notching than those of adult trees and each leaf is considerably larger than leaves found in the canopy. They are shaped in such a way that light absorption is maximized, even given the low light that they receive. Monolayered trees have leaves so closely aligned that usually less than 10% of the light striking a branch will penetrate below the branch. In Flowering Dogwood, the figure is closer to 5%. These trees are adapted to present a single, uninterrupted layer of leaf surface, rather like the shape of an umbrella, to capture both low levels of sunlight (seedlings, saplings, young trees) or canopy sunlight (mature trees).

Leaf Angle

OBSERVATIONS: Most leaves are angled toward sunlight rather than horizontal. This is especially true of grasses, whose blades point nearly vertically, but it can also be seen in trees such as Weeping Willow. Why doesn't grass grow with blades that are flat against the ground rather than upright? Why do willow boughs droop? Leaves of forest wildflowers growing in shade also hang at an angle, drooping from the stem. Why? Angled leaves are particularly evident in species with lanceolate leaf shapes but can be seen in other species as well.

EXPLANATION: An angled leaf receives less concentrated sunlight per unit of leaf area than a leaf that is positioned horizontally and therefore most directly exposed to the sun. This lessens the heat stress on the plant, a definite advantage since water is not always available to cool it. An angled leaf, by spreading out the sunlight, exposes all of the tiny chloroplasts, the factories of photosynthesis, but at the same time it enhances the overall efficiency of the plant by reducing stress caused by heating. Grasses, growing in open fields, are very much exposed to heat stress. Their angled blades generate some shade and diffuse light over a wider surface, lessening heat stress. In spring, forest wildflowers usually have leaves that are oriented horizontally rather than drooping. At this time light is abundant because the canopy has not developed and temperatures are still relatively low, minimizing any possible heat stress. It is then to the plant's advantage to expose its leaves horizontally, since heat stress is not a factor. However, when the canopy has fully developed, the leaf angles of wildflowers change, and the leaves droop. Increasing the surface area over which the already diffused light can fall is more of an advantage at this time.

HABITAT AND GROWTH FORM IN PLANTS (PL. 40)

A major ecological event of summer is the flowering of myriads of herbaceous plants. July, August, and September are when the diverse asters and goldenrods flower along with many others. Flowers are, of course, conspicuous. Their appearance attracts pollinators (see p. 270). However, some plants produce many more flowers than others. Plant species that produce prodigious numbers of flowers often, but not always, are devoting more energy and overall effort toward reproduction than plants that produce fewer flowers. Why do such differences exist? What we find is that the type of habitat in which a plant lives seems to influence the degree to which the plant "spends energy" on flowering and reproduction. In other words, *different habitats* select for *different growth forms* of plants. A comparison of herbaceous plants from woodlands, open dry fields, and wet habitats, such as meadows, will demonstrate many examples of how plant growth forms vary with habitat.

Forests and Woodlands

To a herbaceous plant, forests and woodlands are stable habitats. Most herbaceous species are perennial and devote the lion's share of their energy to leaf, stem, and root growth. Taproots are common and taproots and related root fibers often far outweigh above-ground growth. This can easily be seen by digging up these plants but such a practice is not encouraged; many species are rare and should be left unmolested. Leaves are often thick, waxy, unlobed, or moderately lobed. Flowers are produced yearly but tend to be small and few in number. Many species flower in the early spring, when sunlight is most available before canopy leaves have opened (see p. 10). Spring flowering is much more common in woodland-dwelling species than in open-habitat species. The pattern exhibited by forest-dwelling wildflowers is called the *stable-habitat strategy.*

The most extreme examples of the stable-habitat strategy occur in species entirely confined to woodlands and forests, such as the trilliums or Indian Cucumber-root. Both flower in spring, but their fruits appear in summer. Fruits are few and obviously represent a low annual energy investment in relation to leaves and roots. One woodland species, the Cut-leaved Toothwort, has been shown to devote merely 1% of its energy annually to reproduction. Woodland herbs, which are perennial, put much of their energy into below-ground tissue, an adaptation that maximizes the chance of surviving winter. Other woodland species showing the stable-habitat strategy include:

- Wood Anemone
- Pipsissewa
- Wintergreen
- Solomon's-seal
- False Solomon's-seal
- Trout-lily
- Wild Oats
- Moccasin-flower and other woodland orchids
- Bloodroot
- Starflower
- Twinleaf
- Hepaticas
- Mayapple

Open Dry Fields

Recently disturbed areas such as open dry fields or roadsides present plants with stresses both in space and time. Physical condi-

Horseweed. This common plant of disturbed early old fields demonstrates the opportunistic growth strategy, by producing hundreds of tiny flowerheads.

tions tend to be variable and often extreme (see Chapter 4). Habitat change is frequently quite rapid as old field succession occurs, and thus the species tend to be opportunistic. Their growth form is reflective of their *opportunistic strategy* (see Pl. 40). Large amounts of energy are spent on reproduction. Although several flowering patterns are evident, all involve large commitments of energy for flower production. Many species possess large numbers of small flowers in various arrangements. Examples include:

Horseweed
Ragweed
Yarrow
Daisy Fleabane
Steeplebush
Butterflyweed
Common Plantain
Bushy Aster

Other species have smaller numbers of flowers, but each flower is relatively large, and the total number of flowers still exceeds that seen in most woodland species:

Common Mullein
Moth Mullein
Butter-and-eggs
Evening-primrose
Hedge-bindweed

Composites (see p. 176) are extremely common:

Common Sunflower
Black-eyed Susan
Tansy
various daisies, thistles, asters, goldenrods, and others

Yellow Bedstraw. An alien species of disturbed areas, it devotes prodigious amounts of energy to flowerheads, producing thousands of seeds.

Leaves are not usually very waxy or thick. Stems may be long if moisture is readily available (see following section on wet habitats) but more often are moderate to short. Many species, such as Sheep Sorrel and the hawkweeds, have basal leaves that are flattened against the ground in a rosette pattern. Leaves are often very deeply lobed or toothed. Highly complex leaf shapes (termed "dissected") are common, as in Ragweed. Many species, such as Hedge-bindweed, proliferate asexually by runners. Taproots occur in many but by no means all species (examples: Common Dandelion, Queen Anne's Lace or Wild Carrot, and Butterflyweed). Root growth often is less than above-ground growth, in marked contrast with stable-habitat strategy plants.

Wet Habitats and Meadows

Wet habitats such as meadows and streamsides are neither as stable as woodlands nor as disturbed and environmentally challenging as open dry fields. Moisture and light, the two environmental agents most responsible for lush plant growth, tend to be abundant in wet areas and plants are characteristically tall. Plant species of wetlands have an *intermediate strategy* of growth. These species tend to resemble open-area species because they produce an abundance of flowers and thus put much energy into reproduction, but they also grow taller, putting energy into stems and leaves. Cardinal-flower may top 4 ft. and Jewelweed reaches 5 ft. in wet habitats. Jewelweed also grows in dry habitats, however, where it is much shorter—see below. New York Ironweed reaches 7 ft. in wet areas and Tall Ironweed tops 10 ft. As with Jewelweed, the ironweeds are at their tallest in rich, moist soils, and grow considerably shorter in drier soils. Wet-habitat species differ from open field–roadside species and woodland species in the appreciable amount of energy they devote toward *stem growth*. Leaf and root characteristics are intermediate between those of woodland species and old field species and flowers tend to be abundant. Other examples include:

- Sweet Pepperbush
- Yellow Loosestrife
- Purple Loosestrife
- Joe-pye-weed
- Irises
- Cattails
- Golden Ragwort

Growth Strategy of Jewelweed in Wet and Dry Habitats

Jewelweed is commonly associated with stream banks and shaded woodland swamps. Its delicate orange (or yellow, depending on species) flowers are produced in abundance from about midsummer through early autumn. The flowers develop into seeds contained in seed capsules. Each capsule houses 3–5 seeds and, once the seeds ripen, the capsule is very sensitive to being touched. Even the slightest impact will cause the capsule to burst, scattering the seeds up to 5 ft. away. Since Jewelweed grows in wet areas, many seeds are distributed by water. Because of its peculiar seed-dispersal method, Jewelweed's other common name is Touch-me-not.

Jewelweed (Spotted Touch-me-not) grows in dense patches, often along a stream bank. Populations of seedlings sometimes number 1 000 per square meter. The plants compete significantly among themselves and natural thinning of stands occurs. The conspicuous, saclike flowers are held horizontally at the branch tips. Each flower has a long sepal ending in a curved spur, where the nectar is stored. Jewelweed is pollinated by long-tongued bees such as Bumble Bees, as well as by hawk moths and hummingbirds. Short-tongued bees, such as Honey Bees and short-tongued Bumble Bees, often feed by piercing the nectar spur to reach the nectar. They are not pollinators, but "nectar thieves."

Jewelweed flowers cannot fertilize themselves. Flowers begin as males, each flower containing 5 anthers coated with sticky pollen. Anthers remain on the flower for a day, then are dropped. The flower is now female, and the stigma (with egg) is exposed within the flower for the first time. All Jewelweed flowers on a given plant always are first male, then female, and thus a flower can never fertilize itself.

At the axils of the branches, where branch meets stem, you may see tiny green, budlike structures hanging by short stems. These are also flowers, but, unlike the showy, insect-attracting colorful flowers at the branch tips, these axil flowers are self-fertilizing. Jewelweed can both cross-fertilize and self-fertilize, and it does so with different flowers.

Jewelweed also grows in dry upland habitats with poor soils. A Jewelweed in such a habitat is shorter than the typical wetland plants and has fewer flowers. However, in dry areas Jewelweed plants will contain many of the tiny self-fertilizing flowers. Jewelweed is largely cross-fertilized when on lush wetland sites and largely self-fertilized when on poor dry sites.

Why Jewelweed Hedges Its Bets

Cross-fertilization ensures a strong measure of genetic diversity among the offspring. New combinations of genes come together and their diversity probably significantly enhances the ability of producing plants to cope with a range of environments (see p. 268). Genetic recombination is the ultimate function of sexual reproduction.

But cross-fertilization has its risks. Suppose an insect fails to visit the flower, or suppose it rains and no insects are about when the flowers are sexually mature? Jewelweed is an annual, and, if it doesn't produce viable seed, the plant is genetically dead. Annuals have but one opportunity to reproduce. Having some flowers that self-fertilize is a means of ensuring that some seeds are produced. Annuals are commonly self-fertilizing (see p. 170).

In rich mesic and hydric habitats, Jewelweed puts large amounts of energy into stem growth and flower numbers. Cross-fertilizing flowers outnumber the tiny self-fertilizing flowers. But in dry areas with poorer soils, a different pattern is apparent. There the plants are shorter and have fewer out-crossing flowers. Studies have shown that in these more stressful environments, it is cheaper (in terms of energy) for the plant to focus its genetic efforts on self-fertilizing flowers. Seeds of these flowers weigh only about two-thirds as much as those of cross-fertilizing flowers. All told, seeds from self-fertilizing flowers "cost" the plant only about one-third to one-half as much as those from cross-fertilizing flowers. This strategy, whereby a species concentrates on out-crossing in one environment (hydric-mesic) and on self-fertilization in another (xeric), is called *environmental cleistogamy* (out-crossing). Jewelweed survives because it "has it both ways."

Growth Strategies of Goldenrods and Asters

OBSERVATIONS: Various goldenrods (Pl. 41) and asters demonstrate all three growth strategies. Although all goldenrod species have conspicuous clusters of yellow flowers, the amount of flowers differs considerably even within a single species, depending upon habitat. Showy Goldenrod occurs in both woods and old fields. Field populations commit up to 30% of their total above-ground tissue to flowers and flower stalks. Woodland populations commit a maximum of 20% of their above-ground tissue to reproduction and many commit as little as 8%. Woodland Showy Goldenrod populations employ a more stable reproductive strategy than old field populations. Such a trend is evident among, as well as within, goldenrod species. Blue-stemmed Goldenrod, a woodland species, has far fewer flowers than Canada Goldenrod, an old field species. You can easily observe these subtle but important differences in flower density as you look at goldenrods in different habitats.

Asters reveal similar patterns. White Wood-aster and Large-leaved Aster have rather small flower clusters in relation to their leaf and stem sizes. Both species occur in woodlands and forests. In contrast, species found in old fields, meadows, and roadsides, such as Small White Aster (see Pl. 40), have many more flowerheads per plant than the woodland species.

In wet habitats and meadows, goldenrods and asters are characteristically quite tall, having invested much of their growth energy into strong, tall stems. Flat-topped Aster and Purple-stemmed Aster both reach heights of 8 ft. and Rough-leaved Goldenrod, a goldenrod species of swamps and wet meadows, reaches 7 ft. Rough-leaved Goldenrod may be no taller than 2 ft. in its driest habitats.

Early Goldenrod. An old field goldenrod, this species demonstrates the opportunistic growth form.

EXPLANATION: An herbaceous plant can devote more energy to its own growth, by adding root, stem, and leaf tissue, or to its reproductive efforts, by growing many small flowers or fewer but larger flowers. This, of course, is in no way a conscious decision by the plant but instead reflects the growth form most successful for a particular environment. Growth form is adaptive, and is selected for by the conditions imposed on plants by characteristics of the environment. The proportion of total energy devoted to each possible growth form varies among habitats. Should a plant be rooted in a stable habitat such as a forest, where dramatic changes from one growing season to the next are unlikely, it will probably be to the plant's ultimate benefit to invest its captured energy heavily in itself. A deep taproot and wide thick leaves aid the plant in competing successfully for water, soil nutrients, and light, as well as in replacing insect-damaged tissue. Such a plant may not use a high proportion of its energy to make flowers and fruits and, in fact, its flowers may be relatively insignificant in relation to the rest of the plant. However, its total lifespan can be many years in a stable habitat. Modest to low reproductive effort, if continued over long timespans, results in very high reproductive success. The use of large amounts of energy for self-maintenance with modest to low annual reproductive effort may be quite adaptive in stable habitats such as forests, where the opportunity for reproduction may be expected to persist for many years. This is the stable-habitat strategy.

In contrast, disturbed areas such as roadsides, old fields, wet meadows, and pastures are habitats where vegetation change can be very rapid (see Chapter 4). Plants need to literally "make hay while the sun shines." It is more adaptive for plants to put as much effort as possible into reproduction, to spread their seeds while they can. It is not an advantage to be long-lived if the environment is basically transient. In old fields, planned obsolescence is probably a blessing rather than a curse. Plants have evolved the opportunistic strategy, which emphasizes the investment of large amounts of energy into reproductive effort and the production of huge numbers of propagules. It is with the opportunistic species that we include the annuals and biennials, but many perennials are opportunistic as well.

Some species fall very clearly into either the stable-habitat strategy or the opportunistic strategy, but many fit somewhere in between. The intermediate strategy, which is seen most easily in wetland and meadow species, demonstrates that the growth form of a species may fit somewhere on a continuum between the two extremes. Even within a species, as in the goldenrods discussed above, individuals vary in growth form depending upon habitat.

PATTERNS OF FRUITING AND SEED DISPERSAL

As summer matures and autumn approaches, eastern forests bear fruit in the most literal sense. Many species of trees, shrubs, and wildflowers enclose their seeds in fleshy fruits. Fruits are usually brightly colored and easily visible. The function of a fruit is to be eaten, so that the enclosed seeds are dispersed. Just as flowers function to attract animals for pollination, so fruits function to attract animals for seed dispersal.

Careful observation of fruiting species and the animals attracted to them will illustrate much about the interactions among plants and animals. Beginning in midsummer, follow the development of fruiting among local species through autumn. The following questions should guide your observations:

1. For each fruiting species, when does the fruit develop? Is it in midsummer, late summer, early autumn, or late autumn? Through what range of time does a given species hold its fruit? Are there still plants with fruit in winter?

2. What color is the fruit and what color changes does it go through as it ripens?

3. When do leaves turn color in fruiting species? Do they turn early in autumn or late? How obvious are fruiting species among the other plants of the forest or field?

4. Open fruits and look at the seeds. Are the seeds large or small? How many in each fruit?

5. What birds and mammals feed on the fruit? To what extent does fruiting coincide with fall migration of birds?

Caution: Though many fruits are edible for humans, you should *never taste fruit or any wild plant unless you are certain of its identity.* Some fruits are poisonous. To learn the edible fruits, see *A Field Guide to Edible Wild Plants of Eastern and Central North America.*

In general, there are three patterns of fruiting evident in eastern forests. First, there are the species that fruit in mid- to late summer and produce fruits high in sugars but low in fats. Secondly, there are the species that fruit in fall, making fruits of low sugar but high fat content. Thirdly, there are the species that fruit in fall but make fruit low in both sugar and fat.

Mid- to Late-summer Fruiting Species (Pl. 48)

Most species that fruit in mid- to late summer produce small berries enclosing small seeds. These include the blueberries, huckleberries, strawberries, mulberries, and blackberries. A few species, however, produce large-seeded fruits. These include the plums and cherries. Mid- to late-summer fruiting species are characterized by having fruits of high sugar content, and thus they taste good to humans. Indeed, many of our most popular jams and jellies are made from these species and a late summer's walk in a woodland or field often brings a tasty reward. Though these fruits are high in sugars (carbohydrates), they are low in fats (lipids). This difference is significant for migrating birds, since carbohydrates yield only about half as much energy per unit weight as lipids (see below).

Many species are also characterized by *pre-ripening fruit flags,* meaning that the fruit undergoes a series of color changes as it matures. A typical berry begins an inconspicuous green color, then turns a bright pink or red, finally turning a deep purple or blue. This pattern is typical of blueberries, blackberries, mulberries, and black cherries. The color changes may serve to signal potential fruit-dispersing animals that fruit is about to ripen, ensuring that these animals will remain in the area and feed on the ripened fruit. Doing so maximizes the probability of successful seed dispersal.

Many of the small-seeded species bear fruit within 2 meters (6½ ft.) of the ground, where both mammals and birds can reach it. The sweet taste and odor of many of these small-seeded fruits may be adaptive in attracting mammals. Birds in general have relatively poor senses of smell and taste and the sweetness of these berries may not be noticeable to birds, though it would be to mammals. Among mammals that feed on small-seeded berries are Red

Ripening Pokeberries. This species is heavily fed upon (and seeds disseminated) by birds, who can see the color change from unripened green to ripened purple.

and Gray foxes, Raccoon, Black Bear, Fox Squirrel, and White-footed Mouse. The small seeds usually pass easily through the digestive systems of these mammals without being digested.

Timing is important to small-seeded species. They fruit after the main flush of insects hatches in spring and early summer. Birds are very attracted to the fruit, as insects are more scarce. This maximizes the chance that a bird will feed on the fruit and disperse the seeds.

Species that produce large-seeded fruits in summer are all in the genus *Prunus,* a large group that includes the wild cherries and plums. Fruits are usually above 2 meters (6½ ft.) from the ground and these species are among the most abundant fruiting species in northern forests. Like small-seeded species, they are attractive to mammals, and common dispersers include the Striped Skunk, Virginia Opossum, Raccoon, Red Fox, and Gray Fox. Seeds ingested by mammals are dispersed by being passed through the intestine. Birds also eat large-seeded fruits and regurgitate the seeds.

The early summer fruiting means that these plants rely on *resident* birds and mammals to disperse their fruits, and do not rely so much on migrant birds.

Fall High-quality Fruiting Species (Pls. 49, 50)

In late summer and fall a group of species produces fruit that is high in fat (lipid) content and contains large seeds. This group includes Spicebush, Flowering and Gray-stemmed dogwoods, Southern Magnolia, Southern Arrowwood, Sassafras, Virginia Creeper, and Black Tupelo (Black Gum). Most of the fruits are red or violet. They do not taste sweet, but are pungent or sour. Not attractive to mammals, they are fed upon voraciously by long-distance migrant birds such as thrushes, catbirds, and waxwings. The high fat content is an ideal food for migrating birds. Since there is roughly twice the energy per unit of weight in lipid (fat) as in carbohydrate, high-lipid fruits are highly economical for birds to eat. They can store a lot of fuel with relatively little weight cost compared with that of more sugar-laden fruits.

High-quality fruits are extremely popular with birds and they mature just as waves of autumn migrants are heading south. The timing is ideal to attract these migrants. Spicebush fruits ripen in New Jersey by the first 2 weeks of September. In one year, 77% of the fruits had been dispersed by 31 October and in another year, the figure was 90%. Birds do not leave these fruits on the shrubs for very long. Seeds pass quickly through a bird's intestine (usually within 30 minutes to 1 hour), and most seeds are deposited relatively near the feeding site.

Fall Low-quality Fruiting Species (Pl. 50)

The majority of plant species that flower in the fall are called low-quality fruiting species because the lipid (fat) content of their fruits is much lower than in the high-quality species. Common representatives of this group include hawthorns, sumacs, Choke-cherry, greenbriers, Mountain-ash, roses, Mapleleaf Viburnum, Fox Grape, Poison-ivy, hollies such as Winterberry, Red Cedar, and Common Juniper. Fruits are usually red or violet. The seeds of these species are also disseminated by birds but are not consumed by birds nearly as quickly as the high-quality species. Instead, many remain on the plant for the entire autumn and well into winter. They are consumed after the high-quality fruits have been taken and are often consumed during periods of severe winter weather. Loading fruits with lipid is expensive to the plant. The high energy content of lipid places a heavy cost on the metabolism of the plant, accounting for why there are fewer high-quality fruit-producing species.

In a New Jersey study of Mapleleaf Viburnum, more than 70% of the fruits were still on the plants on 1 January, though they had been mature since late August. Of the missing fruits, about 20% had dropped off the plant and the remaining 10% had been dispersed by birds.

Low-quality fruits have less than 10% lipid (fat) by weight. For instance, hawthorns typi-

Winterberry. This low-quality fruit remains on the plant over the winter and is fed upon by birds returning north in the spring.

cally have from 1% to 2%, greenbriers from 0% to 1%, and Chokecherry from 0% to 2%. Compare this with the high-quality species such as Spicebush (35%), magnolias (33%–62%), and Flowering Dogwood (24%).

The sumacs and bayberries represent an interesting situation. These species have small fruits, which are densely clustered in the sumacs, less so in the bayberries. But, though the fruits are small, the seeds are relatively large. This means the amount of pericarp or fleshy fruit relative to the indigestible seed is low. Though these fruits are relatively high in lipid (over 20% fat), the energy gain per fruit is quite low, so a bird must eat many of these fruits in order to really profit. For this reason, the fruits are considered low quality and, indeed, you will see sumac and bayberry fruits present on the plants well into the winter. Sumac fruits last very long, some remaining on the plant for several years. They are dry, oily fruits with high tannin content. White-tailed Deer are known to feed on sumac in many areas and both bayberries and sumacs are fed upon in spring by birds migrating north. They also form an emergency food supply for birds in winter.

It is not possible to separate high- from low-quality fruits on the basis of any consistent characteristic. They are colored alike, and many display foliar fruit flags (see below). There has been a general convergence among fruit color in all species. Red, blue, and violet are virtually the only colors represented. The way to best separate high- from low-quality fruits is to watch what gets eaten. Low-quality fruits are still around after migrating birds have passed. High-quality fruits are usually gone.

Foliar Fruit Flags

Though most fruits are red, deep blue, or violet, they still may not be easily discovered by migrating birds, most of which will be unfamiliar with the specific field or forest in which they search for food. Many fruiting species undergo a *very early leaf color change* in fall. In these species leaves change from green to brilliant orange-red or yellow. This early flush of color, called *foliar fruit flags,* may aid in attracting birds to the plant. Vines such as Poison-ivy and Virginia Creeper undergo early color changes and become highly visible as they entwine around tree trunks. Foliar fruit flags also occur on Black Tupelo (Black Gum), Sassafras, Spicebush, sumacs, wild grapes, and dogwoods. Many plants with foliar fruit flags are high-quality fruiting species, or have widely dispersed rather than clumped obvious fruits. High-quality fruits rot much more quickly than low-quality fruits, so they must be eaten quickly after ripening. Foliar fruit flags may be the long-dis-

Poison-ivy. This vine exhibits the characteristic of foliar fruit flags, with leaves brightly colored when fruit is ripe.

tance signal that enables the migrating birds to locate and consume the fruit quickly.

Fruit-consuming Birds and Mammals

MAJOR FRUIT-CONSUMING BIRDS

Cedar Waxwing
American Robin
Eastern Bluebird
Hermit Thrush
Wood Thrush
Gray-cheeked Thrush
Swainson's Thrush
Veery
Gray Catbird

Except for the waxwing and catbird, the species listed above are thrushes. All nest from northern forests southward. Most of the thrushes, as well as the Gray Catbird, migrate to the tropics. Waxwings are common wintering species throughout eastern North America and consume many of the low-quality fruits that are present in winter. Robins often overwinter in northern states when fruits are abundant, but Robins prefer high-quality fruits.

OTHER FRUIT-CONSUMING BIRDS

Wild Turkey
Ruffed Grouse
Northern Bobwhite (quail)
Pileated Woodpecker
Downy Woodpecker
Red-bellied Woodpecker
Yellow-breasted Chat
Northern Oriole
Orchard Oriole
Scarlet Tanager
Summer Tanager
Rose-breasted Grosbeak

Common Flicker
Tree Swallow
Northern Mockingbird
Brown Thrasher
Yellow-rumped Warbler
Evening Grosbeak
Pine Grosbeak
Rufous-sided Towhee
Purple Finch
House Finch

This group contains many permanent resident species as well as long-distance migrants. The chickenlike birds (turkey, quail, and grouse) are resident species that feed heavily on fruit. Woodpeckers are not normally considered fruit-eaters, but they often turn to fruits in the fall. The crow-sized Pileated Woodpecker can readily perform the acrobatics necessary to cling to a Virginia Creeper vine and pick berries. The orioles and tanagers are long-distance migrants and rank close to the thrushes in their importance as fruit-consumers. Tree Swallow and Yellow-rumped Warbler show a particular fondness for Northern Bayberry. Huge flocks of Tree Swallows gather along the coast in autumn and can be observed feeding on ripened bayberries. Yellow-rumped Warblers often overwinter when bayberries are abundant.

COMMONEST FRUIT-CONSUMING MAMMALS

White-footed Mouse
Fox Squirrel
Eastern Gray Squirrel
Eastern Chipmunk
Red Fox
Gray Fox
Raccoon
Striped Skunk
Virginia Opossum
Black Bear
White-tailed Deer

Mammals defecate seeds, leaving them in large piles, which may attract seed predators and may result in competition among germinating seedlings. As mentioned above, only mid- to late-summer fruiting species seem to be adapted to attract resident mammals, principally through sweet taste and odor (mammals have very keen senses of smell and taste), and by having fruits low on the plant, and often dropped on the ground.

Adaptations of Fruit for Seed Dispersal

Fruit is an adaptation to attract seed-dispersing animals, mostly birds but also mammals. But, why is this form of dispersal adaptive? The reason, quite simply, is that plants can't move, and the worst place a seed can fall is right under the parent plant. Seeds must get away from the parent plant because it is highly doubtful that a seedling could become established when in intensive competition with a mature member of the same species. The parent plant has a hold on the soil, light, and other resources of the particular spot it occupies. A young seedling would begin life with a very low probability of success, simply because of the presence of the plant that made it! As if parental competition were not enough to be up against, there is yet another reason for dispersal. When many seeds drop beneath a plant, their very concentration serves to attract seed predators as well as dispersers. Seed predators range from insects to mammals. Rather than disperse the seed, they digest it or otherwise damage it to the extent that it cannot germinate. When seeds are widely dispersed rather than concentrated, each seed has less of a chance of being discovered and destroyed by a seed predator.

Mammals, especially mice, can be seed predators as well as seed dispersers. If a seed is digested in the gut of a mammal, it cannot germinate and grow a new plant. The size extremes evident in mid- to late-summer fruits may help reduce the probability of digestion by mammals. The very small seeds characteristic of many species pass quickly through the gut of mammals and emerge intact. Also, many small seeds are discarded as the mammals chew the fruit. Intermediate-sized seeds are at risk, which may be one reason there are virtually no intermediate-sized seeds among the plant species whose fruits attract mammals.

There are more than 300 species of fleshy fruiting plants in eastern North America. The fruits are an extremely important energy source for birds and mammals. Fruit fuels migrating birds and sustains birds and mammals through the rigors of winter.

Why is fruit only produced during the end of summer and fall?

Why don't some plants fruit earlier? Fruits are fertilized mature ovaries. In most fruiting plants, insects are the pollinators. Before the ovaries can be fertilized, there must be insects to serve as pollinators, and so the warmth of spring, which brings a swarm of insects, initiates the process that will end with the mature fruit. Birds are the major fruit-eaters and seed-spreaders. Birds nest in spring and early summer, and at this time insects abound. Birds feed nestlings almost exclusively an insect-worm diet rich in protein necessary for the rapid growth of baby birds. Fruit contains only 3–13% protein, compared with up to 70% protein in insects; it would likely be at a disadvantage for dispersal should it mature while birds are nesting. Insects would still be both favored and abundant and nesting birds might easily pass up available fruit. Indeed, any bird feeding its young mostly fruit (if it were available) could be less successful rearing its young, because the nestlings would grow more slowly and thus be exposed to predation in the nest for a longer time period.

The correlation between fruit maturation and bird migration is very striking. As waves of migrants move southward, fruit-laden vines, shrubs, and trees supply the fuel needed for long migratory flights. The tens of thousands of migrants, many of which were hatched in Canada and northern states, find plentiful fruit as they migrate south. Clearly, the plants have evolved the timing of fruit production to attain the maximum probability of being eaten by a bird. Late summer and fall is when bird numbers are maximized in forests and when birds are moving. There is no finer animal to disperse a seed than a migratory bird.

Mammals are less able to reach high into trees (though many do climb), so mammal-dispersed fruit must be lower on the plant or must drop from it. Once fruit is on the ground, it is more subject to rot. Overall, it is not surprising that relatively few temperate-zone plants attract large numbers of mammals to their fruits. Only the mid- to late-summer species, which produce their crops early (before migrating birds come through in numbers), are adapted to attract mammals as seed-dispersers. Interestingly, in the tropics, where there is a year-round supply of fruit, many bat species have adapted to feed on fruit.

At first glance you might be surprised that there are so few high-quality fruits in fall compared with low-quality fruits. There is a significant trade-off involved here, however. To produce high-quality fruits, a plant must contribute much energy to fruit production. High-fat fruits do not last long. They are very subject to rotting. High-quality fruiting species essentially "put all their eggs in one basket," relying on the first waves of fall migrant birds to consume all of the fruit and disperse the seed crop. High-quality species tend not to fruit simultaneously. For instance, in New Jersey, Flowering Dogwood fruiting peaks after that of Spicebush. This separation in peak fruiting times avoids intense competition between these plants for birds as seed-dispersers.

Low-quality fruiting species play the game in a more conservative manner. Though their fruit is less attractive to birds and will be passed over in favor of the high-quality fruit, once the high-quality fruit is consumed, the low-quality is the only choice. Because of the low fat content, these fruits are slow to rot and can last a long time on the plant. Some even last throughout the winter and are present on the plant to be eaten by migrants returning north in the spring! Low-quality fruits rely on time and the harsh weather of winter to "force" birds into choosing them.

It is not easy to generalize as to which fall-fruiting plants produce high-quality fruits and which ones produce low-quality fruits. In general, many of the dogwoods produce high-quality fruits as do Spicebush and Sassafras. Most other species, however, produce low-quality fruits that remain on the plants well into the winter.

MILKWEED ECOLOGY (PL. 42)

OBSERVATIONS: Milkweeds are abundant in old fields throughout eastern North America. They are opposite-leaved plants, with rounded clusters of five-petaled, uniquely shaped, white, reddish, or orange flowers (depending on the species). Flowers develop into conspicuous seedpods and seeds are attached to feathery, parachute-like structures. Milkweed is named for its sap, a milky, sticky juice. Eleven milkweed species occur commonly in eastern North America, though the family (which is mostly tropical) is represented globally by more than 250 genera and 2000 species.

Locate a milkweed clump and observe its development and animal visitors over the course of a summer. Common Milkweed, distributed abundantly from northern Canada through Georgia and the Midwest, should be easy to find.

Common Milkweed tends to occur in dense clumps often separated by considerable distances. A large milkweed patch in a field may be a mile or more from a second patch. This patchiness is caused by the reproductive pattern of the plant (see below).

Common Milkweed flowers from early June throughout the summer. Flowers are in rounded clusters and each individual flower consists of 5 petals, 5 sepals, 5 stamens, and 2 ovaries (best seen with a hand lens). The 5 petals are united at the base and pollen is packaged in sacs, called pollinia, which occur in pairs, attached to a wishbone-like structure called a corpusculum. A careful dissection of a milkweed flower will reveal the corpusculum-pollinaria unit, called a pollinarium. Insect pollinators (see below) remove the entire pollinarium.

Insects abound around milkweed flowerheads. You should see Bumble Bees, Honey Bees, ants, wasps, and butterflies of several species, including the brilliant orange and black Monarch and probably the less conspicuous hairstreaks. Close inspection should reveal well-camouflaged crab spiders lurking within the flowerheads. Smaller insects, such as flies and Honey Bees, may become entrapped in the flowers and be struggling to escape. Predatory insects, such as colorful Yellow Jackets and hairy black tachinid flies are usually in the vicinity.

A visit to a milkweed clump at dusk can also be rewarding, as you might see hawk or sphinx moths, geometer moths, and underwing moths (see p. 349).

Common Milkweed. The large, conspicuous, ball-like flower clusters attract numerous insects. Milkweed Beetle, on leaf, occurs only on milkweeds.

Milkweed leaves are usually relatively undamaged, but large, yellow-, white-, and black-striped Monarch caterpillars may be munching on them. By midsummer, boldly patterned, orange-red Milkweed Beetles, often mounted one upon another in copulation, are common on the leaves. Equally boldly patterned, red and black Milkweed Bugs are roaming about the developing seedpods.

Both the Monarch butterfly adult and caterpillar are conspicuous, along with the Milkweed Beetle and Milkweed Bug. These latter two insects do not occur on plants other than Common Milkweed. Monarch adults visit a wide variety of flowers in quest of nectar but Monarch caterpillars, like the Milkweed Beetle and Milkweed Bug, feed only on milkweed.

EXPLANATION: The natural history of milkweed illustrates in microcosm many of the interactions that affect larger ecological systems. Milkweed is adapted for long-distance pollination, dispersal, and defense. Its fauna demonstrate adaptation and specialization.

Milkweed is clumped because, following its establishment in a field, it rapidly reproduces asexually, forming a clone or genet (see p. 32). Such a pattern is characteristic of many pioneering plant species in old fields (see Chapter 4). All milkweed stalks in a given clump are genetically alike, united by an underground rhizome system.

To establish itself in newly opened areas, milkweed must produce fertilized seeds, and for this, the action of pollinating insects is essential. Because milkweed clumps are widely scattered, for cross-pollination to occur, long-distance flying insects make ideal pollinators. Bumble Bees, wasps, butterflies, and nocturnal moths, all powerful, long-distance fliers, are the principal pollinators. Bumble Bees can account for 75–90% of all daytime pollinations and moths can contribute up to 25% of the overall pollinations. Bumble Bees are effective pollinators of milkweed because they are abundant, large enough not to get trapped in the flowers (as many smaller insects do), and they are sufficiently strong to remove the pollinarium, which attaches to their bristly legs. Big insects require large food rewards and milkweed is a prolific nectar-producer. It also becomes quite fragrant at night, an enhancement in attracting moths. The large, pale flower clumps are easily visible to both day- and night-flying insects.

Not all insects are pollinators. Ants are common milkweed visitors and walk from flower to flower on the same plant. However, they rarely walk to neighboring clumps, because the distances are great. Ants, therefore, do not successfully cross-pollinate and are actually nectar parasites.

The presence of an abundance of nectar-feeding insects provides a concentrated food source for predators such as crab spiders and Yellow Jackets. The pale spiders are well concealed among the flowers. They lie in wait, ambushing pollinators, including the large Bumble Bees as well as the luckless small creatures that have become entrapped among the network of flowers. Yellow Jackets attack from above.

Milkweed Butterflies

OBSERVATIONS: A milkweed, over the course of the summer, will host eggs, larvae, pupae, and adult Monarch butterflies. The Monarch depends entirely on milkweed for its life cycle. Females attach cone-shaped green eggs to the undersides of milkweed leaves. The eggs are ⅛ in. long, have vertical ridges, and are attached either singly or in small clusters. You must check milkweeds frequently to spot Monarch eggs, as they hatch in only 3–4 days.

The tiny Monarch caterpillars have yellowish green bodies and shiny ebony heads. Newly hatched caterpillars first eat their eggshells, then begin on the leaves. Within a week the caterpillar has grown substantially, shedding its exoskeleton and taking on the yellow, white, and black stripes that characterize its remaining week before metamorphosis.

Watch caterpillars (of any species) eat. They chew the edges of the leaves, taking large chunks with great efficiency. Leaves may be chewed away entirely, right to the stem. An easy way to locate a milkweed with Monarch caterpillars is to look for leaf damage.

By the end of 2 weeks the Monarch caterpillar undergoes its final molt and begins metamorphosis to an adult. The pupa stage of butterfly development is called the *chrysalis.* Monarch chrysalises hang, sometimes several to a plant, dangling beneath branches and on the undersides of leaves. The cocoon is spun by silk from the caterpillar. It is attached by an adhesive pad also secreted by the caterpillar. The cocoon is initially green but later becomes transparent, revealing the developing butterfly inside. After about 10 days as a chrysalis, the adult butterfly emerges.

It is unusual to be present when an adult Monarch emerges from its cocoon but, if you are fortunate enough to do so, you will see that it takes about an hour for the butterfly to pump fluid into its wing veins, stiffening and expanding the wings so that it can fly. At this time, the insect is highly vulnerable to predation. Once the wings are stiffened and fully expanded, the bright orange and black butterfly takes to the air, perhaps feeding on the milkweed flowers for its first meal as an adult.

EXPLANATION: Why does the Monarch depend entirely on the milkweed as a host for its life cycle? Members of this butterfly family are called "milkweed butterflies" because as caterpillars they feed exclusively on plants in the milkweed family. Such a characteristic is not unusual among butterflies and moths—many butterfly and moth larvae feed exclusively on plants of a single family (see p. 261). Adult Monarchs feed on milkweed nectar, but not exclusively. Like the milkweeds, milkweed butterflies occur mostly in the world's tropical regions. The most common milkweed butterfly in North America is the Monarch.

The question of why these caterpillars eat only milkweed leaves is related to the question of why other families of butterflies and moths are not found on milkweed. Also, why are both Monarch caterpillars and adult butterflies so boldly patterned and so colorful? Not only are adults bright and obvious, but they fly lazily and are seemingly easy targets for avian predators.

The answers to all of the above questions center on the

Monarch. The bold coloration serves as a warning to potential predators. The butterfly is feeding on goldenrod. Note long tongue.

fact that milkweed produces an abundance of a chemical called cardiac glycoside (see p. 260). The effects of consuming cardiac glycosides, of which digitalis (a drug obtained from Foxglove) is a common example, range from mild to severe poisoning. Cattle can easily become ill by eating milkweed. Cardiac glycosides are *defense compounds* that aid in protecting the plant from herbivores. Like cattle, most plant-eating insects cannot tolerate cardiac glycosides, but the milkweed butterflies (monarchs and their relatives) have evolved the ability both to tolerate and to store the toxic chemical in their tissues. Monarch caterpillars are as poisonous as the leaves they eat, protected by the same chemical defenses. The milkweed butterflies have turned an evolutionary trick on the milkweed—they can eat the plant without ill effects and then use its defense compound for their own defense! No other group of butterflies or moths has conquered this chemical defense so efficiently, so milkweed butterflies have exclusive access to milkweeds. Their tolerance to cardiac glycosides permits them to become milkweed specialists.

It is not clear exactly how these butterflies store cardiac glycoside, but what is clear is that the chemical remains in the adult butterfly after its metamorphosis from a caterpillar. Birds will readily attempt to eat adult Monarchs, but those that do will become ill from the glycoside. Once a bird is exposed to the nauseating experience of eating a Monarch, it will avoid future encounters. The Monarch's bold, orange and black pattern serves as an easily remembered warning to avoid future encounters with this toxic butterfly; birds ignore this *warning coloration* at their peril. The Monarch caterpillar and adult also have tough, rather leathery skin. This adds an additional element of protection for surviving the attacks of birds slow to "remember" what they are attacking.

Monarch natural history is intimately related to the adaptation for eating milkweeds and storing the cardiac glycoside, using it for protection. Other milkweed specialists include the red Milkweed Beetle, which feeds on milkweed stems and roots, and the small Milkweed Bug, which eats milkweed seeds. Both insects are quite colorful and both are toxic to predators for the same reason as the Monarch—cardiac glycoside is stored in their tissues.

Mimicry and the Monarch (Pl. 42)

OBSERVATIONS: Look very carefully at adult Monarchs. Occasionally a smaller "Monarch," which has 2 *white* spots near the tip of the upper wings, may appear. This insect is not a real Monarch, which is larger and has orange spots in this position. This butterfly is a Viceroy. Viceroys bear an uncanny resemblance to Monarchs, a trait that helps them survive. Though Monarchs may be quite common, Viceroys, in comparison, are uncommon (see below). You will have to search to find a Viceroy.

EXPLANATION: Though the Viceroy looks extremely similar to a Monarch, it is not even in the same family! It belongs to the brush-footed butterfly family. Viceroy caterpillars never feed on milkweed, but rather on leaves of trees in the poplar and willow families. Neither the Viceroy caterpillar nor the adult butterfly is toxic to birds. The Viceroy is a *mimic* of the Monarch. It gains protection because it so closely resembles the unpalatable Monarch that predatory birds presumably take it to be a Monarch and avoid it. The Viceroy caterpillar does not occur on milkweed and does not look like a Monarch caterpillar. There would be no advantage to such a resemblance. Only the adults look alike.

Why are Viceroys less common than Monarchs? Viceroys would be their own worst enemies if they were abundant. Were Viceroys to outnumber Monarchs, birds would encounter palatable Viceroys so frequently that they would not be effectively "trained" to avoid orange and black butterflies. Viceroys are only protected when well outnumbered by Monarchs.

PATTERNS OF INSECT OR SPIDER CAMOUFLAGE AND MIMICRY

OBSERVATIONS: There is a very clear correlation between the color patterns of many insect and spider species and the background pattern of their habitats. One of the most widespread adaptations in nature is *camouflage* (also called *crypsis*), the mimicking of background patterns. Another common pattern is mimicry of other species, such as the Viceroy butterfly mimicking the unpalatable Monarch. Mimicry and camouflage take many forms and you can spend numerous pleasant hours unraveling mimicries in your area. To get started, I suggest focusing on the following categories:

1. **LEAF AND TWIG MIMICS.** Search the foliage and branches for insects and spiders. You should find quite an impressive variety of mimics that resemble the leaves and twigs they frequent.

2. **BARK MIMICS.** Take a very careful look at the bark of various trees. You might be surprised by the number of well-camouflaged arthropods you encounter.

3. **FLOWER MIMICS.** Look into a flower, especially a large species like a Queen Anne's Lace (Wild Carrot) or goldenrod. Careful searching will usually reveal ambush bugs and spiders colored very much like the flower itself (see p. 350).

4. BEE AND WASP MIMICS. Not every "bee" is a bee, nor every "wasp" a wasp. Many kinds of insects look a lot like bees or wasps and are often, if not usually, mistaken for them. That's the idea. Search for bee and wasp mimics near flowers.

5. ANT MIMICS. Many insects and some spiders have evolved body forms that mimic ants. These creatures often live among ant colonies, deriving food, protection, or both, from their formicine hosts.

6. BUTTERFLY MIMICS. As discussed above, some palatable butterflies (such as Viceroys) mimic unpalatable species (such as Monarchs). There are other examples of this pattern that you can identify.

Bear in mind that the whole object of mimicry, whether the animal is mimicking background patterns such as leaves or another animal such as a bee, is to *avoid detection*. Therefore, you'll have to search diligently and observe carefully to appreciate the effectiveness of mimicry and camouflage. In the case of bee and butterfly mimics, the trick will be separating the mimic from the model.

In this guide there is only space to cite a few examples in each category. You should be able to find many more, and should refer to the Peterson Field Guides to insects, moths, and butterflies for identifications.

Leaf and Twig Mimics (Pl. 43)

The following are among the commonest and most widespread leaf-mimic insects:

The **TRUE** or **NORTHERN KATYDID** has forewings colored and shaped in such a way as to closely resemble leaves, including wing veins that resemble leaf veins. Katydids live among foliage in treetops and are known for their endless calls on hot summer nights. They feed on the very leaves they resemble.

The **BUFFALO TREEHOPPER** lives on low vegetation of forests, meadows, and old fields. Look for it on willow, elm, locust, cherry, and various fruit trees as well as goldenrod, aster, and clover. This small green insect has a humped upper back, hence the name "buffalo." Its green, leaflike wings make it difficult to spot. It feeds on plants.

The spectacular **LUNA MOTH** is common but highly nocturnal, and is therefore often overlooked. A member of the moth family Saturniidae, along with the equally impressive Io, Promethea, and Cecropia moths, the Luna Moth is a giant silkworm moth. The Luna Moth frequents deciduous forests and hangs leaflike from the foliage. The Luna closely resembles leaves, but notice its long, streamerlike tail. The streamers are more conspicuous than the rest of the wing and may act to misdirect attacks of predators. Though part of the streamer is lost, the moth escapes. Hence this species "hedges its bets" by being both camouflaged and having a *distraction signal* in the event that it is discovered. For more on distraction signals, see p. 354.

The following insects are twig mimics:

The **NORTHERN WALKINGSTICK** feeds on foliage of deciduous forests. This remarkable-looking insect is very easy to miss since it so closely resembles a woody twig. Males are brown and females greenish brown. Each sex tends to orient to the branches most similar in color to the insect. You may have some difficulty finding walkingsticks but you should search for them on oaks and hazelnuts.

The **PRAYING MANTIS** and **CAROLINA MANTIS** are the tigers of the foliage. Their slender bodies effectively camouflage them as they lurk among old-field shrubs and wildflowers. The Praying Mantis ranges in color from green to tan and the Carolina is green to brownish gray. Mantises are voracious predators. They have sharp, serrated forelegs, useful for capturing and butchering a great variety of butterfly adults and caterpillars, moths, flies, and various other insects. These insects are also known for their risky sex lives. Some females kill and eat the males after copulation. (This is adaptive, at least for the female—she acquires a large protein meal at a time when the eggs are being developed. Though the protein benefits the male's future offspring, it is nonetheless more difficult to argue that being devoured is adaptive for the male.) In the case of mantis camouflage, the insect's ability to mimic the background twigs protects it from predatory birds and aids it in lying in ambush for potential victims. Incidentally, the Praying Mantis is not a native North American insect, but was introduced in 1899. However, it is so effectively colored to blend in among virtually any shrubby background that it is adequately camouflaged in its alien country.

The **LOCUST TREEHOPPER** is neither a twig nor a leaf mimic, but is a thorn mimic! Living in forests and old fields, this insect is well named—look for it in locust trees. Like the large thorns that adorn locusts, this insect rests on twigs looking just like one of the thorns. To find this creature, touch the thorns until one moves. It feeds on juices which it sucks from the plant.

Bark Mimics (Pl. 43)

Many species of moths are active at night and rest during the day on exposed bark. They are well camouflaged, blending with the

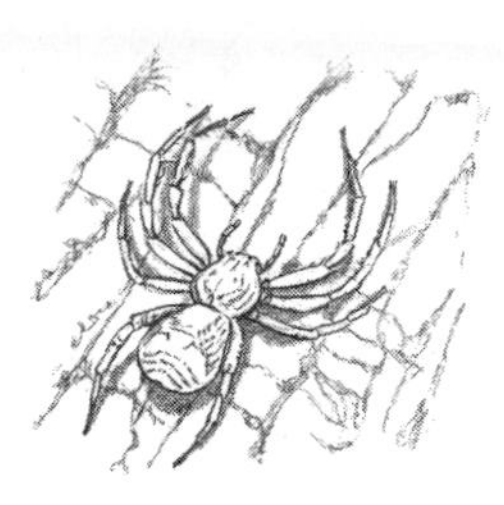

Fig. 42. Inconspicuous Crab Spider on bark.

bark pattern. Indeed, the first unambiguous case of natural selection involved the Peppered Moth of Great Britain, a bark mimic (see p. 266). It is impossible to include a comprehensive list of even the most common bark mimics among the moths, but one group, the underwing moths, will serve well to illustrate the principle of bark camouflage. You will find many other examples.

UNDERWING MOTHS number 104 species in the United States, approximately 70 of which are found in our region. These insects characteristically rest head downward on bark, with their forewing patterning almost precisely in alignment with the bark patterning. Different species are found on different species of trees, and their wing patterns precisely mimic bark patterns. For instance, one of the so-called "white underwing" species rests on white birch bark. A "dark underwing" species is found on trees with blackened stumps, like those in burned-over tracts. It is the forewing patterning of these moths that is patterned like bark. The hind wings, generally concealed at rest, exhibit very bold, often bright patterns. As you might suspect, these patterns can be very distracting when suddenly flashed at a predator.

The **HORNED FUNGUS BEETLE** is one of many beetles that closely resembles the rough-textured bark where it commonly rests. This forest-dwelling insect feeds mostly on bracket fungi. In addition to being protected by its resemblance to bark, it feigns death, which makes it look even more like a bark fragment as it lies among the forest litter.

Many spiders, including the very well-named Inconspicuous Crab Spiders, hug the bark, blending remarkably with bark patterns (see Fig. 42). Though common, these spiders are so cryptic that they are usually overlooked, except by birds such as the Brown Creeper, which feed by carefully spiraling around a tree trunk, searching the bark crevices.

Flower Mimics (Pl. 43)

Nectar-rich flowers attract numerous hungry insects. As such, these flowers present opportunities for predatory animals, including many insects and spiders. Flowers are distinctly and brightly colored, but predatory species have evolved to be cryptically colored when lying in wait inside or on the flower.

Two examples, both of which are common on goldenrod, are the Yellow Ambush Bug and the Goldenrod Spider.

The **YELLOW AMBUSH BUG** feeds on any of the small goldenrod-feeders, including butterflies, bees, flies, and other true bugs. The Ambush Bug does exactly that, tucking itself among the dense yellow goldenrod flowers and grabbing prey using its thick, strong first pair of legs. With its piercing mouthparts, the Ambush Bug stabs its prey and injects toxin in its saliva to immobilize it. Look carefully for Ambush Bugs. Though they are not small, they are amazingly well camouflaged within the spray of tiny goldenrod blossoms.

The **GOLDENROD SPIDER** is one of the more than 200 species of crab spiders. These arachnids do not spin webs to capture prey but, like the Ambush and Assassin bugs, they lie in wait, cryptically blending with the background. Many species inhabit flowers and one of the most common is the Goldenrod Spider. Colored pale yellow, usually with reddish streaks, this predator feeds on Bumble Bees and syrphid flies, grabbing them and paralyzing them with its bite.

In any clump of goldenrod, you should be able to find both the Yellow Ambush Bug and Goldenrod Spider.

Bee and Wasp Mimics (Pl. 43)

Bees and wasps sting. Though bees are nectar-feeders, they are very good at defending themselves, affording them considerable protection. Many predaceous species, including insects and vertebrates, avoid bees. Wasps are often aggressive and many are predators of insects and spiders. By virtue of their aggressive behavior and potent stinging abilities, bees and wasps are superb examples of the axiom, "the best defense is a strong offense."

When observing bees and wasps buzzing about flowers, look very carefully. Some may not be what they appear to be. In fact, a bee is not always a bee.

If you notice a Honey Bee without a waistline, it is probably not a bee at all, but a **DRONE FLY**. The Drone Fly looks extremely similar to a Honey Bee, and flies like one. However, it lacks the narrow, pinched-in waist of the Honey Bee. Drone Flies, which neither bite nor sting, are essentially defenseless. They feed on nectar and pollen, mingling among the bees, gaining protection from predators by looking like bees.

Other flies are bee mimics, including the Progressive Bee Flies and the Large Bee Flies.

The **LARGE BEE FLY** feeds on nectar. It follows solitary bees, laying its eggs in nests excavated by the solitary bee female. The fly larvae then feed on the bee larvae. This unique species not only derives protection from mimicking bees, but parasitizes them as well. Large Bee Flies are often seen hovering.

The **PROGRESSIVE BEE FLIES** also feed as larvae on the larvae of other insects, including bees. Adults are prone to hovering, and they frequent open fields and low vegetation.

The **AMERICAN HOVER FLY** looks very much like the aggressive Yellow Jackets. The name "hover fly" is descriptive. To identify a hover fly, look closely at any Yellow Jacket that seems to hover motionless over a flower, but be careful. Remember that female Yellow Jackets can sting more than once, and their stings are painful. Hover flies, on the contrary, are stingless and generally defenseless. The American Hover Fly is nonetheless well protected by its resemblance to a potent Yellow Jacket. As adults, hover flies feed on nectar from flowers in old fields, but larvae feed on scale insects and aphids.

Some hover flies (sometimes called **"FLOWER FLIES"**) look like dark bees. They are mostly blackish but have bee-like, yellowish-striped abdomens and a bee-like shape. Like the American Hover Fly, these insects feed on nectar from flowers in old fields as adults and on aphids as larvae. They neither bite nor sting.

The **FRINGE-LEGGED TACHINID FLY** resembles a bee, even to the pinched-in abdomen. It is striped with black and yellow and frequents old field flowers, where adults feed on nectar. Tachinid

Fig. 43. Drone Fly compared with Honey Bee.

flies lay eggs on living insects, and the larvae invade the host, growing inside their victim. Hosts include a range of insects from caterpillars to aphids, depending upon the species of tachinid. The Fringe-legged Tachinid female lays her eggs on various species of true bugs.

Bee Killers (Pls. 43, 44)

Adaptation does not guarantee success. Although bee and wasp mimics do gain protection by their resemblance to stinging insects, not all predators are intimidated by bees and wasps. One group, the **ROBBER FLIES**, contains several species that specialize in killing bees and bee mimics.

Robber flies are easy to recognize. They are streamlined, sleek-looking insects, larger than their prey. Quick and skillful fliers, robber flies either leap on their prey or catch it in mid-air. One species, called the **BEE KILLER**, specializes in killing Honey Bees. The Bee Killer is very common in old fields, where it ambushes nectar-feeding Honey Bees, killing them by stabbing them with its needle-like mouthparts and sucking the liquid from them. Many bee mimics, as well as true bees, probably fall prey to Bee Killers.

Clear-winged Moths

One family of moths, the clear-wings (Fig. 44), tends to strongly resemble wasps and gains protection from this similarity. One common eastern species, Doll's Clearwing moth, bears a striking similarity to paper wasps. These moths are woodland-dwellers, but tend to frequent forest edges. Their caterpillars (larvae) burrow in aspens, poplar, and willows.

Fig. 44. Doll's Clearwing moth compared with Paper Wasp.

The ant-mimic spiders and the Ant-mimic Jumping Spider both look more like ants than spiders. Ant-mimic spiders live in close proximity to ant hills in the southeastern states. They do not eat ants but gain protection from looking like them. Ants are full of a noxious chemical called formic acid, which discourages most predators from dining on such spicy insects. By looking like ants, these perfectly palatable little spiders gain a measure of protection. The Ant-mimic Jumping Spider is found throughout our region and shares a similar natural history.

Why are mimicry and camouflage so beneficial?
It may seem very obvious that mimicry or camouflage, nature's art of disguise, shields its animal practitioners from discovery, and so it does. But, looking beyond the obvious, why is protection from discovery important? Aside from those animals with warning coloration (Monarch, Coral Snake), nature seems to select widely for animals that go incognito. This is because protection from discovery is valuable for both prey and predators. Prey animals avoid detection by predators but predators, by being cryptic, avoid detection by prey. Thus, predatory species are able to approach prey more closely and are more successful at capture. As explained in Chapter 5, populations can evolve a different appearance if their background changes. This is what occurred in the Peppered Moth, which turned from light to dark as tree trunks accumulated soot from industrial pollution. It is hardly surprising that so many different kinds of animals, ranging from insects and spiders to snakes and grouse, blend so perfectly with their background environments. Indeed, you might keep track of how many different kinds of cryptically patterned animals you find in a half-hour of looking on a summer's day. Most are camouflaged, or cryptic, in some way.

Mimicry of other animals (called models) is also widespread in

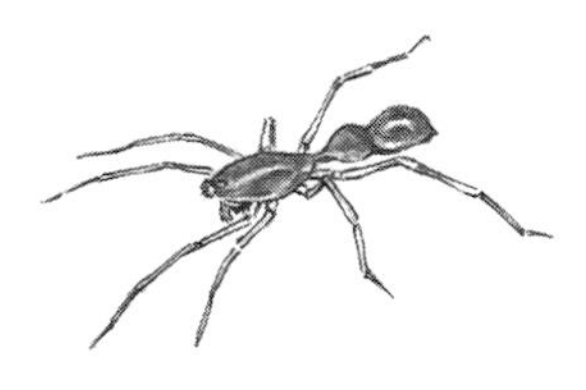

Fig. 45. Ant-mimic spider.

nature. The mimic gains protection by being mistaken for the model, which is usually poisonous or in some way dangerous. Models represent "resources" which have been evolutionarily exploited by the mimics. Mimicry is discussed again on p. 356.

DISTRACTION PATTERNS OF BUTTERFLIES AND MOTHS (PLS. 43, 44)

Often, when a butterfly or moth is at rest, it will be cryptic, blending very well with its background, but many butterflies and moths "hedge their bets" against escaping predation by also employing *distraction patterns* in the event of discovery. Distraction patterns are evident when the insect spreads its wings, and include *large spots resembling eyes, flash coloration, wing-border spots,* and *wing-border lining*. Look at butterflies and moths carefully, noting where, on their wings, their flashiest colors appear. Typically, the colors are most visible when the wings are spread, and are at or near the wing edges.

The **UNDERWING MOTHS,** as mentioned above, are normally extremely cryptic when resting head-downward on bark. When an underwing moth is discovered, however, it is by no means at a loss for survival tricks. The underwings (hind wings) of these moths display brightly colored flash patterns, usually of bright orange or reddish interrupted by black barring. These sudden bursts of color are presumed to work by confusing a predator long enough for the moth to take flight before being grabbed. Many of the sphinx moths have similar patterning on their hind wings.

Once in the air, an underwing moth may do any of several things. It may quickly fly around to the other side of the tree trunk and alight, again becoming cryptic. Or, it may flutter away erratically, zigzagging in such a manner as to make accurate pursuit by a predator less likely. Finally, it may drop to the ground, again becoming cryptic with the background pattern of litter. Close observation of underwing moths has revealed that wings are occasionally torn along the hind edges, evidence of attacks as the moth took flight. However, in some underwings, the wings show impressions made by birds' beaks, but the wing is not torn. Such impressions may indicate that the moth was grabbed while on the bark, but the sudden flash pattern of the hind wing so startled the bird that it momentarily let go, permitting the moth to escape. Such are the survival strategies of the many species of underwing moths.

Another adaptation that serves to startle predators is the presence of eye-like spots on the wings of various butterflies, moths, and a few mantids. Some insects, such as certain species of the big silkworm moths, show very prominent eyespots. The adult (especially the female) **IO MOTH**, for example, has large black spots surrounded by bright yellow-orange on its hind wings (see Pl. 44). Other moths with bold eyespots include the **PROMETHEA MOTH**, which has spots on the outer part of its forewings, the **POLYPHEMUS MOTH** (named, presumably, for the cyclops vanquished by Odysseus), which has small eyespots on the outer part of both fore- and hind wings, and several of the **SPHINX MOTHS**.

Though several of the moths and butterflies with bold eyespots superficially resemble owls, it is doubtful that predators mistake these insects for birds. More likely, the sudden appearance of the eyespot, revealed only when the wings are spread, serves to startle the predator and misdirect its attack, perhaps even to delay the attack, providing precious escape time. The "eyes" are probably targets to draw the attention of the predator from the more vulnerable parts of the insect.

The position of the wing spots is important. If these markings are adaptive in that they misdirect the predator's attack, then the farther from the potential victim's body they are, the better. A survey of butterflies and moths will show that, almost invariably, the spots are near the outer part of the wings, where a bird's beak would do the least damage to the insect.

Examples include the colorful **BUCKEYE**, with large spots on the outer edges of both fore- and hind wings, and several other satyrs and nymphs, including the **LITTLE WOOD SATYR**, which has various-sized black spots lining the outer parts of both fore- and hind wings.

The **WOOD NYMPH** has two black spots on its outer forewings, emphasized because they are surrounded by a bold splash of yellow.

The **MOURNING CLOAK**, one of the earliest butterflies to appear in the spring, has the outer edges of its wings lined with bright pale

Fig. 46. Butterfly patterning.

yellow. A predator's attention is very much focused, at least initially, on the least vulnerable part of the butterfly.

EXPLANATION: The prevalent pattern that emerges in the coloration of these moths and butterflies is a combination of camouflage and distraction patterning. Eyespots, wingspots, colorful wing lining, and color splashes all function in a similar manner—to misdirect a predator's attack. Close inspection of butterflies and moths often reveals damage done by bird beaks to the outside parts of the wings. The insect pays a very small price to escape. As you examine the butterflies and moths in your area, note how many are patterned in ways similar to the examples provided here.

A Complex Butterfly Mimicry System (Pl. 45)

The Monarch and Viceroy (see Pl. 42) represent but one example of how a palatable butterfly species may evolve mimicry with an unpalatable model. Such examples are multiplied many times in tropical areas, but other examples also exist in North America. Perhaps the most complex mimicry pattern involves the Pipevine Swallowtail butterfly, an unpalatable species with 5 different mimics!

OBSERVATIONS: The **PIPEVINE SWALLOWTAIL** occurs throughout eastern North America, and ranges westward as well. Adults feed on many species of flowers, including honeysuckles, milkweeds, thistles, lilacs, and azaleas. The caterpillar, however, feeds only on plants of the birthwort family, such as Dutchman's-pipe and Virginia Snakeroot. These species contain aristolochic acid, which Pipevine caterpillars can store, just as Monarch caterpillars store cardiac glycosides. The result is that Pipevine adults are unpalatable.

Pipevine Swallowtails are black, with shiny metallic bluish purple on the hind wings. Bright orange dots surrounded by black rings highlight the outer hind wings. Five species of butterflies, none of which dines on birthworts as a caterpillar, mimic the Pipevine Swallowtail. All of these butterflies are perfectly palatable to predators.

The **EASTERN BLACK SWALLOWTAIL** caterpillar is partial to parsley and carrots, such as Queen Anne's Lace, but the adults look very much like Pipevine Swallowtails, even down to the orange spotting on the wings.

The **TIGER SWALLOWTAIL** is usually quite unlike the Pipevine Swallowtail, being bright yellow with bold black linings and wing veins. However, where the Pipevine Swallowtail occurs along with the Tiger, *female* Tigers often are mimics of the Pipevine. They are black on the forewings and metallic blue on the hind wings. The males never mimic and many non-mimetic females remain in the population. Though the mimicry protects the female Tigers, they are at a bit of a disadvantage nonetheless, since males seem to prefer mating with non-mimetic females. Where Tiger Swallowtail mimicry has been studied, the abundance of the mimetic females varies with the abundance of Pipevine Swallowtails. The more Pipevines there are, the more Tiger females mimic them.

The **SPICEBUSH SWALLOWTAIL** is named for the principal food of the caterpillar (which also eats Sassafras). Adults feed on nectar of Jewelweeds, honeysuckle, and Joe-pye-weed. These palatable butterflies, especially the females, closely resemble the Pipevine Swallowtail. Like the Tiger Swallowtails, Spicebush females most closely resemble Pipevines in areas where Pipevines are very abundant. As one would expect, the mimicry pattern clearly correlates with the abundance of the model.

The **RED-SPOTTED PURPLE** male and female both closely resemble the Pipevine Swallowtail, but, as with the others, the resemblance only occurs in the part of their range where Pipevines are common. North of the Pipevine's range, it hybridizes with the **WHITE ADMIRAL** and becomes banded. This species is closely related to that other famous mimic, the Viceroy, and eats many of the same plants (e.g., poplars).

Finally, the **DIANA FRITILLARY** female represents one of the most dramatic examples of the Pipevine mimicry complex. Fritillaries are normally brilliant combinations of orange and black spotting or banding and the male Diana of this species fits this pattern. Its outer wing linings are bright orange. The female, however, looks very like the Pipevine Swallowtail, and is completely different in appearance from the male.

EXPLANATION: The Pipevine Swallowtail mimicry complex illustrates how species from several butterfly families have converged to look, to varying degrees, like a model. All gain protection from avian predators by this similarity. When a palatable species evolves a close resemblance to an unpalatable species, the association is termed *Batesian mimicry,* after its discoverer, Henry Walter Bates. The Monarch (model) and the Viceroy (mimic) offer another example.

A second kind of mimicry, termed *Müllerian mimicry,* occurs when two *unpalatable* species share a close resemblance. Müllerian mimicry is very common in the tropics, where there are so many species of plants with noxious compounds. It is less common in North America, though one clear example exists. The Queen butterfly of the southeastern states is closely related to the Monarch, and feeds on milkweeds with cardiac glycosides. Like the Monarch, the Queen is unpalatable. The Queen is also a strikingly beautiful and obvious orange and black butterfly, and the Queen and the Monarch represent a case of Müllerian mimicry. Müllerian mimicry is beneficial to both models, since only half as many individuals of either species have to perish in order to establish the unpalatability.

You might wonder why usually only females mimic the Pipevine Swallowtail. A possible explanation is that in these species, it takes two doses of a certain gene to produce the mimicry, and the gene is located on the female sex chromosome, called the X chromosome. Males have only one X chromosome and one tiny Y chromosome. Males, therefore, could not acquire a double dose of the gene and become mimics.

Caterpillar Defenses

OBSERVATIONS: Caterpillars are generally large, slow moving, and juicy. In short, they ought to make nice meals for bird and insect predators alike, and many do. There are many caterpillars present in any summer field or woodlot, ranging from the obvious colonial tent caterpillars, which make their silky tents in cherry trees, to the small loopers, or "inchworms," which creep along, arching their wormlike bodies as they consume the edges of leaves.

Many caterpillars are cryptic. Green is the prevalent color for most species, but there are many exceptions. The Monarch caterpillar is boldly striped. Caterpillars of many fritillaries are patterned black and red. The Pipevine Swallowtail caterpillar is black with red spots.

One common characteristic is the presence of hairiness or spines along the caterpillar. In the huge family of brush-footed butterflies, which includes such notable species as the many fritillaries, checkerspots, tortoiseshells, commas, painted ladies, buckeyes, and admirals, caterpillars typically are lined with clumps of short, sharp spines. **TENT CATERPILLARS** and the familiar "weather predictor," the **WOOLLY BEAR,** are other examples of hairy caterpillars. Among the large silkworm moths, the **IO MOTH** caterpil-

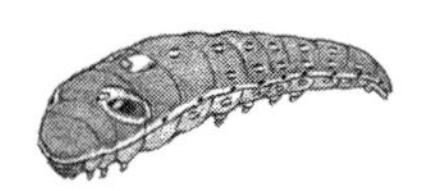

Fig. 47. Spicebush Swallowtail caterpillar.

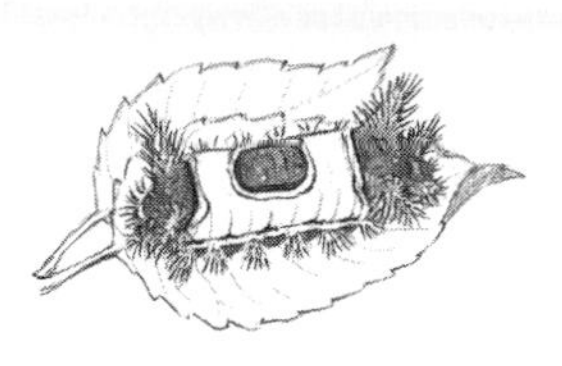

Fig. 48. Saddleback caterpillar.

lar is lined with thick clumps of bristles all along its upper surface. **GYPSY MOTH** caterpillars are quite bristly.

Some caterpillars appear to have huge "eyes," except that, upon close inspection, the eyes are merely markings on the skin. The caterpillars of the **TIGER SWALLOWTAIL** and the **SPICEBUSH SWALLOWTAIL** both show this pattern.

Finally, some caterpillars are toxic, at least to many would-be predators (see above).

EXPLANATION: Caterpillars are bird food. They are produced in huge numbers, but their mortality rates are staggering. Caterpillars form a very important component of avian food during breeding season. Still, many have evolved adaptations that provide at least some chance of avoiding falling prey to hungry thrushes and warblers. That these adaptations are effective is evidenced by the continued abundance of many butterfly and moth species.

Camouflage helps. Caterpillars dine on green leaves and it is therefore hardly surprising that so many species are green. However, they have other evolutionary tricks.

Many, like the Monarch, Queen, and Pipevine Swallowtail, are unpalatable, and these caterpillars are often boldly patterned, somewhat like the adult butterflies. Being indigestible affords considerable protection if the predators can be so educated. Large eyespots in Tiger and Spicebush Swallowtail caterpillars probably serve nearly the same function as in adults, namely to distract or confuse potential avian predators.

Finally, there are the bristly, spiny caterpillars, the porcupines of the insect world. These prickly insects are, to a large extent, protected from predation by their needle-like exteriors. Relatively few species of birds, the cuckoos being notable exceptions, dine readily on hairy caterpillars. A further adaptation is the fact that many spiny caterpillars are *urticating,* meaning that the spines are mildly poisonous, sufficiently so to irritate skin and mucous membranes. The **IO MOTH** caterpillar (Pl. 44) is among the most urticating of North American species, so be careful not to handle it.

The well-named **SADDLEBACK CATERPILLAR** (a moth larva) is boldly patterned and also urticating. Urticating hairs provide considerable protection for thick, slow-moving, otherwise vulnerable caterpillars.

LEAF ROLLERS, FOLDERS, TIERS, AND MINERS

A close look at a variety of trees, shrubs, and herbs will reveal that some have leaves with edges rolled together, edges oddly curled, one edge folded over the other, or even several leaves seemingly tied together. Some leaves have tunnels, where the chlorophyll has been removed. All of these deformed leaves are the work of various insects, the rollers, folders, tiers, and miners.

Thousands of species of insects are known to roll, fold, tie, or mine. It is impossible to list species here, but it is possible to understand why insects do what they do to leaves, and which major insect groups are the most common leaf-deformers.

Leaf Rollers and Folders

OBSERVATIONS: Look closely at any leaves you find with curled or folded edges. The folding is usually supported by silk, and there is usually an insect, most often a larva, inside the fold. The leaf may be rolled lengthwise (parallel to the midrib) or crosswise.

Rolled or folded leaves are found on basswoods, dogwoods, willows, cherries, oaks, birches, alders, poplars, maples (especially Box-elder), grapes, apples, Witch-hazel, sumacs, Poison-ivy, and many other species, including garden species such as roses and beans.

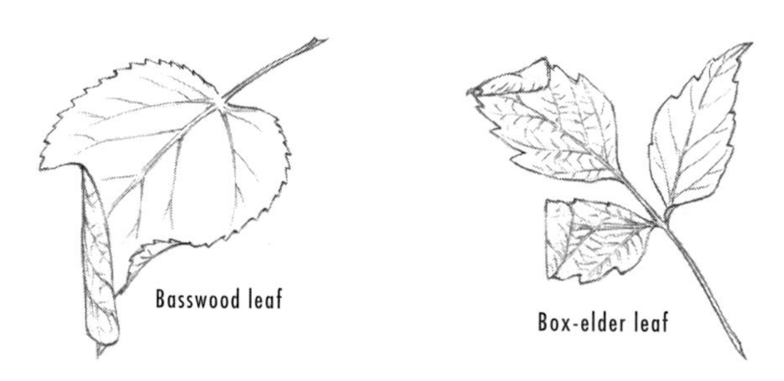

Fig. 49. Leaf rollers.

EXPLANATION: Leaf rolling or folding is usually done by larval insects spinning silk to curl the leaf over, enclosing the larva. Most rollers are larvae of butterflies and moths. However, other insect groups, including beetles, sawflies, and grasshoppers, also contain leaf-rolling and -folding species.

The leaf is rolled as the insect secretes sticky silk that attaches firmly to the leaf edge. As the insect weaves its silk back and forth across the leaf, between edges, the silk shortens and dries, rolling the edges of the leaf together. The final product is a tube-like leaf, which serves various functions, depending upon the type of insect that engineered it (see below).

A leaf is folded by bending the edge and sewing it with secreted silk. Sometimes much silk is involved in the process but the leaf is not tube-shaped, as with a rolled leaf. Rather, one leaf edge is folded over and tied firmly, making a permanent curl in which the larval insect is housed.

For butterfly or moth larvae, rolled or folded leaves are protective, housing the larva in some cases while they feed, in some cases through the pupation period, and in some cases, both. The beautiful **PROMETHEA MOTH** caterpillar rolls a leaf as a pupation site. Some species of tiny moths in the family Gelechiidae roll leaves of oaks, birches, and alder. Many other gelechiid species fold leaves. Other moth families that include leaf rollers and folders are Tortricidae (of which the famous **SPRUCE BUDWORM** is a member), Pyralidae, and Oecophoridae. In this latter family of small brown moths, the caterpillars tie flowerheads, feed inside them, and then burrow into the stem to complete their development. One pyralid that is quite common is the **GRAPE-LEAF FOLDER.** Look for grape leaves with folded edges and inside these leaves will likely be the pupa of this moth. Another common species is the **APPLE-LEAF SEWER,** found in orchards.

The **BASSWOOD LEAF ROLLER** is very common. It actually makes two rolls during its life cycle. First, it cuts a section of the large basswood leaf and rolls it into a tight feeding tube, where it feeds on the rolled leaf. When ready to pupate, the green caterpillar (with a black head and thorax) makes another roll and becomes a pupa. Look for rolled basswood leaves beginning in midsummer.

The **BOX-ELDER ROLLER** mines the leaves of Box-elder before rolling them. You can recognize the work of this insect by noting pale blotches on the leaflets (Box-elder, also known as Ashleaf Maple, is a maple with compound leaves). Usually, where these blotches occur, the edges of the leaflets will be rolled, protecting the feeding caterpillars that began life by mining inside the leaflets.

Other rollers include beetles and web-spinning sawflies, plus one grasshopper species. One group of weevils rolls a small section of a leaf into a thimble-shaped tube where an egg is laid. The newly hatched larva is protected as it feeds both in and on the rolled portion of the leaf.

It has often been said that nature abhors a vacuum. Abandoned leaf rolls and folds are used extensively by various scavenger beetles and other insects as well as mites and spiders.

Leaf Tiers

Many lepidopterans (butterflies and moths) habitually tie needles and leaves together. They live inside, using the shelter for feeding, egg-laying, or pupation. The common **EVERGREEN BAGWORM MOTH** weaves a bag of cedar needles in which the female lays her eggs. The **PINE-TUBE MOTH,** as the name implies, ties pine needles together as protection. The caterpillar pokes its head out of its pine-needle enclosure to feed. As it grows, the caterpillar may make several tubes, pupating in the last one. Locust trees often host the **SILVER-SPOTTED SKIPPER,** a lepidopteran that ties leaflets of locusts and other legumes together as protection for the larvae.

Cherry leaves are frequently tied by the **SCALLOP-SHELL MOTH.** Several leaves are tied together by larvae to make an elongate tube. The leaves are killed in this process and turn brown, making these tubes easy to spot. The leaf tube usually drops to the ground in fall, but the larvae pupate within the fallen tubes until emergence the following spring.

The **EUROPEAN CORN BORER** sometimes ties milkweed leaves into a tentlike structure. This pest species is best known for its attacks on corn stems.

The most obvious of the leaf tiers are the **FALL WEBWORM MOTH,** a member of the tiger moth family, and the **EASTERN TENT CATERPILLAR,** which belongs to a different family of moths. The Webworm ties leaves using much silk, forming the web which gives the insect its name. Though the web is very conspicuous, the leaves are hardly rolled or folded. The webworm feeds on the leaves within the webs, and webs often cover whole branches. Tent caterpillars live colonially inside the large tents, emerging to feed. These insects favor cherries and other related species.

Leaf tying, like leaf rolling and folding, is adaptive as protection for feeding and pupating larvae. Both larvae and pupae are highly vulnerable to predators. The silk which attaches the pupae to a branch or leaf is used to mold leaves together, making a shelter for the larva and/or pupa.

A leaf may seem to be a flat, almost two-dimensional object to you, but to a tiny insect, it is very much a three-dimensional entity. Many butterflies and moths, as well as species from a few other insect groups, such as the beetles, flies, and sawflies, have larvae that burrow inside leaves and feed on the chlorophyll-rich cells, mining tunnels through the leaf as they eat their ways through.

Recognize mined leaves by pale, blotchy, irregularly shaped tunnels. Sometimes the tunnels go across the leaf, but usually they are either parallel to or radiate out from the midrib. Mines have been described as blotch, linear, linear blotch, trumpet-shaped, digitate (fingerlike), or tentiform. This latter form overlaps with the technique used by leaf tiers, and is represented by such species as the Box-elder Roller (see above).

If you very carefully cut open a mined leaf, using a single-edge razor blade, you should be able to locate tiny insect larvae. You should also see the remains of digested leaf tissue, called frass. Larvae are tiny and wormlike, with legs that are reduced (very small) or even absent. The head is large, with eyes and antennae reduced or absent, but with strong jaws and teeth for chewing. Some larvae have wedge-shaped heads, adaptive as they mine their way through the leaf.

Leaf-mining larvae are most common among the butterflies and moths, where over 400 North American leaf-mining species occur. About 200 fly species, 50 beetle species, and 15 bee and wasp species, mostly sawflies, mine leaves. Only larvae mine, hatching from eggs and burrowing into the leaf. Larval habits vary according to species, with some larvae spending all their time inside the leaf and others spending part of their time inside, part outside. Pupation may occur either inside or outside the leaf, again depending on the species.

Leaf miners feed on rich organic compounds manufactured by the leaf and consume the food-producing cells as they mine. Severe mining damages leaves, leaving the pale brownish blotch indicative of the presence of the leaf miner.

Miners eat so much that frass disposal can pose problems. Some miners leave the frass inside the mine, basically ignoring it. Others habitually change mines, and still others push the frass from the mine.

Miners are most conspicuous from mid- to late summer. They infect most species of plants and you should have little difficulty finding evidence of their presence. For instance, oaks host many species.

Galls are tumorous growths on plants, usually on stems, but also on leaves. They vary in shape: many are ball-like, others elliptical, others spiny or hairy, and still others funnel-shaped. Some plants have only one or two galls, some have many. They are caused by either insects, mites, or roundworms, which somehow induce the plant to form tumorlike tissue. You will find galls on many species, but common examples include goldenrods (which harbor several types of galls), oaks (which harbor many gall types), roses, legumes, willows, composites, maples, birches, beeches, hickories, spruces, pines, elms, and Witch-hazel. In all, over half of all plant families are routinely parasitized by gall-causing animals.

Galls are easy to find, but it is not so easy to see the gall-causing animals. To do so, carefully open the gall with a single-edge razor blade. Many gall insects are very tiny and dwell in the gall only for a relatively short time. Many fall prey to gall parasites. Empty galls are common. Sometimes more than one kind of insect can be found in a gall. Some galls are used by insects that have nothing to do with the gall formation, but only take advantage of the gall for protection. These insects, which use galls but do not induce them, are termed *inquilines*. Galls are caused by beetles, butterflies and moths, aphids, thrips, flies, and wasps, as well as one mite family, and nematode worms (roundworms—see Chapter 8). One comprehensive list of gall-makers totalled 1440 species.

Goldenrod Galls

A very common and easily observed gall occurs on the leaves of goldenrods. Odd-shaped and dense clusters of small leaves form a ball-like mass near the top of a goldenrod plant, just below the flower stalks. A single plant may have several galls. Each gall is caused by a tiny midge that is found only on goldenrods. The larval midge completes its development inside the gall.

Look for galls on the stem as well as leaves of goldenrod. Some stem galls are elliptical; others are rounded and ball-like. These are caused by two different types of insects. The elliptical gall is induced by the caterpillar of a moth in the family Gelechiidae, the same family of small moths that contains many leaf rollers and folders. Eggs are laid inside goldenrod stems during summer and larvae hatch the following spring. Each caterpillar burrows inside a goldenrod stem and feeds, an event which inspires the plant to make the gall. Pupation occurs inside the gall, and the moth hatches in late summer. Before pupating, the caterpillar bores a small exit hole in the gall, and closes it with easily remov-

Fig. 50. Goldenrod galls.

able silk fibers. The adult moth, which cannot bore, can then easily escape.

More rounded, ball-like galls also occur on goldenrod stems. These are the result of a fly that lays its eggs on goldenrods; the gall serves as an overwintering home for the maggot (larva) when it hatches. The fly larva normally feeds by tunneling through the goldenrod stem, but it completes its development in the gall.

Goldenrod-gall inhabitants are subject to predation and parasitism. The **ICHNEUMON WASP,** using a long, needle-like ovipositor, "injects" its egg through the elliptical goldenrod gall and lays it next to the pupating moth caterpillar. The larval wasp eats the moth pupa and then makes a cocoon, emerging in late summer. Beetle larvae, along with other insects, commonly prey on the fly larvae inside goldenrod ball galls. Downy Woodpeckers are often seen hammering on galls.

Witch-hazel Galls

Witch-hazel galls are small but distinctive cone-shaped galls, several of which can be seen on a single leaf. They are caused by a tiny aphid. The cones are hollow inside and each contains several young aphids. There is an opening for each gall on the underside of the leaf. Many galls, called "open galls," have such an opening; "closed galls," also very common, lack openings.

The life cycle of this aphid species is complex. Females lay eggs inside the galls and young aphids hatch and leave the galls and feed on birches, going through several generations before returning to Witch-hazel. Each generation has a different morphology (body structure). The insects show progressive reductions in legs, antennae, and mouthparts and their bodies become oval, coated with a white, shiny, waxlike substance. After completing six generations on birch, winged aphids hatch and return to Witch-hazel,

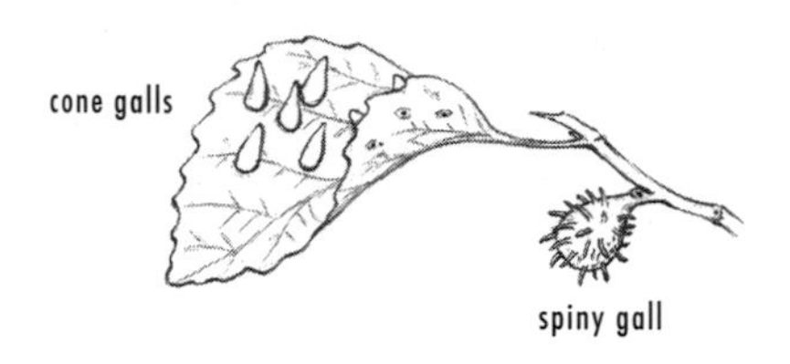

Fig. 51. Witch-hazel galls.

reproducing a seventh generation, which then lay eggs inside the cone-shaped Witch-hazel galls.

Yet another aphid also makes Witch-hazel galls and it also alternates between birches and Witch-hazel. The name Spiny Witch-hazel Gall describes the gall well, because it is oval with an abundance of blunt spines protruding from it. Unlike the cone-shaped galls, it is attached to the stem, not the leaf. Young aphids leave the gall through an opening (this is also an open gall) and migrate to birch. There they complete six generations, returning finally to Witch-hazel. However, while on birch, the larval aphids make "pseudo-galls," which are really folded upper leaf surfaces. Larvae complete their metamorphosis inside the folded leaves. Look for them on various birch species.

Oak Galls

Oaks support a remarkable gall-making fauna, with approximately 800 species. This high diversity is even more remarkable since virtually all oak gall-makers are in but one family of wasps, the Cynipidae. Many cynipid wasps make their own galls but some use the galls of other gall-makers. Cynipids are very common and you should have little difficulty finding oak galls of various kinds. Galls can be found on leaves, twigs, branches, roots, and even flowers and acorns. The adult wasps are small, shiny black insects with oval abdomens.

Oak leaves often have "Button Galls" caused by a cynipid wasp. These tiny galls, which do indeed look like buttons, can number over 100 on a single leaf. Another very common oak gall is the so-called "Oak-apple," a large (1–2 in.), rounded gall, light brown in color, attached to a leaf, branch, or twig. Apple galls formed in spring and early summer are softer in texture than those of late summer and fall. Many cynipid wasps cause these galls, and the galls themselves are large enough to house many different inquilines (gall-using animals). Apple-gall wasp larvae are frequently preyed upon by other insects invading the gall. The "Hedgehog Gall" is compound, containing several larval wasps each housed in a separate chamber. The Hedgehog Gall is, like its namesake, covered with prickly spines, and is located on the midrib of a White Oak leaf. Each gall is thick and hard and contains many tiny larval wasps that hatch in spring. The newly hatched wasps lay their eggs, not on leaves (which have not yet even opened) but on buds. The eggs cause tiny, red, thin-walled galls to form on the buds. Each gall contains a single egg, which hatches into a tiny larva. This wasp can harm the oak by interfering with bud opening. Another type of gall, the Bullet Gall, is a grape-like cluster of brownish galls found on oak twigs. Many other oak galls, some quite irregular in shape, can be found on twigs.

XPLANATION: Galls protect developing larvae, though many do fall prey to gall intruders. Gall formation is caused by the insect (or mite or roundworm) inducing the plant to form a tumorous tissue. The larval insect, not the egg, seems to cause gall formation. As it feeds, the insect secretes an enzyme that breaks plant starch into simple sugar. This seems to be a key element in stimulating gall formation. Plant cells begin to multiply, forming the gall, which may be hard or soft. Gall formation is most active during spring, when plant tissue is still growing.

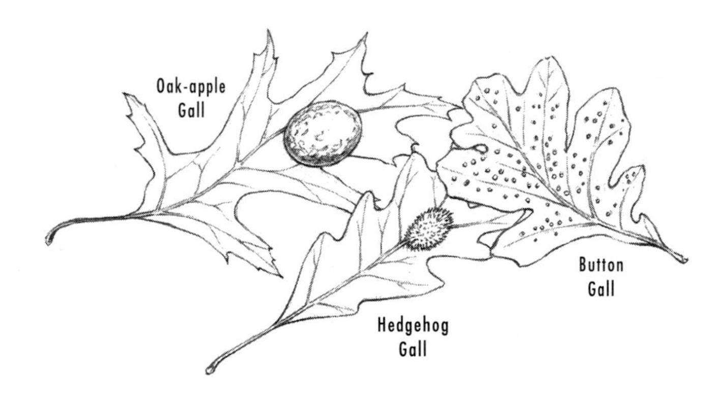

Fig. 52. Oak galls.

The most abundant gall-causing insects are the flies. Nearly all of the approximately 700 gall-causing flies belong to one family, popularly called gall gnats, of which there are 682 species. The Hessian Fly derives its scientific name *(Mayetiola destructor)* from the depredations it causes to wheat fields. Other gall gnats attack legumes, especially clovers. Another family of flies includes goldenrod gall-makers.

The bees and wasps, especially wasps, are also prominent gall-makers. Most are in the cynipid wasp family and concentrate heavily on oaks.

There are a few gall-causing species of beetles, and many beetle gall-makers also are borers. Depending upon the preference of the species, beetle galls appear on maples, poplars, willows, grapes, Virginia Creeper, pines, and birches.

Among the butterflies and moths, gall-causers are mostly in the genus *Gnorimoschema* (a common group of moths), generally found on goldenrod but with one species on asters. Homopteran gall-causers, mostly aphids, make galls on many species.

The above survey shows that though there are many gall-causing insects, they tend to be confined to single families or even genera (smaller groups). The gall-causing characteristic has not evolved repeatedly but has evolved independently in several widely different groups of insects. Once gall-causing evolved, further speciation within each of the gall-causing groups led to the present diversity, such as is seen in the cynipid wasps.

Chemical irritation produced by boring larvae was probably the initial stimulation leading to gall evolution. Plants developed "tumors" in response to the irritation caused by insects, and the insects adapted to use the galls to protect developing larvae. Insects with gall-causing characteristics were more fit (in an evolutionary sense), and their genes were passed on to future generations.

DRAGONFLY MATING PATTERNS

OBSERVATIONS: Spend any amount of time around a pond or quiet stream in summer and you will see dragonflies. They zoom along, alighting briefly, then continue to sweep about the pond. Dragonflies are easily differentiated from damselflies by their swifter, more determined flight, compared with the fluttery, butterfly-like flight of damselflies. Both dragonflies and damselflies have 2 pairs of transparent wings, but dragonflies hold their wings stiffly horizontal, while damselflies fold their wings in an upward angle above their abdomen.

There are approximately 450 species of dragonfly and damselfly in North America, but you can learn much about their natural history without identifying any of them to the species level. Just learn the major families and then watch them go about the business of mating, which they do in the open, around ponds, during daylight hours. Dragonflies are territorial and males are aggressive, both with each other and with females.

There are several major groups of dragonflies.

DARNERS are large, metallically colored dragonflies, many with shiny blue and green bodies. They are very swift fliers. The **GREEN DARNER** is a common and widespread species. Female Green Darners lay single eggs in hollowed-out stems just beneath the surface of the water. Females are relatively secretive and, though the males are obvious, you probably won't see females, at least not while they are depositing eggs.

GOMPHIDS are called **CLUB-TAILED DRAGONFLIES,** and, as that name implies, they have swollen tips at the end of their abdomens. Gomphids tend to hover a lot. You won't find too many around ponds —they favor streams.

SKIMMERS consist of many easy-to-observe species. They abound around ponds. The wings are usually spotted or barred, and species can be identified by wing pattern. Some, the **CLEARWINGS,** have clear wings. Body colors differ according to species, and, in some species, the sexes are colored differently. Red, blue, black, and green are the prevalent body colors. Common skimmers include the **GREEN CLEARWING,** the **WIDOW,** the **RED SKIMMER,** and the **BLUEBELL.** Skimmers are highly territorial. Species like the **BLUE PIRATE** and the **WHITE-TAILED SKIMMER,** both of which are common and widely

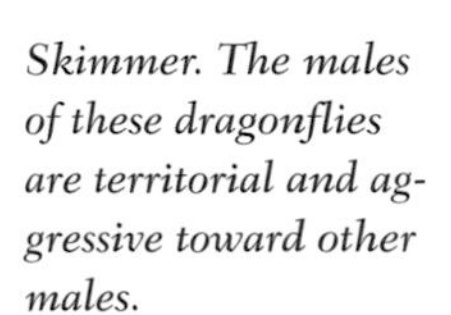

Skimmer. The males of these dragonflies are territorial and aggressive toward other males.

distributed, are among the most aggressive, but other species are strongly territorial as well. The behavior descriptions that follow deal with skimmers.

Dragonfly eggs are laid in water. Dragonfly nymphs spend the first part of their lives as underwater predators. Following metamorphosis to an adult, the animal becomes an airborne predator.

Territoriality

Watch what happens when two skimmers meet. They often engage each other in brief but intensive aerial combat. Many males may converge at a single pond and clash in mid-air. The actual territory of a given male is over a part of a large pond or all of a small pond. Territories are defended for mating and mate-guarding only, and they can change size from one day to the next, depending upon how many males are competing. Male territorial behavior is variable and is strongly affected by the presence of other males. A solitary male or just a few males on a large pond are less territorial and do not act as aggressively. On the other hand, large numbers of males inspire a sort of insect war. Try to ascertain the number of territories and degree of aggressiveness of male skimmers at a pond in your area.

Mating

When a female visits the pond, the males become very active in pursuit. Usually several males will pursue a single female, which may be unreceptive to all of them, merely flying away from the pond. When the female is receptive, mating occurs, usually on a

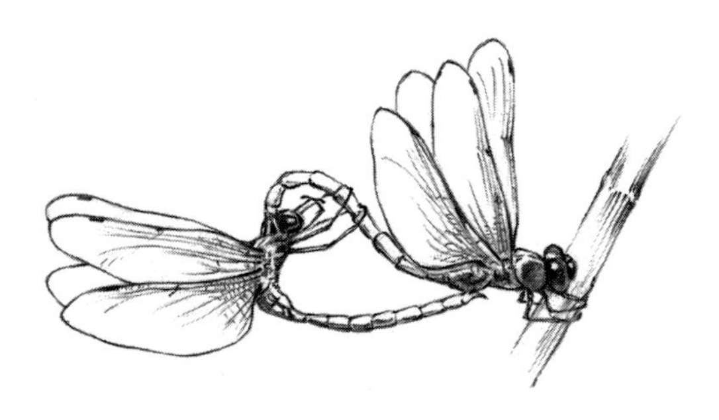

Fig. 53. Wheel position in mating dragonflies.

Fig. 54. Damselfly depositing eggs.

reed stalk or some other vegetation along the pond edge. The male holds the female behind her head and the two bend their abdomens to form a "wheel," a position that enables the receptive female to take sperm from the front of the male's abdomen. Dragonfly copulation is very quick, requiring only a few seconds.

Female dragonflies can store sperm for a considerable period of time and males compete vigorously for access to females. Any female dragonfly can mate with several males, but usually, the last male to mate before the female lays eggs will contribute the most sperm to actual fertilization. It is very much in the male's best genetic interest not only to mate, but to prevent other males from doing so after him. Males therefore practice *mate-guarding*, staying as close as possible to a female with which they have just mated.

Mate-guarding and Egg Deposition

OBSERVATIONS: You may not see an actual mating, but you should have little difficulty observing mate-guarding. Following copulation, the female needs to locate a suitable spot along the water's edge in which to lay her eggs. Females shed eggs into the water, where they sink and hatch into larvae. However, other males are still quite interested in mating with this recently mated female. Remember, the one who gets to her last is usually the one whose sperm does most of the fertilizing.

Look for a pair of dragonflies, one hovering above the other as the lower one flies about the pond edge. The male is hovering above the female, guarding her from other males, and thus protecting his genetic interest in the future of dragonflydom. The female is searching for a site to lay her eggs. Other males may approach, but the guarding male will swoop at them and try to drive them off. Even as the female lands to lay her eggs, other males may zoom in. Sometimes an interloper will succeed, and another mating will occur. Sometimes other males so harass the female that she flies off before laying eggs, almost undoubtedly lowering the reproducing success of the male that she just mated with, since she will in all probability now mate with another male at another pond.

Two dragonflies may literally become attached to each other. Some males guard their mates by holding the female after mating, preventing any other male from access to her. This behavior is called *contact-guarding* and, though it prevents the male from mating with another female, it does help ensure that the female with which he just mated will oviposit (lay eggs) without harassment by other males.

The best way to observe a female in the act of ovipositing is to follow as she flies along the pond edge or else locate a pair where the male is contact-guarding the female. Watch until they land and she oviposits. Egg-laying requires only about 4–5 minutes in skimmers.

EXPLANATION: It is easy to understand why it is ultimately adaptive for a male dragonfly to be territorial. Ponds and lakes offer limited areas for egg deposition and females must deposit eggs under water. To gain access to females, the male must inseminate her just as she is about to oviposit (lay eggs). The male has little choice but to defend an area around a pond or lake and intercept whichever female comes along.

It is also clear why mate-guarding is adaptive. To abandon a female after copulation would make a male less "fit," since other males might copulate with her and their sperm would outcompete his. What is less clear is why all males don't practice contact-guarding. Being literally attached to the mated female is a strong guarantee against another male adding his sperm to her collection. One possible advantage of noncontact-guarding focuses on the very brief copulation time required for dragonfly mating. A male who has mated may temporarily "abandon" the female and quickly mate with another female, then return to guard the first or remain with the second. Males are sometimes seen attempting to fly above and prevent access to two females simultaneously.

Why is mate-guarding of ultimate evolutionary advantage to females? Perhaps because it enhances the probability of successful, uninterrupted egg-laying. A female that must frequently return to a pond wastes energy and is exposed to predation by birds. Getting egg-laying accomplished as efficiently as possible holds substantial advantage for females as well as males.

DEFENSE ADAPTATIONS OF SNAKES AND OTHER REPTILES (PL. 46)

Reptiles are most easily seen in summer, as they require warm weather to be active. All reptiles have dry scaly skin, and those with legs have toes with claws.

Summer is an ideal time to observe snakes, which, along with bats and spiders, serve to sort out the faint-hearted from the strong among natural history addicts. All snakes are predators—there is not a herbivore among them. All can bite, some quite painfully, some poisonously. Without question you should exhibit care when seeking out and handling snakes. Though snakes are predators themselves, they nonetheless have many enemies and they display an impressive range of defensive adaptations, including both color-patterning and behavior, and it is well worth your effort to seek out and observe them.

Snakes are active during the summer and will occasionally be encountered along forest trails or darting through fields. They also coil up and rest beneath rocks, logs, and boards. You should take care to stand ***behind*** any rock or log you lift, especially if poisonous snakes are found in your area. In fact, it is more instructive to merely observe snakes rather than handle them. For obvious reasons, ***never attempt to pick up a snake that may be poisonous.*** Should you decide, however, to pick up a snake, grasp it firmly ***behind the head***; otherwise you will in all likelihood be bitten. For more information on handling snakes, as well as identification of the many species found in the East, refer to *A Field Guide to Reptiles and Amphibians.*

Terrestrial Turtles

Snakes are not the only reptiles you will find in summer sojourns to fields and forests. The **EASTERN BOX TURTLE** is found throughout most of our area (it is absent largely from the Northern Hardwood Forest and all of the Boreal Forest). Box Turtles are terrestrial, and are well adapted for defense against predators. The dome-shaped carapace, or upper shell, is thick and the plastron, or lower shell, has a hinge both in the front and rear. The turtle can therefore close itself up entirely, efficiently protecting its vulnerable head, feet, and tail. It then patiently waits out the danger.

Should it be turned upside down, it has little difficulty righting itself.

In the northeastern states and southern Canada, look for the **WOOD TURTLE.** This turtle is considerably less terrestrial than the Box Turtle and lacks the domed carapace and hinged plastron. Named for its tendency to wander about in mesic (rich, moist) woodlands, it is nonetheless rarely far from water. The scientific name of this turtle *(insculpta)* refers to the ornate ridges radiating from the carapace scales.

Lizards

Primarily animals of the deserts and arid lands, there are few lizard species in eastern North America. Most eastern species are confined to southern states. These include the sleek, glossy skinks; the snake-like, legless lizards; and the vocal, wall-clinging geckos. This latter group is mostly confined to southern Florida. Two widespread species are the **GREEN ANOLE** and the **EASTERN FENCE LIZARD.** The former is common in moist forests and swamps throughout the southern states and the latter is found in dry pine-oak woodland, reaching the mid-Atlantic states and southern New England.

Lizards rely on camouflage and speed to escape predators. Both the anole and the fence lizard are able climbers, literally capable of running up vertical tree trunks and walls. The anole blends well with its background, as it can change pigmentation from green to brown. The fence lizard is very agile and quick and can scurry out of harm's way. Should either an anole or fence lizard be grabbed, however, it can lose a section of tail and regenerate the lost piece. The severed section of tail keeps thrashing, occupying the predator's attention while the lizard makes its getaway. This characteristic, called *autotomization,* is adaptive because the loss of part of a tail permits the rest of the lizard to escape unscathed. If you see a lizard with a tail stub, you know it had a close call.

Camouflage Patterning and Defensive Behaviors in Snakes

Snakes represent excellent examples of cryptically colored animals. Many species blend quite closely with the background pattern of their habitats. Such camouflage both increases their effectiveness as predators, and helps conceal them from their own potential enemies. Included below are a few of the most common snakes that show cryptic coloration. There are many more, not included here. It will not be easy to find all of these snakes, though they are all relatively common. Their populations are not very dense in most areas—predatory animals in general are far less common than plant-eaters. Secondly, their effective camouflage obviously adds substantially to the difficulty of finding them!

Snakes have evolved a diversity of behaviors that they use when threatened. Though you may feel intimidated when in the company of a snake, bear in mind that (1) you are bigger than the snake and (2) you can probably outrun it! Snakes are routinely attacked, killed, and eaten by mammalian predators, as well as by hawks, owls, and often other snakes. Defensive behaviors are essential for snake survival and each snake has its own repertoire. Don't be afraid to *mildly* intimidate ***non-poisonous*** snakes to observe various defensive behaviors. ***Do not, under any circumstances, attempt to intimidate a poisonous species.*** Its "court of last resort" can be fatal to humans. Remember, ***if you have any doubt about the identity of a snake, leave it alone, and do not try to touch it.***

Non-poisonous Snakes

GREEN SNAKES (1–2 ft.) are diurnal snakes inhabiting branches of shrubs and trees, especially along woodland edges and streamsides. They feed on large arthropods such as caterpillars, grasshoppers, and spiders. Two species occur, the **SMOOTH** and **ROUGH GREEN SNAKES.** These species are well named—their greenish yellow bodies camouflage them very effectively. Both species, the Smooth and the Rough (named for "smooth" scaling on the former and the ridged or "keeled" scales on the latter), are similar in appearance, size, and food preferences. They differ somewhat both in range and habits. The Smooth is more northern but also occurs throughout much of the Appalachians, where the Rough is absent. The Smooth is less arboreal (tree dwelling) than the Rough. Both species are gentle and usually tolerate handling without biting. Their camouflage is their defense, followed by impressive speed. They can dart quickly away from would-be attackers.

The **COMMON GARTER SNAKE** (1–4 ft.) is probably the most frequently encountered eastern serpent. It is most easily identified by its prominent yellow side stripes. A very similar species, also cryptically colored, is the **EASTERN RIBBON SNAKE** (1–3 ft.). Both snakes occur in old fields, wet meadows, and moist woodlands. The yellow-brown striping pattern effectively camouflages the snakes against a background of grass and herbs or dried leaves. The Ribbon Snake is more aquatic than the Garter Snake, preferring streamsides, ponds, and other wetland habitats. Garter snakes, like green snakes, are very swift and dart purposefully away from perceived danger. When trapped, however, they flatten their bodies, the effect of which is to make them appear larger, emphasizing their color pattern. If you pick one up, it will usually emit a rancid-smelling musk from a gland at the tail base.

The **BROWN SNAKE,** a small (1 ft.), yellowish brown recluse found beneath logs, boards, and rocks in moist woodlands and meadows, uses virtually the same tactics when threatened, including the rancid musk. The tactic of secreting foul-smelling (and tasting) fluid is likely to be much more effective against mammalian predators rather than owls or hawks. Birds have far poorer senses of smell and taste than mammals.

The **EASTERN BLACK RACER** is named for its best defensive weapon, its speed. This 5-ft. snake with a thin, whiplike tail is mostly active in daytime and you will probably encounter it as it glides with impressive speed through grasses in old fields. Adults are all-black except for some white on the chin, though juveniles are mottled with browns. Speed is both a defensive and offensive weapon for racers. These snakes pursue their small rodent prey very effectively. In spite of the scientific name *(constrictor),* racers do not constrict their prey, as rat snakes do (see below). They grab their prey with a quick strike.

The **NORTHERN WATER SNAKE** (2–4 ft.) is a dark brown, blotchy inhabitant of streams, river banks, and wetlands. Though non-poisonous, it is one of the most aggressive snakes in defending itself. It will strike repeatedly and bite hard. The wound is often slow to stop bleeding, as this snake can inject a mild anti-coagulant.

The **BULLSNAKE** is part of a large complex of various races that includes the pine snakes and gopher snakes. These snakes feed almost exclusively on rodents, killing them by constriction. Constrictors do not crush their prey. Their steady constriction prevents the rat or mouse from expanding its chest to breathe, soon suffocating the animal. Bullsnakes vary in appearance, with several common races ranging from speckled blackish to tan with brown blotches. Each is well camouflaged. These snakes are relatively large (4–8 ft. long) and their defensive behavior is often very aggressive. A threatened Bullsnake will hiss forcefully, vibrate its tail with great vigor (note the similarity to a rattlesnake), flatten its head (again, note the similarity to a rattlesnake), and, when all else fails, strike in its own defense.

Similar in size and behavior to the Bullsnake are the various **RAT SNAKES,** which include the **COMMON RAT SNAKE,** the **CORN SNAKE,** and the **FOX SNAKE.** The Corn Snake is the most intricately patterned of the rat snakes, with colorful reddish blotches on a yellowish tan body. Though strikingly colored when in the hand, the Corn Snake can be hard to see in dried leaves beneath forest shade. Rat and Fox snakes are cryptically colored and coloration varies among the several races, ranging from black to yellowish to gray. Fox Snakes are pale brownish with black blotches. Rat snakes vigorously vibrate their tails when in danger. The tail vibrations make a very audible sound when the snake is in dry leaves. If tail vibrations fail to impress its tormentor, a rat snake will often erect itself, in cobra fashion, and hiss. If this fails, it will strike. Like Bullsnakes, rat snakes are constrictors.

The virtuoso actor among the snakes is the 3-ft. **EASTERN HOGNOSE,** which occurs in both light and dark (melanistic) forms. The light form is more widespread and is the most cryptically colored. It is yellowish brown with varying-sized blotches of dark brown and tan on its upper surface. Like the Timber Rattlesnake and Copperhead (see next section, below), which it somewhat resembles, it is well adapted to blend in among the dried leaves of the forest floor, though it also frequents old fields and meadows. When threatened, the Hognose usually begins by puffing its head to resemble a cobra's hood. The snake can seem very intimidating, but, should you not be fooled, its next strategy is to go into convulsions, finally turning belly up with its mouth open, feigning death. Interestingly, opossums go through a very similar series of defensive behaviors, culminating also in feigning death. Hognoses thus "play possum," while possums "play hognose."

Poisonous Snakes

Both the **TIMBER RATTLESNAKE** (3–6 ft.) and the **COPPERHEAD** (2–4 ft.) are members of the highly poisonous pit viper family and both are extremely well-camouflaged dwellers of the shady forest floor. They also rest between rocks of stone walls and among the boards in debris piles. These potentially dangerous snakes coil characteristically when at rest and usually warn intruders by rapid tail vibrations. In rattlesnakes, the tail vibrations are greatly enhanced by the rattle, a unique structure of highly modified scales. They are remarkable for the rapidity with which they strike, both when capturing prey and when defending themselves. Another common southern species, the **COTTONMOUTH,** or **WATER MOCCASIN** (2–6 ft.), is aquatic, dwelling in cypress swamps and other wetlands.

Do not approach any pit viper.

Pit vipers are named for their unusual sense organ, the pit, which is located below and a bit to the rear of their nostrils. The pits are sensitive to heat, and aid the snake in detecting warm-blooded prey such as mammals and birds. The toxin is delivered by injection through hypodermic fangs, which normally are ensheathed and folded against the upper jaw, but which swing out when the snake is biting. The venom injures both blood cells and nerves, resulting in hemorrhage and suffocation.

EXPLANATION: The above examples illustrate the overall pattern of *crypsis* or camouflage in snakes. Just as in predatory insects (see p. 350), crypsis functions in two ways: First, it permits the animal to be inconspicuous. Snakes have a lot of meat on their serpentine bodies and make very satisfactory meals for many mammalian and avian predators. Being incognito is a definite advantage. Second, crypsis permits the snake closer access to its prey. When prey come too close to a waiting snake, they become snake food. If snakes were easy for prey to see, dinner would be much more difficult to come by.

All cryptically colored animals, whether snakes or insects, must be able somehow to recognize the "correct" background. A green leafhopper on dark bark is not cryptically colored and would not last long. Green leafhoppers must instinctively know to land on green leaves, where their crypsis is an advantage. Likewise, snakes must recognize appropriate background. Hunting at night or at dawn or dusk permits the snake to venture into less cryptic backgrounds with darkness providing added protection. Finally, snakes are rather reclusive, coiling beneath rocks and boards. Not only are they cryptically colored, but their behavior also helps them avoid discovery.

More than one function is often accomplished by a given adaptation. When a snake retreats to spend the day under a rock, it not only is inconspicuous, but it also stays cool. Snakes, like all modern reptiles, have no way to control their body temperatures by regulating their metabolism and must therefore seek out cool areas in which to pass the warmest hours; otherwise, they would overheat.

Defensive behavior in snakes can be roughly divided into two patterns, as the examples above show. Small snakes attempt to slither away rapidly, opting to "run" from the danger. If they cannot escape, they have a series of back-up defenses, including body-flattening, musk emission, thrashing, and biting. Large snakes seem to follow the adage that the best defense is a strong offense. These serpents vibrate their tails, raise their heads, and "hiss with purpose." The message they send is clear—"bother me and I'll prove dangerous." They *bite hard* when handled carelessly. The Eastern Hognose Snake, intermediate in size, is also intermediate in defense strategy. It huffs and puffs and spreads its "cobra hood," only to roll over and feign death if bluffing fails.

One defense pattern evident in virtually all snakes is to make themselves appear larger. Body and head flattening and head raising make these animals appear larger and increase their similarity to poisonous species (a possible behavioral mimicry—see next section).

A final speculation concerns the origin of the rattlesnake's rattle. The widespread use of tail-vibrating by many snake species suggests that it is adaptive to warn approaching animals that the snake is prepared to defend itself. Rattlesnakes have tails with scales modified to emit a loud rattle when the tail is vibrated. Rattlers need not rely on dried leaves to make their vibrations audible, as other species do. Rattlesnakes apparently represent the only snakes where the behavior of tail-vibrating has resulted in the evolution of an anatomical specialization. Such an event is the normal direction evolution takes, with *function preceding form*. Once a function is established as advantageous, such as tail-vibrating in the case of the rattlesnakes, form sometimes follows, as with the evolution of the rattle. The presence of so many other species that tail-vibrate shows that the rattle probably came second, and the tail vibration first—behavior and function preceded form. Why don't other snake species evolve rattles? These species may not yet have the genetic mutations that occurred at least once in rattlesnakes to permit the formation of rattles. Before a trait can evolve, there must be genes to make it! Mutations occur by chance alone, and may simply never have occurred in pine snakes or rat snakes. Secondly, though rattlesnakes enjoy the advantage of the rattle, they back it up by being very dangerous due to their venom. Non-poisonous species may be better off relying on somewhat less bravado.

Warning Coloration and Mimicry in Snakes (Pl. 45)

OBSERVATIONS: Not all snakes are camouflaged. Another pattern, less common, is warning coloration. In our area there is only one clear example, the potentially deadly **EASTERN CORAL SNAKE,** and, fortunately (!), it is usually hard to find. Nonetheless, it is interesting to consider that this snake evolved a different pattern of protection from the usual cryptic coloration and also that other snake species occurring in the range of the Coral Snake look remarkably like it.

Caution: Do not ever attempt to handle a Coral Snake or any of its mimics (described below) unless you are both a skilled snake handler and utterly certain of your identification of the snake.

The **EASTERN CORAL SNAKE** (2–4 ft.) is related to the Old World cobras. Its venom is extremely toxic—if a victim is bitten, death is a very likely result. Fortunately, Coral Snakes tend to be secretive, crepuscular (active at dawn and dusk), and rarely encountered. They habitually coil beneath rocks and logs in southeastern woodlands. Coral Snakes are brightly colored, with alternating *wide red, narrow yellow, and wide black bands,* surrounding the belly as well as the upper body. In no way is the Coral Snake camouflaged. Quite the contrary, it is very obvious (though it usually will not be out in the open). The Coral Snake color pattern appears to represent an example of warning coloration, such as was described for the milkweed butterflies above (see p. 345). In this case, however, the warning is not based on unpalatability but on danger.

There is another species that occupies much of the southeastern range of the Eastern Coral Snake, the **SCARLET SNAKE** (1–2 ft.). This non-venomous species is also a crepuscular snake and is fond of burrowing, with many of the same habits as the Coral Snake. It certainly suggests a Coral Snake when you first see it. Like the Coral Snake, it is red, yellow, and black, but both the *band width* and the *order* of the banding are different. In the Scarlet Snake, *yellow* bands are wide, black *narrow,* and *red touches black* (whereas in the Coral Snake *red touches yellow*). Also, in the Scarlet Snake, the bands are really blotches—they do not encircle the belly as they do in the Eastern Coral Snake. Though these differences clearly differentiate between the two species when one looks closely, the initial impression of a Scarlet Snake is very similar to that of an Eastern Coral Snake. The Scarlet Snake is a mimic.

Yet another species, the non-venomous **SCARLET KING SNAKE** (3–6 ft.), has a color pattern similar to that of the Eastern Coral Snake. Like the Scarlet Snake, the Scarlet King Snake ranges throughout the southeastern states. It is also banded bright red, yellow, and black, but the black bands are *narrower* than those of a Coral Snake, and the *red and black bands touch,* rather than red touching yellow. The bands encircle the snake, as they do in the Coral Snake.

Finally, consider the **EASTERN MILK SNAKE** (2–3 ft.), another member of the king snake family. You may see this common species over much of the range of the Copperhead. The Eastern Milk Snake bears a fairly close resemblance to the poisonous Copperhead (see above). Though the Copperhead is not an example of warning coloration, it does seem to represent a "model" for the non-venomous Milk Snake, which may benefit from the resemblance.

EXPLANATION: Why should Eastern Coral Snakes be patterned to warn creatures of danger? Such a characteristic seems non-adaptive because the Coral Snake, after all, has venom to protect itself. However, it takes energy to make venom and the snake must use its venom to secure food. It would not be very economical to use it principally for protection. Not only that, but the Coral Snake lacks the long sharp fangs of pit vipers (see above). Instead, its teeth are small and it must bite hard and chew a bit to inject the maximum amount of venom. A large mammal or predatory bird could easily do serious damage to the snake before being bitten. In fact, it may avoid being bitten altogether and kill the serpent. Because of its relatively small size and consequent vulnerability to predation, it is clearly to the Coral Snake's advantage to send a message of warning to the effect, "Don't tread on me!" Many species of Coral Snakes occur in the tropics and experiments with several species of large tropical birds have shown that these birds instinctively avoid wooden dowels painted with the bright red, yellow, and black patterns of Coral Snakes. Whether this instinctive avoidance of Coral Snakes is true of North American mammals and birds awaits experimentation, but indications are that this colorful snake is a reptilian example of warning coloration.

The Scarlet Snake and Scarlet King Snake, and perhaps the Eastern Milk Snake, like the Viceroy butterfly, have evolved color patterns that mimic another unrelated species. In all cases, the snake mimics resemble venomous species that are very dangerous. The mimics' defense is based on the presumed knowledge that mammal and bird predators have of poisonous species. There is little doubt that the Scarlet Snake and the Scarlet King Snake are Eastern Coral Snake mimics. The Coral Snake is very distinctive and the fact that both non-venomous species so closely resemble it over exactly that part of their ranges where Coral Snakes also occur argues forcefully for mimicry.

The mimicry argument is least certain for the Eastern Milk Snake. Though it resembles a Copperhead, both species are well camouflaged; thus the Milk Snake may have evolved camouflage patterning independently of the presence of the Copperhead. However, any Milk Snake most closely resembling a Copperhead would probably enjoy an additional evolutionary advantage and thus leave more offspring. In other words, what began as merely camouflage patterning could have been further refined by the process of natural selection to become mimicry.

PATTERNS OF BIRD NESTING (PL. 47)

OBSERVATIONS: From early spring throughout midsummer birds are engaged in reproductive activity, centering around establishing territories, courtship (see Chapter 6), nesting, and fledging young. It is a real challenge to find birds' nests, which are usually well concealed, and many species are secretive when attending young. You will need sharp eyes and patience to become a nest watcher. Use care to avoid disturbing nesting birds or inadvertently providing a trail to the nest that could be followed by foxes or other predators. The best way to observe nesting behavior is from a distance with binoculars. To find birds' nests, the best method is to watch for birds bringing material for nest-building or to watch parent birds returning with food for the young. Birds' nests vary considerably from species to species (see below) and this guide only contains an overview of the general characteristics of nesting. For specific information and illustrations of the nests of each species, see *A Field Guide to Birds' Nests.*

There are several stages of nesting behavior: nest location, nest-building, egg-laying, incubation, feeding young, and fledging. Many common dooryard species, especially the American Robin, make excellent subjects for observing.

Nest Locations

There are three major kinds of nests:

1. OPEN NESTS IN SHRUBS AND TREES: Thrushes, mimic thrushes, vireos, orioles, tanagers, kinglets and gnatcatchers, crows, jays, blackbirds, grosbeaks, buntings, finches, some sparrows (Chipping, Field, Clay-colored), most flycatchers, most wood warblers, most hawks, some owls (Great Horned, Long-eared), Mourning Dove, Ruby-throated Hummingbird. Most forest bird species are in this category.

2. CAVITY NESTS IN TREE SNAGS OR BIRD BOXES: Woodpeckers, wrens, titmice and chickadees, nuthatches, Tree Swallow, Purple Martin, Great Crested Flycatcher, Eastern Bluebird, European Starling, Prothonotary Warbler, House Sparrow, some owls (Screech, Sawwhet, Barred, Barn), American Kestrel, Wood Duck, American Goldeneye, Common and Hooded mergansers.

3. GROUND NESTS: Woodcock, Ruffed Grouse, Northern Bobwhite, Wild Turkey, Ring-necked Pheasant, goatsuckers (e.g., the Whip-poor-will), Killdeer, Spotted Sandpiper, many warblers (Black-and-white, Blue-winged, Golden-winged, Tennessee, Nashville, Palm, Kentucky, Ovenbird, waterthrushes, Wilson's, Canada), most sparrows, Eastern Meadowlark, Horned Lark, several sparrows (Savannah, Grasshopper, Vesper, White-throated, Lincoln's, and Song sparrows, Northern Junco).

Some species fit more than one of the above categories. The Rufous-sided Towhee usually nests on the ground but may nest in a low shrub.

In addition there are a few bird species that do not fit easily into any of the above categories. Belted Kingfishers and Bank Swallows excavate cavities in embankments. Cliff Swallows and Barn Swallows nest in barns along rafters. Chimney Swifts nest in chimneys and hollow trees. The Brown Creeper places its nest beneath a strip of loose bark on a tree trunk. Brown-headed Cowbirds are nest parasites—they lay their eggs in other birds' nests (see p. 386).

EXPLANATION: The diversity of nest locations displayed by bird species is impressive. It is difficult to imagine any sort of possible nest site that some bird has not adapted to use. Within a forest, virtually all possible nest locations are utilized, from canopy (orioles and vireos), to understory (thrushes and grosbeaks), to tree cavities (woodpeckers and chickadees), to the litter layer (American Woodcock and Ovenbird). Nest sites are resources and birds compete both within and between species to obtain suitable nest sites. Each species has a genetically programmed pattern of nest-site recognition, as well as genetic programs for nest "blueprints," and parental behavior.

Tree cavities are in particular demand. Unfortunately, many forests are managed in such a way that tree snags are systematically removed. Doing so eliminates possible nest sites for cavity-nesters. Many cavity-nesters depend on woodpeckers to excavate nest holes. Nest cavities are not only in demand by many bird species but are also used by such mammals as flying squirrels and Gray Squirrels. Birds must compete with these animals as well.

Canopy nests are often placed well away from the trunk, on thin branches that will not support the weight of large predators. Oriole nests are excellent examples of this practice. Some nests, like those of vireos, hang from a branch fork in the canopy shade. Nests of gnatcatchers and hummingbirds, though on a branch in the open, are cryptically colored and tiny, and thus are very easily overlooked.

Eggs also tend to be cryptically colored, except those of many cavity-nesters. If the eggs are in a dark cavity, there is no adaptive advantage in having cryptically colored eggs. In fields, a high percentage of species nest on the ground (not surprisingly) but many other species utilize dense shrubs as nest sites. Nests are generally well concealed, especially those located on the ground.

Nest-building, Eggs, Incubation, and Defense (Pl. 47)

Nest Materials

A survey of birds' nests reveals that birds are opportunistic in nest construction as well as location. Many birds use one basic material, such as grass, twigs, or mud, but most combine these materials. Great Crested Flycatchers use shed snake skin, and many species use twine, string, or hair if they can find it. Gnatcatchers and hummingbirds use spider silk, and kinglets, hummingbirds, and other birds use lichens and mosses as nest materials. Swallows make heavy use of mud and the American Robin mixes mud with twigs and grass fibers. Feathers appear in the nests of many birds, especially swallows.

Birds building nests are often observed flying with nest materials in their beaks, returning to the same spot over and over. Swallows gather about pond edges and puddles taking billfuls of mud. Robins have bills full of grasses, and chickadees patiently peck away at dead snags, excavating their cavities. Woodpeckers accomplish the same purpose by drilling, but note that when a woodpecker is excavating a nest cavity, it does not drum, as it does when establishing territory. Rather it hammers from side to side, earnestly chipping away at the bark, a practice that is much more quiet than the loud tattoo of a drum.

Some Examples

It usually takes about a week for a bird to construct a nest. In some species both male and female take part in the nest-building effort but in most species the bulk of the job, if not the entire job, goes to the female. Some common species that are (relatively) easily observed nesting include:

AMERICAN ROBIN: Open cup mud nest of grasses. This bird is so abundant that it should be easy to find a nest. Robins tend to nest low and have no aversion to nesting near human dwellings. The nest is made by the female and she uses a wide variety of materials ranging from string to fine or coarse grasses. Robins are well known for their habit of returning yearly to nest in the same (or nearly the same) spot. This behavior, called *nest-site fidelity,* occurs in many species. Only the female incubates and the normal clutch size is 4 eggs. Eggs are easy to recognize, since they are the well-known color of "robin's-egg blue." If you approach the nest closely, the female will stay put, becoming very rigid. If she is forced to flush (leave the nest), both male and female will become agitated, flying about the nest site and calling.

NORTHERN ORIOLE: This species makes a pendulous nest of grasses, vines, string, and hair, hanging in a canopy tree, often fairly high. The nest tends to be near a branch tip, affording some protection from tree-climbing predators. The female constructs the nest but both male and female, because of their bright colors and vocalizations, are usually easy to locate as they fly back and forth to the nest site. The female incubates the clutch of 4–5 eggs. Eggs are pale gray with thin, streaky brown blotches. The male remains close by and can become very aggressive, loudly chattering at intruders such as grackles and Blue Jays.

HOUSE WREN: The tiny, feisty House Wren takes readily to all manner of bird houses (as well as natural cavities), stuffing the cavity with twigs and grasses. Look closely at House Wren nesting behavior. You might notice that the wrens seem to build more than one nest. This is because the male arrives first and builds one or more "dummy nests," a behavior which presumably signals the female that the territory is acceptable. The female may then use one of the male's nests or, more often, build her own nest. The female alone incubates and clutch size is 5–8 eggs. Eggs are shiny white with reddish brown speckles.

COMMON FLICKER: Cavity-nesters, flickers are easily located by their loud *wicker* call and their preference for relatively open areas. You might see flickers in contention for nest sites with European Starlings, also cavity-nesters. Competition among species for cavities can be quite severe. Both male and female flickers excavate the nest hole, and both sexes incubate. Clutch size is usually 6–8 eggs but is variable—3–10 eggs can appear in a clutch. Eggs are glossy white.

CHICKADEES: Both the Carolina Chickadee and Black-capped Chickadee (and the northern Boreal Chickadee) will either nest in old woodpecker holes or excavate their own holes. They usually nest fairly low in a tree and are easy to watch while excavating. They persistently peck at the bark and wood, removing flakes until the cavity is satisfactory. They usually excavate in a dead pine or other softwood tree. Only the female builds the nest, which is lined with moss, feathers, and all manner of fluffy, "cottony" material. The female does all the incubation and the male brings her food while she is on the eggs. Clutch size is usually 6–8, and the eggs are white with small rufous spots.

OVENBIRD: You may have to do considerable searching to find an Ovenbird's nest but it is well worth the effort. Located on the ground in a forest, the nest resembles a tiny old-fashioned bread oven. It is an open cup, but it opens on the side rather than the top. It is made of plant fibers and leaves, and is cryptically placed on the forest floor. To make it more difficult to locate, females tend to stay put and do not flush from the nest until you are about to step on it. One of the best ways to locate an Ovenbird's nest is to search a territory until the bird flies up from near your feet. Search diligently or patiently wait until the female returns. Only the female builds the nest and only the female incubates the 3–6 eggs. Eggs are white with reddish brown spots.

EASTERN MEADOWLARK: Grassy fields are where the Eastern Meadowlark makes its ground nest of dried grasses and hair from livestock. Watch where the birds tend to land in the field and walk the area until the female flushes from the nest. Many nest locations have tiny trails made by the female as she moves to and from the nest. Females begin several nests before selecting a final location. Only the female incubates and she may keep adding to the nest during incubation. Clutch size is 3–5, and eggs are white with brown blotches. Meadowlarks are polygamous; with careful observation, you may see more than one female associated with a territorial male. Other open-field nesters, such as the Bobolink and Red-winged Blackbird, are also polygamous (see below).

KILLDEER: This species, a plover, has adapted to nesting on gravelly fields, parking lots, and golf courses. The loud call, which gives the species its name, will help you locate it. The nest, made by the male, is nothing more than a scrape, but the eggs are well camouflaged to blend into the background. If you come upon a Killdeer's nest with an incubating female, she will probably put on a "broken-wing display." This behavior, a form of *distraction display,* mimics an injured bird attempting to escape on foot. The display even includes a distress call as the female, wings and tail spread, lamely pulls herself along the ground just beyond your reach. Eventually, presumably when she is satisfied that she has lured you far enough from the nest, she "recovers," and flies away, eventually to return to the nest. The mother bird's behavior is obviously adaptive in that the most common predator threats to an open-area ground-nesting bird are from cats, foxes, raccoons, and other prowling mammals. The broken-wing act is a complex behavior that diverts the predator's attention from the vulnerable nest. Killdeer nestlings are extremely well camouflaged and sit quite still when danger threatens.

BROWN-HEADED COWBIRD: A brood parasite, the Brown-headed Cowbird is unlike any other bird in our region. Females have territories and are courted by groups of nomadic males. The dominant male of the group usually does the actual mating. Females typically perch atop trees, and they are skillful at finding nests. The female Cowbird then lays a single egg in the nest of another species, often a warbler, like the Ovenbird or Yellow Warbler, but many other species are parasitized as well. From that point on, the female Cowbird has nothing more to do with the egg. She may then search out another nest and lay another egg. The young Cowbird is raised by the parasitized species. Baby Cowbirds tend to be fast-growing and aggressive and often crowd out nest mates, monopolizing the foster parents. Cowbirds have had a strong negative impact on certain species like the Kirtland's Warbler. Not every parasitized species is so vulnerable, however. You might notice that parasitized birds do sometimes recognize the foreign egg. American Robins remove it, and Yellow Warblers often build a second nest atop the old nest containing the Cowbird egg.

Observing Nestlings and Fledging

The number of eggs per clutch varies somewhat both within and between species. As the examples above show, clutch size may be as low as 2–3 eggs or up to a dozen or more. Each species has an *optimal clutch size* for its particular environment. Most birds lay 1 egg per day but do not begin incubation until all eggs are laid. Since rate of development is dependent on incubation (see below) all eggs hatch at about the same time (*synchronous hatching*). However, some species, like raptors and also the Eastern Meadowlark, begin brooding before all the eggs are laid, resulting in *asynchronous hatching*.

Try to observe parent birds returning to the nest with food for the young. Keep your observations of baby birds to a minimum, because you may disturb the parents and risk endangering the young. Depending on the species, nestlings may hatch blind, naked, and essentially helpless, a condition called *altricial*, or be covered with down, with eyes open and alert, in which case they are called *precocial*. Most forest and old-field nesters are altricial. Ruffed Grouse, Northern Bobwhite, Ring-necked Pheasant, Woodcock, and Killdeer are precocial.

Nestlings are vulnerable and it is certainly to their advantage to grow quickly. Daily observations of a Robin's nest will show you how rapidly helpless, naked nestlings become alert, feathered fledglings. The entire process requires only about 10–14 days. Altricial birds are *ectothermic* when hatched, meaning that they cannot maintain a warm body temperature. Of course, the same is true of the bird while in the egg. That's why parent birds incubate, both before the egg hatches, and for some time thereafter. Nestlings must grow in size, grow feathers, and become *endothermic*, a condition sometimes called "warm-blooded" because of the high constant body temperature that is physiologically maintained. As the birds are growing, they aggressively demand food from the parents. Baby birds *gape*, opening their mouths widely, exposing brightly colored throats. This behavior signals parent birds to deposit the food morsel into the nestling's throat.

Insects, especially caterpillars, make up the bulk of most of the food parent birds bring to the young. Robins are known for their fondness for earthworms. Animal food, high in protein, is the best diet for nestlings. As they grow, nestlings are converting worm and caterpillar protein into bird protein. Even seed-eating species like Towhees and Cardinals feed nestlings animal food. Also, animal food is abundant during nesting season. Parent birds can make literally dozens of foraging trips daily, finding food on each occasion.

A pair of Robins nesting in your backyard may begin another nest cycle while the fledglings from their first brood are still soliciting them for food. Many species have 2 broods (sometimes 3) in the course of a summer. In areas with longer growing seasons, like the Southeast, there are more species that are multibrooded. Blue Jays, for instance, occasionally have 3 broods per summer in the South but are often only single-brooded in the North. Most boreal species are only single-brooded, and most southern species are multibrooded. Those of mid-latitudes are variable. After a nest cycle has begun, if something happens to the nest, most species will start over.

A bird that is ready to leave the nest is called a *fledgling*. By the time the juveniles become fledglings they are very active and alert and have their flight feathers, though they are not yet good fliers. You can observe fledglings hopping about branches, calling vigorously to their parents for food, and "testing" their wings. Many people think fledglings have fallen from the nest or are otherwise in difficulty. Sometimes this is true, but more often parents are attending them. You should definitely leave fledglings alone. The fledgling period of the life cycle is also a vulnerable time, but many fledglings do survive. As noted above, you might see fledglings begging for food even as the parents are beginning to raise another brood.

Bird Nesting as an Adaptation

Birds have highly developed parental behavior. After establishing a territory (see p. 298), birds go through *courtship*, followed by *pair-bonding*. Most bird species are *monogamous*, meaning that a single male and single female pair for the duration of the breeding cycle. Pair bonds may change, with new mates chosen for a second brood, a practice common to House Wrens. On the other hand, some species, like the Tufted Titmouse and Northern Cardinal, often pair-bond for life, the same two birds mating year after year. Some birds, including the House Wren, Red-winged Blackbird, Eastern Meadowlark, Bobolink, Indigo Bunting, Common Yellowthroat, and Yellow Warbler are *polygamous*. One male mates with several females, all of which nest within his territory. This mating system is less common than monogamy and tends to occur mostly in birds that nest in open habitats like grasslands and marshes. A few species of birds practice *polyandry*, where one female mates with several males. Phalaropes are an example, but no species in our area does this. In our area, the Ruby-throated Hummingbird, Ruffed Grouse, American Woodcock, and Brown-headed Cowbird are *promiscuous* because essentially no pair bond is established. The female builds a nest and mates with the first male to come along.

Both male and female have an equal genetic investment in the eggs but what, exactly, is adaptive about parental behavior? Birds are endothermic, having high body temperatures and high metabolic rates. Endothermy (the ability to regulate body temperature by adjusting metabolism) develops slowly, however, and eggs must be warmed as the embryo develops. Birds also have very complex brains, and require developmental time for their nervous systems to grow. Parental behavior accomplishes both protection and warming of the developing eggs, insuring the shortest possible development time. Eggs, nestlings, and fledglings are all vulnerable. Since two birds can feed and defend young better than one, monogamy is adaptive in many situations. Monogamy helps insure that juveniles grow as quickly as possible, escaping from their most vulnerable period.

Why are some species polygamous?

Birds such as Ring-necked Pheasants, Red-winged Blackbirds, Indigo Buntings, and Bobolinks, all polygamous, nest in open areas. One male can defend a large territory but there is often enough food for more than one pair and their young. Each of these species exhibits sexual dimorphism (see p. 300), and strong sexual selection has probably occurred in each. Males compete against one another for territories and those with the largest and best territories attract the most females. In some species, the females select territories, not males, so the critical period for the male is during territory establishment. Sexual selection is not the only force responsible for polygamy, however, since species such as Eastern Meadowlarks, House Wren, and Marsh Wren, which exhibit essentially no plumage sexual dimorphism, are also polygamous. The type of habitat is likely to be the major selection pressure for the evolution of polygamy in these species.

In some habitats, like insect-rich fields and marshes, males can "afford" not to participate in parental care. If the male can accomplish additional matings, and the female alone can successfully raise the young, then it is in the male's evolutionary best interest not to help the female raise the young following mating, a behavior called *desertion*. He will get more of his genes in the next generation if he mates with other females that choose his territory. Males that desert females will be more successful reproductively than "faithful" males. It may also be in the female's best interest that the male desert. Males, especially those with gaudy plumage that live in open areas, could attract unwelcome attention to an already-vulnerable nest, attention which would result in predation. Especially colorful males, like Ring-necked Pheasants, could easily lead a predator to a nest. Females are cryptically colored, probably for good reason. Open-area nesting is risky. The Bobolink seems to choose an intermediate strategy of nesting. The territorial male Bobolink helps feed young hatched by the first female with which he mates, but he will mate with other females as well. He contributes no parental behavior to the young hatched by these other mates.

Why do females do most of the nest-building and incubation?

Again, sexual selection may play a role, with males having to defend territory, and females being the more cryptically colored sex. However, another reason may be that though both male and female have equal genetic investments in the egg, the female actually has the egg! Once copulation has occurred, the female is left to lay the fertilized eggs. She has no options other than nesting. The male still has the option of mating with another female. Evolutionarily, females have a clear-cut "mission" following copulation, and that is to nest. Males have the "choice" of abandoning or parenting. There has probably been stronger selection pressure on females for nest loyalty than on males. Only in brood parasites (such as the Brown-headed Cowbird) and a few other species (such as the phalaropes) have the females managed to avoid nesting responsibilities.

Why don't birds lay more eggs?

Some species, called *determinate* layers, can only lay so many eggs in a given season, but others, called *indeterminate* layers, can produce indefinite numbers. The common domestic chicken is an indeterminate layer, as is the Common Flicker. Clutch size, therefore, is not necessarily limited by physiology—some birds could lay dozens of eggs rather than just a few. Egg-laying is just the beginning, however. Birds must be able to raise their juveniles. Clutch size is really a function of the number of young a bird can successfully raise. It does no good to lay a dozen eggs when you can only find sufficient food to fledge three or four young.

Finally, why does brood parasitism exist? The Brown-headed

Cowbird, like the European Cuckoo and a few other species, has abandoned nesting altogether, to become a parasite. Cowbirds originally may have followed Bison herds, feeding on the insects scattered by the large hoofed animals. Such a nomadic lifestyle may have contributed to selecting for brood parasitism. However, not all brood parasites are nomadic. Brood parasitism is another means of exploiting a resource, in this case another bird's nest and parental behavior.

WILDFLOWERS — HABITAT AND GROWTH FORM

Stable-habitat Growth Form (p. 325)

Plants generally small, long-lived. Usually native. Large taproots common. Flowers often small, few in number, insect-pollinated. Relatively few seeds produced. Less of the plant's annual energy is devoted to making reproductive parts than in opportunistic species. Some spread by rootstocks. Mostly in forests, woodlots.

CUT-LEAVED TOOTHWORT

To 15 in. Leaves in whorls of 3, deeply toothed. Usually a single cluster of 4 white flowers.

RED TRILLIUM

To 16 in. Single large red flower; all parts in threes. Whorl of 3 leaves, roundly oval. Very thick taproot (not shown).

INDIAN CUCUMBER-ROOT

To 3 ft. Leaves oval, untoothed, in 2 whorls. Delicate greenish yellow flowers, hanging bell-like. Very thick taproot.

BLOODROOT

To 12 in. Stem and thick taproot exude orange-red juice when cut. Single large flower. Leaves widely oval, lobed.

Opportunistic Growth Form (p. 326)

Upright, often tall. Often alien. Flowers in clusters, usually atop plant in a stalk, spike, or plume. Many flowers, many seeds. Most insect-pollinated, some wind-pollinated. Often short-lived annuals or biennials. Many have no taproot.

SMALL WHITE ASTER

To 5 ft. Dense clusters of flowerheads on upper branches. Small root system.

STEEPLEBUSH

To 4 ft. Dense flower spike atop plant. Old fields, meadows.

QUEEN ANNE'S LACE or WILD CARROT

To 4 ft. Wide "plate" of tiny flowers. Small taproot smells like a carrot. Old fields, roadsides. See also Pl. 21.

Intermediate Growth Form (p. 327)

Usually tall. Insect-pollinated. Usually perennial. "Intermediate" amount of energy devoted to flowering parts. Open moist areas.

PURPLE LOOSESTRIFE

To 4 ft. Long spike of reddish purple flowers. Opposite leaves. Wet meadows, freshwater marshes.

JEWELWEED or TOUCH-ME-NOT

To 5 ft. Many trumpet-shaped orange flowers, each hanging on a long stalk. Moist sites in forests, edges.

CARDINAL-FLOWER

To 4 ft. Deep scarlet flowers in long spike. Moist woods, swamps.

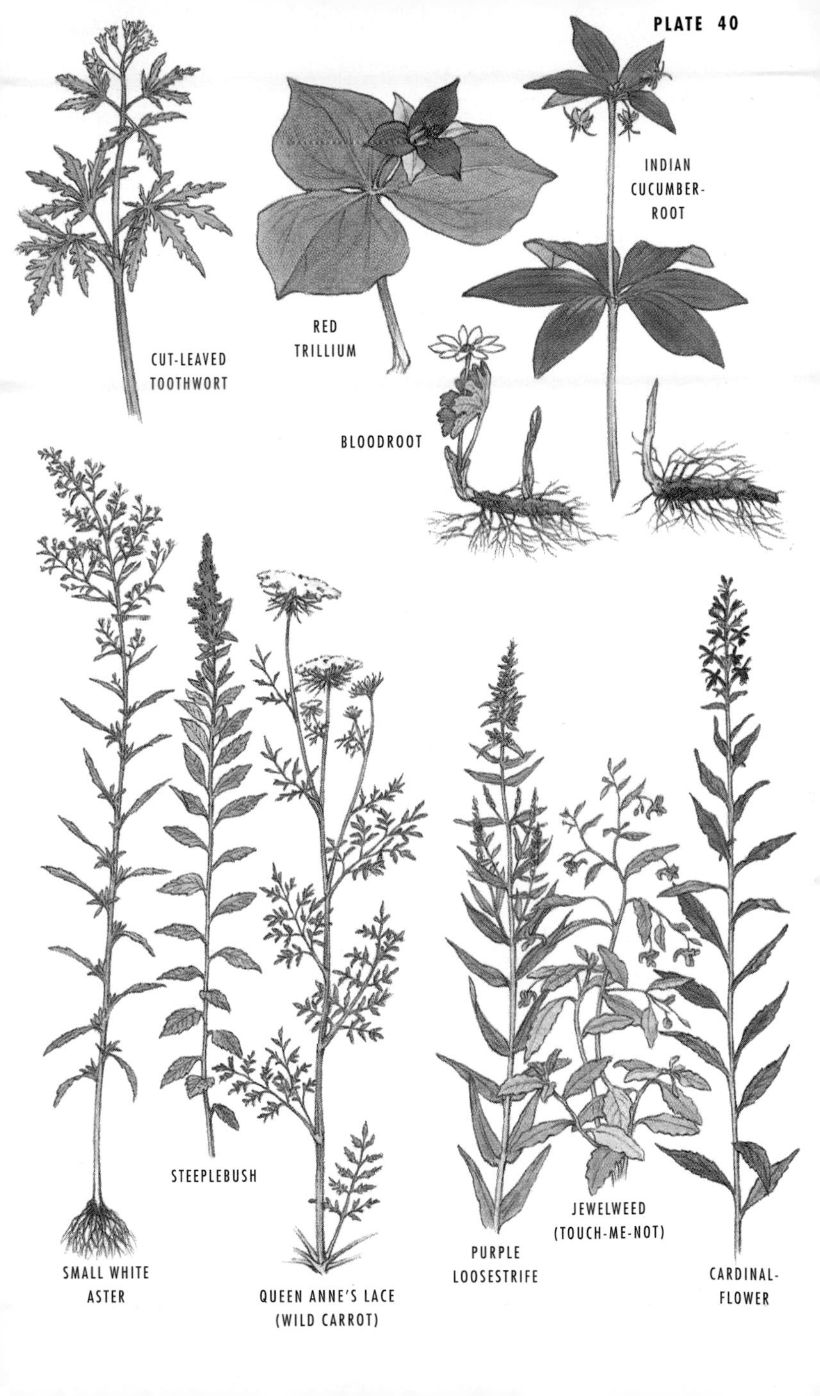
CUT-LEAVED TOOTHWORT
RED TRILLIUM
INDIAN CUCUMBER-ROOT
BLOODROOT
SMALL WHITE ASTER
STEEPLEBUSH
QUEEN ANNE'S LACE (WILD CARROT)
PURPLE LOOSESTRIFE
JEWELWEED (TOUCH-ME-NOT)
CARDINAL-FLOWER

GOLDENRODS — HABITAT AND GROWTH FORM (P. 329)

Within the goldenrod genus (*Solidago*) the stable-habitat, intermediate, and opportunistic growth strategies are all present, sometimes within a single species (e.g., the Rough-leaved Goldenrod, below). See text for details.

CANADA GOLDENROD

Grows along dry roadsides (opportunistic growth form) or moist thickets (intermediate). In a mesic habitat, it grows up to 5 ft. tall. In a more xeric habitat, may reach only 1 ft. Flowerheads in a dense plume atop plant. Dense growth of leaves, each oblong, sharply toothed.

BLUE-STEMMED GOLDENROD

A woodland species with the stable-habitat growth form. Relatively little energy devoted to flowering compared with overall growth. Stem bluish or purplish.

ZIGZAG GOLDENROD

A woodland species with the stable-habitat growth form. Flower clusters at leaf axils. Total amount of flowers small in relation to overall size of plant. Leaves wide and pointed, stem zigzags.

GRAY GOLDENROD

Grows in xeric habitats, such as dry sandy soils. Shows opportunistic growth form. Short, 6–12 in. tall. Flowerheads in a dense, plume-like cluster atop plant. Much energy invested in making flowerheads. Stem grayish, covered with hairs.

ROUGH-LEAVED GOLDENROD

Grows in xeric (dry) and mesic (moist) areas. Shows opportunistic form in xeric areas: short (1–2 ft.) but with densely clustered flowerheads. Flowers take up much of the plant's energy. Shows intermediate form in mesic areas: tall (to 7 ft.) with less overall energy devoted to making flowers. Upper leaf surface is quite rough.

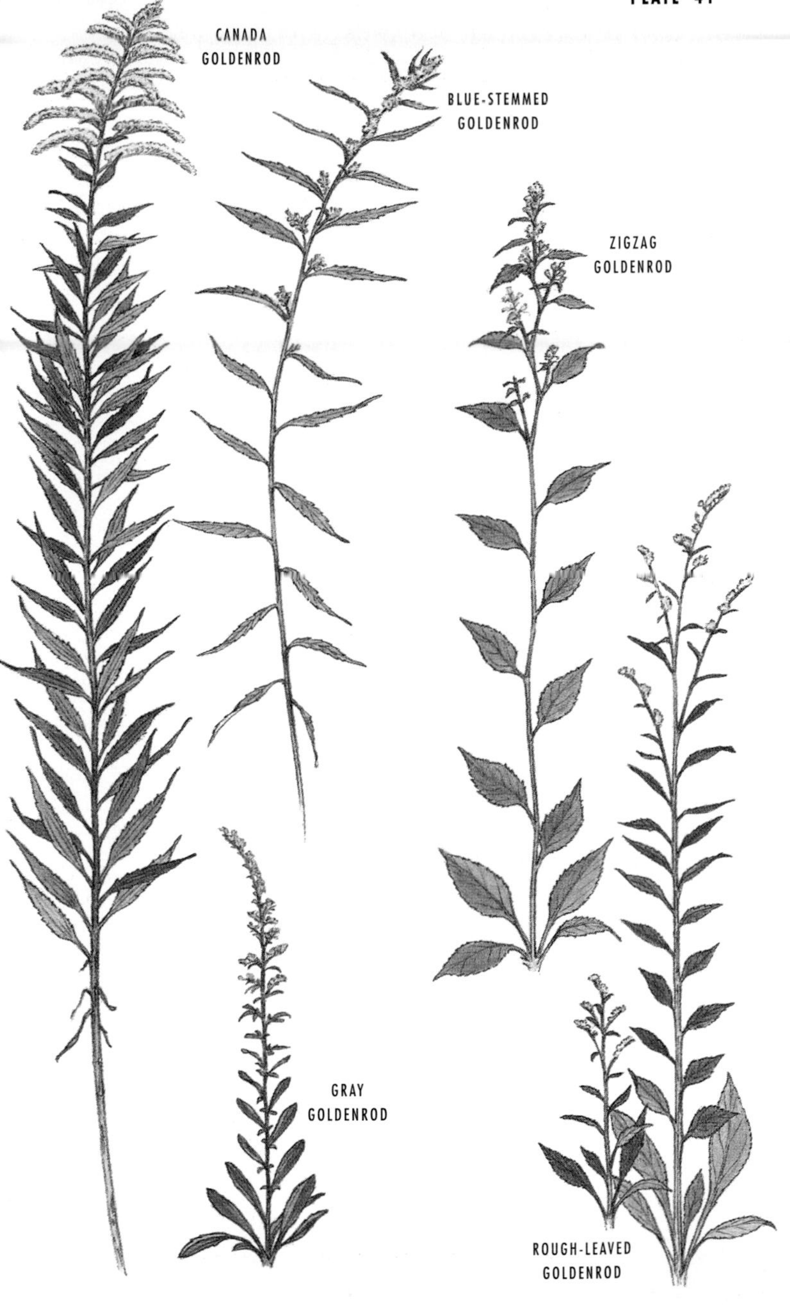
CANADA
GOLDENROD
BLUE-STEMMED
GOLDENROD
ZIGZAG
GOLDENROD
GRAY
GOLDENROD
ROUGH-LEAVED
GOLDENROD

MILKWEED NATURAL HISTORY (P. 341)

COMMON MILKWEED

Tall (to 6 ft.), with opposite, oval, untoothed leaves. Stem and leaves exude a sticky white sap when broken. Flowers small, pink, in dense, rounded clusters. Flowers develop into green pods that rupture and release many tiny seeds, each attached to a feather-like "parachute." Seeds are dispersed by wind.

MONARCH

Adult: Large; bright orange wings with black veins (see Viceroy, below). *Cocoon:* Greenish, like a "wrapped leaf," attached to twig. *Caterpillar:* Black, banded with yellow and white. *Egg:* Small, green, attached to underside of leaf.

VICEROY

Adult is a Monarch mimic. Very similar to Monarch but smaller, with 2 white spots on the front of the forewing (lacking in Monarch).

Many adult butterflies and moths feed on milkweed nectar. The plate illustrates three moths —a large **Sphinx** or **Hawk Moth,** a **Geometer Moth,** and an **Underwing Moth.**

AMERICAN BUMBLE BEE

These insects are principal pollinators of milkweed. Their large, robust bodies have no difficulty moving over the flower clusters. Smaller **Honey Bees** pollinate many species, including milkweed, but their small bodies often become trapped in the flower cluster.

EARLY TACHINID FLY

This fly is a parasite of caterpillars. Female Tachinids lay 1–2 eggs on underside of a caterpillar. Newly hatched larvae burrow through caterpillar's skin and feed by consuming the caterpillar from the inside out.

RED MILKWEED BEETLE

This red beetle with black spots is found exclusively on milkweed, where its entire life cycle occurs (see text). Note the long, black antennae.

EASTERN MILKWEED BUG

Found on milkweed pods. Active, feeds on seeds. Red X on forewings. Life cycle occurs entirely on milkweed (see text).

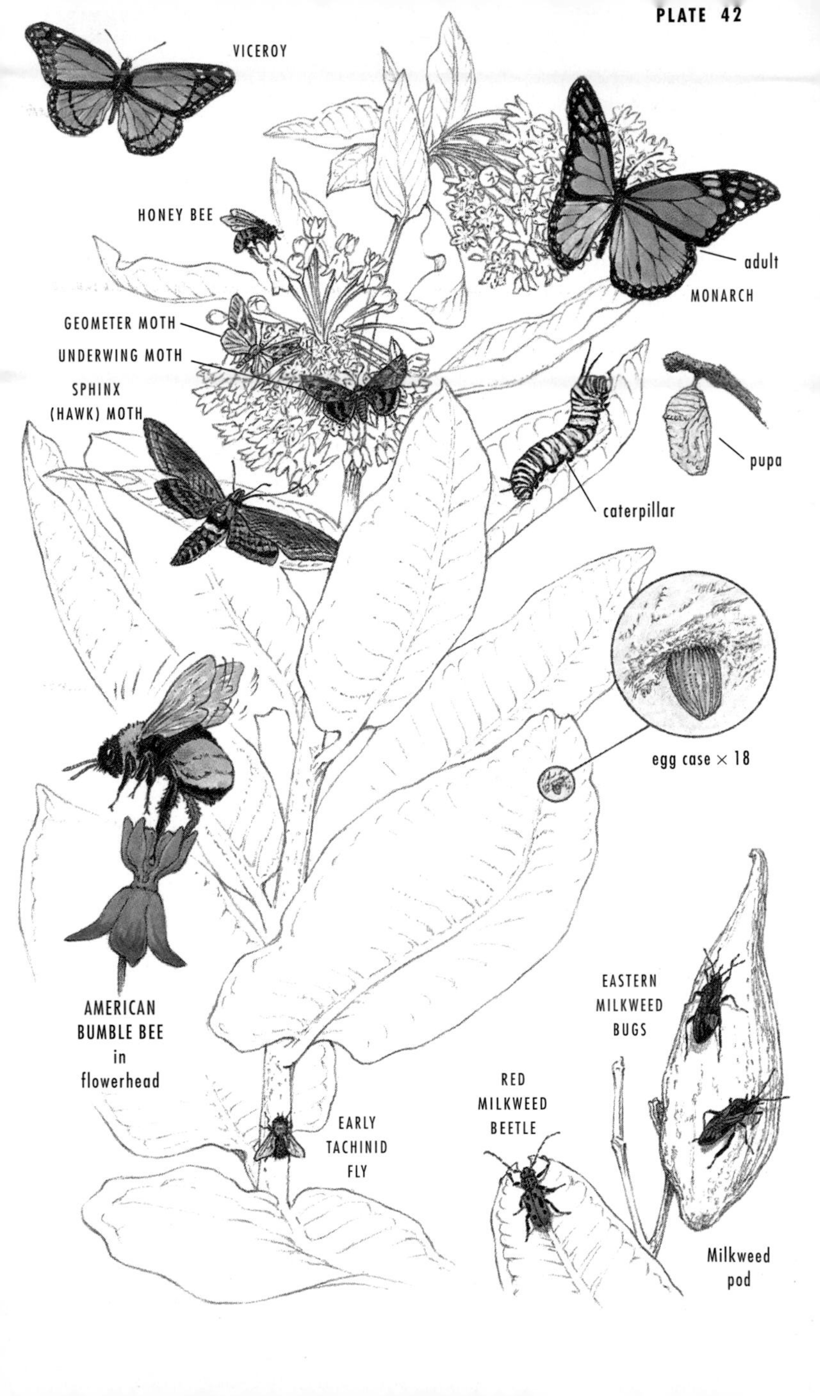
VICEROY
HONEY BEE
adult
MONARCH
GEOMETER MOTH
UNDERWING MOTH
SPHINX (HAWK) MOTH
pupa
caterpillar
egg case × 18
AMERICAN BUMBLE BEE in flowerhead
EASTERN MILKWEED BUGS
RED MILKWEED BEETLE
EARLY TACHINID FLY
Milkweed pod

INSECT AND SPIDER CAMOUFLAGE (CRYPSIS) AND MIMICRY (P. 347)

FOLIAGE (P. 348)

LUNA MOTH

A large, nocturnal silkworm moth. Light green wings resemble leaves as the insect hangs beneath branch during the day.

TRUE KATYDID

Somewhat like a grasshopper, this insect has forewings that closely resemble leaves, even with cross veining. From treetops, it makes a strident monotonous *katydid* sound all summer.

NORTHERN WALKINGSTICK

Very elongated, slender, closely resembling a twig. Male brown, female greenish. Feeds on leaves.

GOLDENRODS (P. 350)

On a single goldenrod plant (such as Canada Goldenrod, shown here) can be found a diverse animal community.

Bee mimics resemble such dangerous insects as **Yellow Jackets, Bumble Bees,** and **Honey Bees.** The non-stinging mimics shown here are the **American Hover Fly** and the **Drone Fly.**

Predators lurk among the dense goldenrod flowerheads. The **Goldenrod Spider** is one of many species of crab spiders that are colored like the flowers they inhabit. The **Ambush Bug** kills bees and other nectar-feeders with its long, stiletto-like mouthparts.

Predators like the **Carolina Mantis** are well hidden among the foliage. The **Buffalo Treehopper** looks like a small leaf. See Pl. 44 for more on mantids and the Buffalo Treehopper.

BARK (P. 348)

WOLF SPIDER

Large, long-legged, rapid-moving spiders, wolf spiders are common in fields and forests. They are well camouflaged both on the ground among the grasses or leaf litter as well as on bark.

LOCUST TREEHOPPER

A hunchbacked, brown treehopper with a projection that resembles a thorn. Feeds on juices of Black Locust trees, which have thorns (Pl. 34).

HORNED FUNGUS BEETLE

A brown, rough-backed beetle with 2 stumpy, hornlike projections on back. Feeds on fungi that grow on bark. If threatened, it feigns death.

UNDERWING MOTH

Many species. When moth rests on bark, wings resemble bark. Open wings reveal bright pattern on hind wing. Also see Pl. 44.

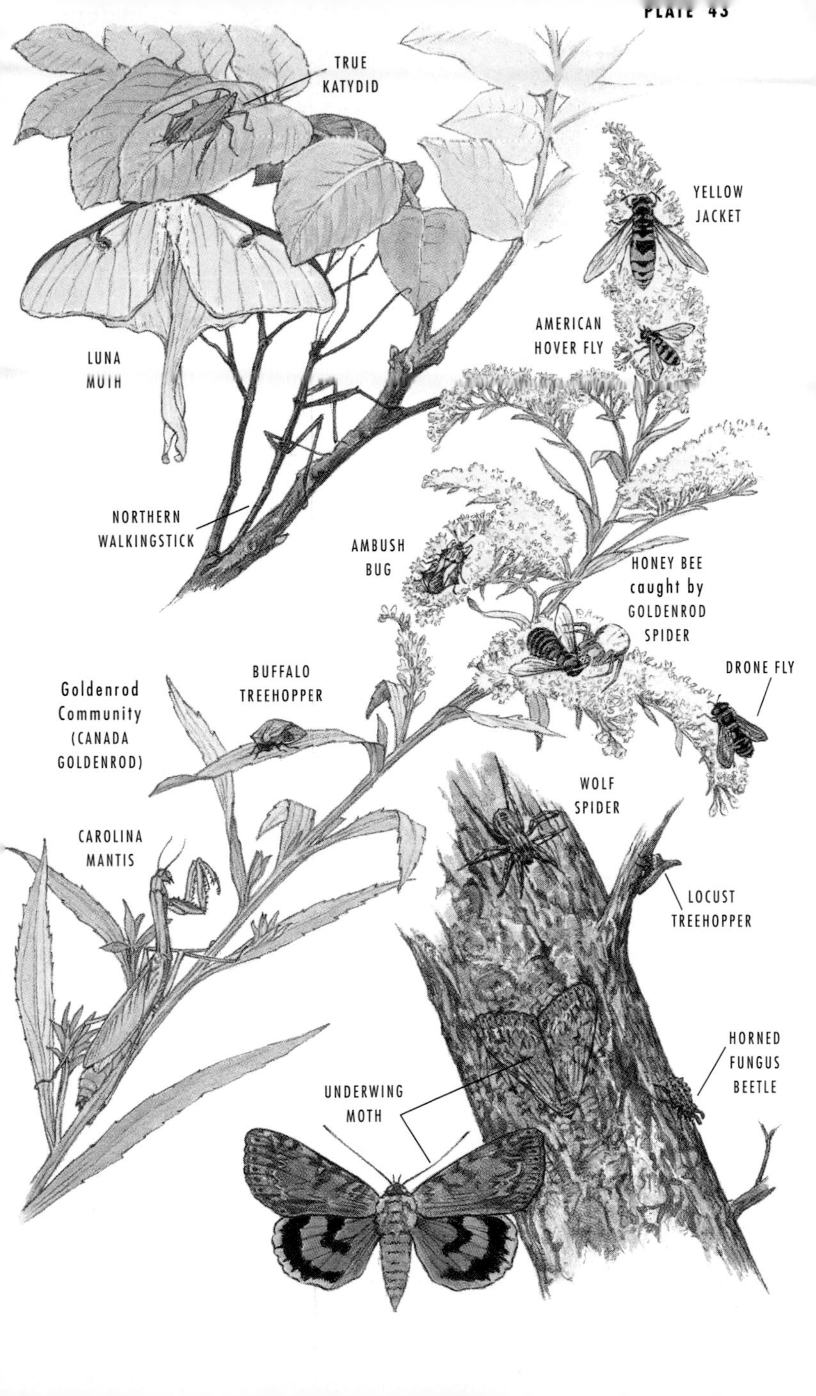
TRUE
KATYDID
YELLOW
JACKET
AMERICAN
HOVER FLY
LUNA
MOTH
NORTHERN
WALKINGSTICK
AMBUSH
BUG
HONEY BEE
caught by
GOLDENROD
SPIDER
DRONE FLY
BUFFALO
TREEHOPPER
Goldenrod
Community
(CANADA
GOLDENROD)
WOLF
SPIDER
CAROLINA
MANTIS
LOCUST
TREEHOPPER
HORNED
FUNGUS
BEETLE
UNDERWING
MOTH

INSECT PREDATION AND DEFENSE BEHAVIOR

These insects show some of the range of adaptations that aid in capturing prey, and that aid in escaping capture. See text.

Prey Capture (p. 352)

ROBBER FLIES

Robber flies attack prey on the wing, dropping on it from above. Swift and skillful fliers, they attack quickly and subdue their prey by using piercing mouthparts. All robber flies suck the prey's body fluids. Most species are slender, but a few are robust. Most frequent old fields, meadows, and other open areas, where they capture pollinating insects. Plate shows a robber fly capturing another fly.

SPHECID WASP

Large wasps; the group includes the digger wasps and mud daubers. These are large, often colorful wasps that have powerful, painful stings. Adults feed on plant juices, but in some species, like the **Black-and-yellow Mud Dauber,** females capture and paralyze a large spider, after having excavated a burrow. The female places her prey in the burrow, then lays a single egg on it; upon hatching the larval wasp feeds on the prey. The closely related **Cicada Killer** (not shown) captures cicadas for the same purpose.

MANTIDS

Voracious insects that are cryptically colored (see Pl. 43). Mantids ensnare and butcher prey using their sharp front legs. Plate shows a **Praying (European) Mantis** devouring a grasshopper.

Prey Defenses (p. 347)

The **Buffalo Treehopper** and **Underwing Moth** (Pl. 43) are both cryptically colored, an adaptation that helps them escape capture. However, if disturbed, the Treehopper can leap quickly out of harm's way. The Underwing Moth flies away erratically, revealing brightly colored hind wings.

IO MOTH

Adult is cryptically colored when at rest but shows bold eyespots on hind wings when in flight. Eyespots on wings may serve to distract predator or direct predator away from the most vulnerable parts of the moth. Caterpillar is lined with clusters of sharp, urticating spines, which are highly irritating.

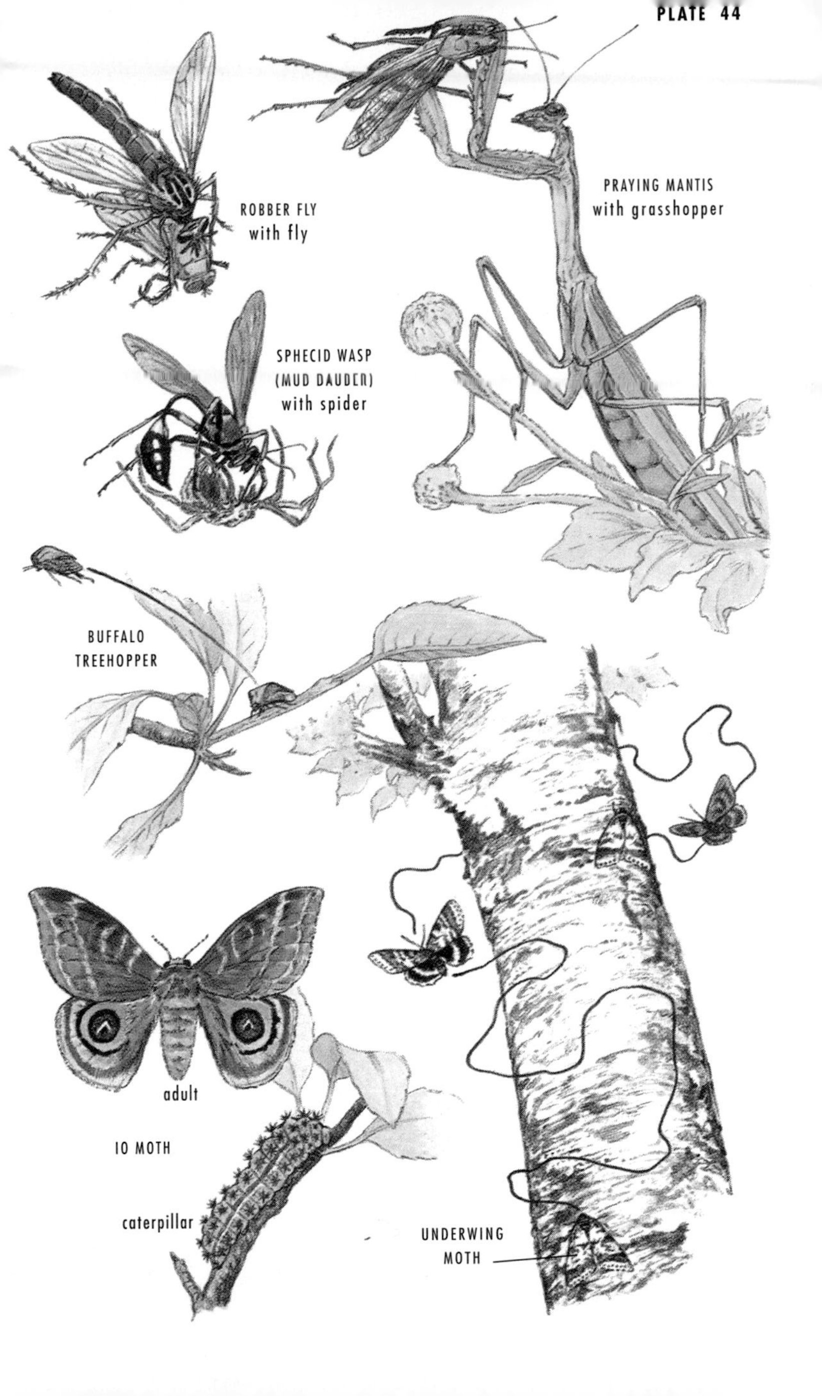
ROBBER FLY
with fly
PRAYING MANTIS
with grasshopper
SPHECID WASP
(MUD DAUBER)
with spider
BUFFALO
TREEHOPPER
adult
IO MOTH
caterpillar
UNDERWING
MOTH

MIMICRY PATTERNS IN SNAKES AND BUTTERFLIES

Mimicry—when one animal closely resembles another, often unrelated species, and derives protection from the resemblance—occurs in many kinds of animals, including snakes and butterflies. See text for details.

Mimicry in Snakes (p. 379)

The non-poisonous **Scarlet King Snake** is a race of Milk Snake that superficially resembles the **Eastern Coral Snake.** Another race of Milk Snake, the Louisiana Milk Snake, also is a Coral Snake mimic, as is the **Scarlet Snake.** See text.

A Butterfly Mimicry Complex (p. 356)

The **Pipevine Swallowtail** is, like the Monarch, unpalatable to birds. As a caterpillar, it feeds on plants in the birthwort family, including Dutchman's-pipe and Virginia Snakeroot, that render it distasteful. It is the model for a mimicry complex that includes 5 species. Each of the 5 species is fully palatable to birds but derives some protection from its resemblance to the distasteful Pipevine Swallowtail.

Only the female **Tiger Swallowtail** is a mimic; the male is not. This is also true of the **Diana Fritillary.** Both the male and female of the **Spicebush Swallowtail,** the **Eastern Black Swallowtail,** and the **Red-spotted Purple** are mimics.

Note that the degree of resemblance of the mimics to the model varies from one species to the next. The resemblance is closest in areas where there are high populations of the model (see text).

SCARLET
KING SNAKE

EASTERN
CORAL SNAKE

TIGER
SWALLOWTAIL

male

female

SPICEBUSH
SWALLOWTAIL

female

PIPEVINE
SWALLOWTAIL

female

female

EASTERN BLACK
SWALLOWTAIL

female

RED-SPOTTED
PURPLE

female

male

DIANA
FRITILLARY

ADAPTATIONS OF REPTILES (P. 373)

COMMON GARTER SNAKE

Non-poisonous. Forests, fields, and suburban yards. Avoids predation by moving quickly and by cryptic coloration—body stripes or blotches break up pattern of the snake against the background.

EASTERN HOGNOSE SNAKE

Non-poisonous. A short, thick snake that superficially resembles some of the poisonous species. If threatened, it will coil, hiss, puff up, and lunge, as poisonous species do. Following this behavior, it will roll over and feign death.

RAT SNAKE

Non-poisonous. To 8 ft. long. Acts aggressively when threatened; will coil, hiss, and vigorously vibrate tail. Captures prey by constriction. Several subspecies, colored differently. Good tree-climbers. Plate illustrates black subspecies capturing food.

BLACK RACER

Non-poisonous. To 6 ft. long. Acts aggressively when threatened, hissing and vibrating tail. Often climbs trees when threatened. Pursues prey and kills by biting, not constriction.

COPPERHEAD

Poisonous—a pit viper. Large, triangular head. Thick body with rufous and tan blotching. Very cryptically colored on forest floor. When threatened, it coils and hisses. Captures food by rapid strike, bite, and injection of venom.

EASTERN FENCE LIZARD

A brown lizard with rough scaling. Found in dry old fields, sand plains, dry woodlands. Fast runner; good climber, cryptically colored against bark. Escapes capture by "sacrificing" tail, which breaks off and continues to wiggle. Tail will regrow.

EASTERN BOX TURTLE

Terrestrial, inhabiting fields and woodlands. Domed, orange and brown blotched carapace (shell). Double-hinged plastron (underside) completely encloses head, legs, and tail. If turned upside down, this turtle can right itself easily when the danger passes.

EASTERN
HOGNOSE SNAKE
COMMON
GARTER SNAKE
RAT SNAKE
constricting
Wood Rat
BLACK RACER
COPPERHEAD
EASTERN
FENCE LIZARD
EASTERN BOX TURTLE

BIRD NESTING BEHAVIOR (P. 381)

The species illustrated are common, and their nesting behavior shows a range of adaptations. See text for details.

COMMON FLICKER and EUROPEAN STARLING

Cavity-nesters. The Flicker is a 12–14-in., pale brown woodpecker with a white rump. Males have a black "mustache." (See Pl. 3.) Starlings are chunky black birds with a greenish and purple sheen, visible only in good light. Both are cavity-nesters and may compete for nest cavities. Plate shows two Flickers being "evicted" as a Starling takes over.

BLACK-CAPPED CHICKADEE

See Pl. 7 for field marks. Chickadees are cavity-nesters, and excavate their own nest holes. Plate shows a Chickadee removing decaying wood from a rotted birch to make its cavity.

HOUSE WREN

Smaller than a sparrow; a perky, brown bird that often cocks its tail up. Very vocal; a rich, loud, gurgling song. Nests in cavities but takes very readily to bird boxes.

BROWN-HEADED COWBIRD

A small blackbird. Males are shiny black with a brown head; females uniformly gray-brown. Brood parasite: plate shows a female laying an egg in the nest of a Yellow Warbler.

KILLDEER

A plover; nests on gravelly, open fields. Brown above with 2 breast bands. Rufous rump. Plate shows female performing distraction display ("broken-wing act") to lure potential predator from nest. Eggs cryptically colored.

OVENBIRD

See Pl. 3 for field marks. A warbler of the forest understory; it walks rather than hops. Nest is located on ground and is shaped like an old-fashioned Dutch oven; very well hidden, but it can be located by observing female returning to it.

female

EUROPEAN STARLING

COMMON FLICKER

male

BLACK-CAPPED CHICKADEE

HOUSE WREN

BROWN-HEADED COWBIRD

nest of Yellow Warbler

KILLDEER

female

OVENBIRD

MID- TO LATE-SUMMER FRUITS (P. 332)

These species develop ripened fruits from mid- to late summer. Fruits are typically berries high in carbohydrates (sugars) with almost no lipids (fats). Most change color twice, beginning as red (pre-ripening fruit flags) and ending as black or blue, an event that may attract the attention of fruit-consumers. Many bird species, both resident and migrant, feed on these fruits and are the principal seed-dispersers. Some mammals also eat them. See text for details.

PIN or FIRE CHERRY

Bright red singular berries on long stalks. Stone (containing seed) in center is quite large. Pulp tastes sour. Plate shows **American Robin.** See Pl. 34 for description of tree.

BLACK CHERRY

Fruits start out red, then turn deep blue-black. Cherries hang in clusters below branch. Taste slightly bitter. Cherry stone large. Plate shows **Eastern Bluebird.**

ALLEGHENY PLUM

Fruits deep reddish, edible. Small tree (to 20 ft.) with alternate, simple, toothed leaves. Grows in thickets in old fields, or in forest understory.

DOWNY SERVICEBERRY

Fruits fleshy, violet; clusters grow upward. Each fruit with several seeds. Fruit tastes sweet. One of the earliest of the summer-fruiting species. See Pl. 32 for description.

FLOWERING RASPBERRY

Fruits red, on stems near branch tips, generally with little taste. Leaves maple-like; twigs and stems very hairy. Fruits appear in early July.

RED MULBERRY

Fruits are tiny and beadlike, in dense, rounded clusters. Sweet-tasting berries are red at first, then turn blue-black. They develop in early to midsummer. Leaves variably lobed.

COMMON HIGHBUSH BLUEBERRY

Fruits start out pinkish, then turn grayish blue to deep blue-black. Berries sweet, each with several small seeds. Plate shows male **Rose-breasted Grosbeak** in changing plumage. See Pl. 33 for description of plant.

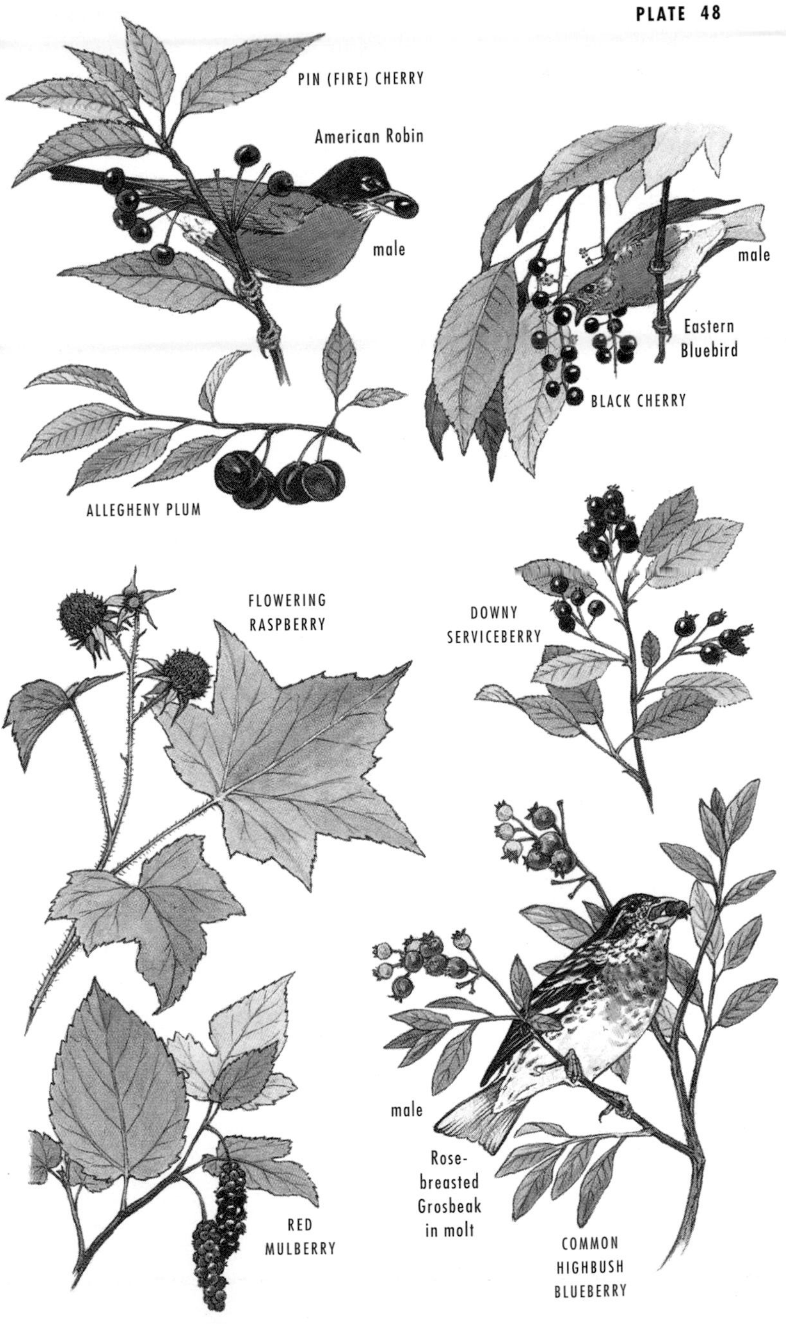
PIN (FIRE) CHERRY
American Robin
male
Eastern
Bluebird
male
BLACK CHERRY
ALLEGHENY PLUM
FLOWERING
RASPBERRY
DOWNY
SERVICEBERRY
male
Rose-
breasted
Grosbeak
in molt
RED
MULBERRY
COMMON
HIGHBUSH
BLUEBERRY

HIGH-QUALITY FRUITS OF LATE SUMMER AND FALL (P. 334)

These species develop fruit during the time when birds are migrating southward; the fruits are very rich in lipids (fat), which birds utilize as fuel for migration. The fruits are not sweet and are often bitter. Birds are the principal seed-dispersers. These fruits are consumed quickly by birds, and rot quickly if not consumed. Some species undergo leaf-color change simultaneously with fruit maturation, perhaps signaling birds of the fruit's presence (foliar fruit flags). See text for details.

SPICEBUSH

Berries become bright red, hanging singly or in twos. Berries, leaves, and twigs very aromatic when crushed. Leaves oval, with smooth margins. Common shrub in mesic forests. See Pl. 10 for description of flower.

FLOWERING DOGWOOD

Cluster of bright red berries at branch tips. Leaves turn orange, then deep red. Plate shows **Swainson's Thrush.** See Pl. 1 for description of plant and Pl. 36 for flower.

SOUTHERN MAGNOLIA

Evergreen. Fruits are bright red drupes, sharply pointed, each with 2 seeds. Fruits are clustered in a reddish brown, cone-like structure. See Pl. 14 for description of plant and Pl. 36 for flower.

SASSAFRAS

Fruits upright, shiny blue-black drupes, each in a red cup on a long red stalk. See Pl. 1 for description of plant and Pl. 36 for flower.

BLACK TUPELO or BLACK GUM

Fruits are large drupes, blue-black, in ones and twos at branch tips. Plate shows **Northern Oriole** in fall plumage (see Pl. 39 for adult male and female in breeding plumage). See Pl. 10 for description of tree.

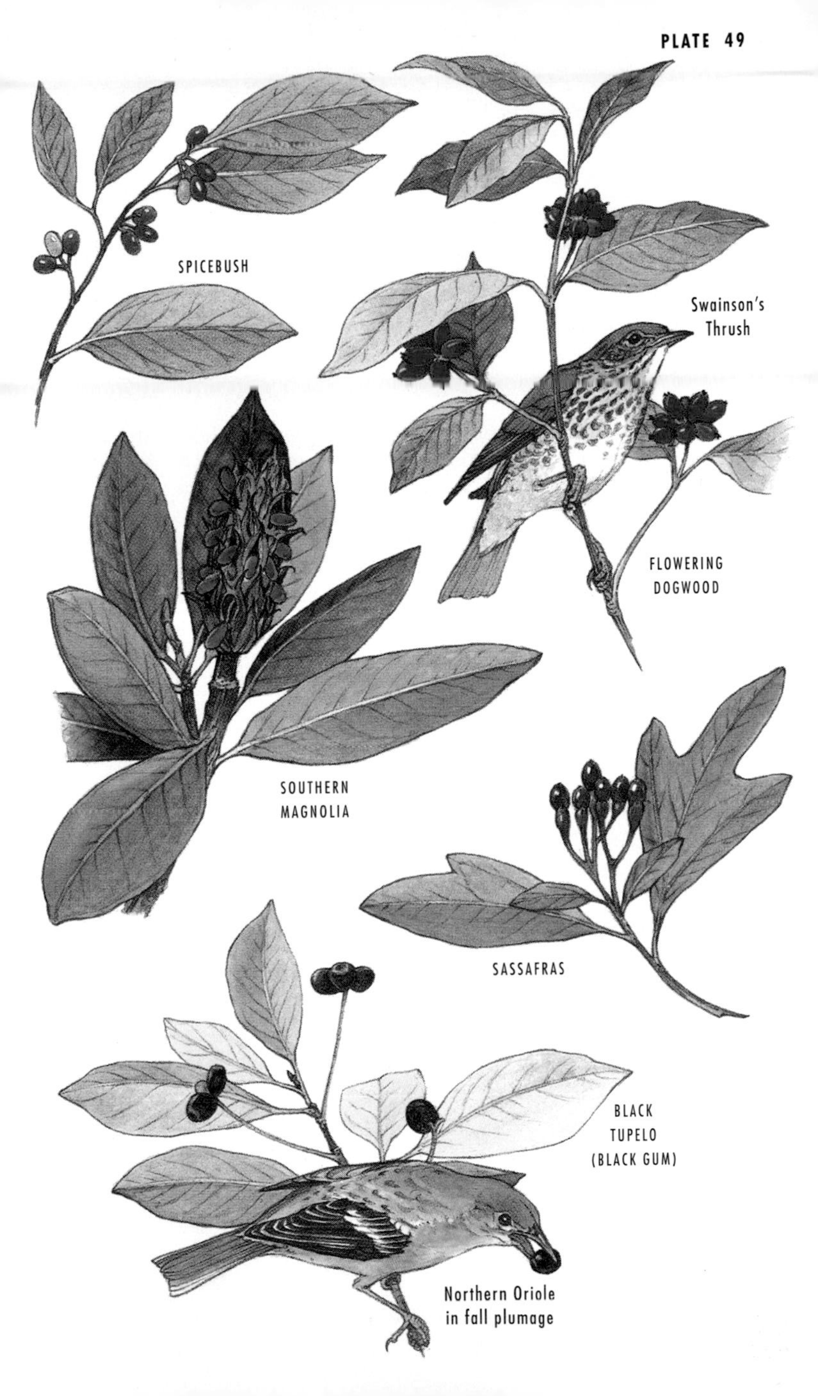
SPICEBUSH
Swainson's Thrush
FLOWERING DOGWOOD
SOUTHERN MAGNOLIA
SASSAFRAS
BLACK TUPELO (BLACK GUM)
Northern Oriole in fall plumage

HIGH/LOW-QUALITY FRUITS OF LATE SUMMER AND FALL (PP. 334–35)

High-quality Fruits

VIRGINIA CREEPER

Berries blue, loosely clustered at branch tips. Grows as a vine; bright orange-red leaves in fall serve well as a foliar fruit flag (see Pl. 49). Plate shows a Red-bellied Woodpecker, which consumes fruits in fall. See Pl. 31 for description of plant.

Low-quality Fruits

These species have fruits that are low in both carbohydrates and lipids (fat), especially the latter. They provide birds with less energy per gram of fruit than the high-quality fruit species and are less favored by migrating birds. They do not rot quickly; many remain on the plant during winter. Birds and mammals eat the fruits in times of food scarcity. Foliar fruit flags (see Pl. 49) are common. Some species—those that keep their fruits all winter—may actually have their seeds dispersed by northward-migrating birds that feed on the fruits in spring. See text for details.

MOUNTAIN-ASH

Bright orange-red, fleshy fruits in dense clusters at branch tips. Leaves compound; leaflets lance-shaped, toothed. Bark grayish. Fruits remain on tree well into winter.

FOX GRAPE

Clusters of fleshy, violet-black grapes hang from vine. High in carbohydrates—sweet-tasting. See Pl. 31 for plant description.

MAPLELEAF VIBURNUM

Fruits are clusters of berries, red at first, later blue-black. Leaves opposite, maple-like. Widespread understory shrub.

COCKSPUR HAWTHORN

Dark red clusters of hard-pulp fruits hang from branch tips, resembling small apples. Fruits persist throughout the winter. Plate shows **Fox Squirrel.**

SMOOTH SUMAC

Fruits densely clustered in a spike-like array; fruits red, with sticky hairs. See Pl. 32 for description of plant.

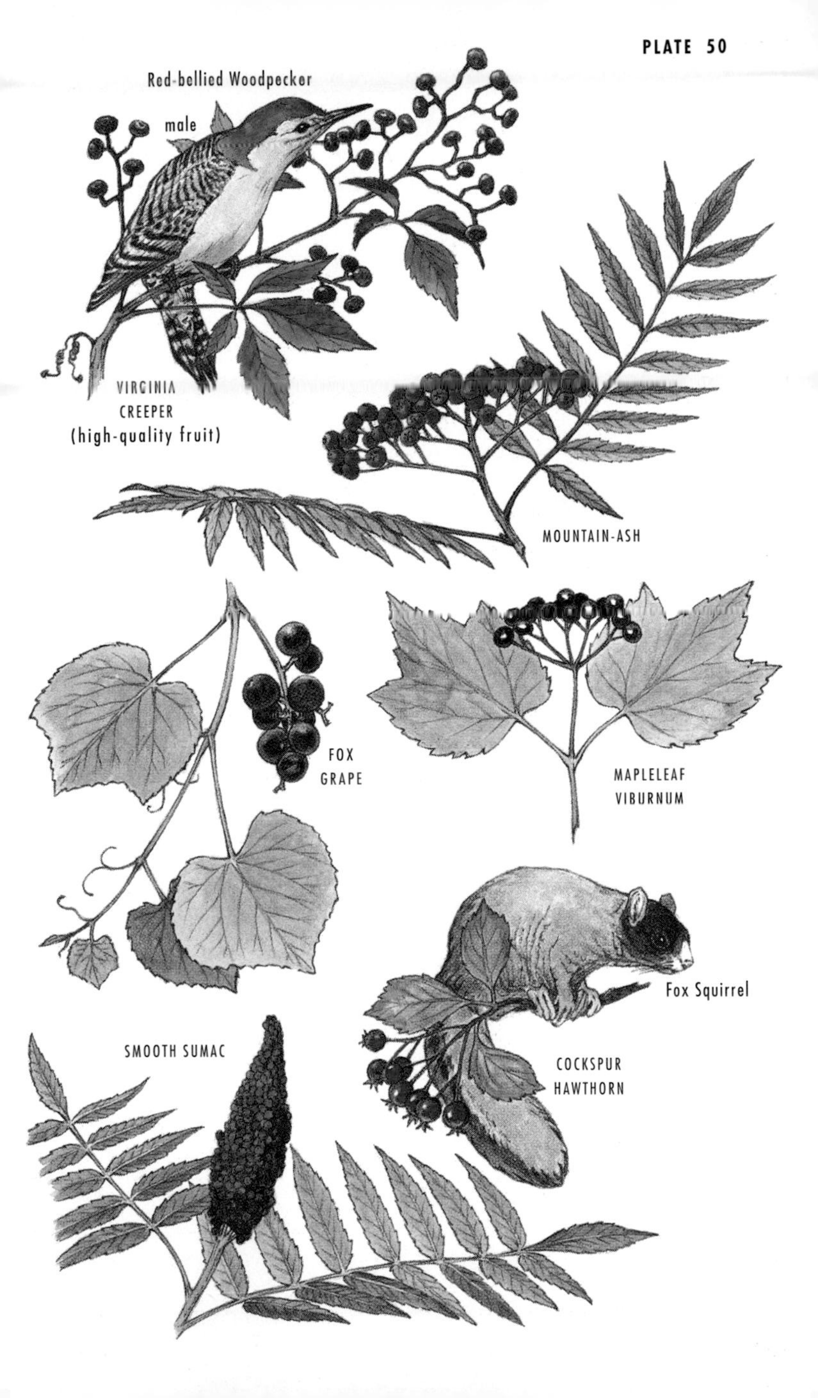
Red-bellied Woodpecker
male
VIRGINIA
CREEPER
(high-quality fruit)
MOUNTAIN-ASH
FOX
GRAPE
MAPLELEAF
VIBURNUM
Fox Squirrel
SMOOTH SUMAC
COCKSPUR
HAWTHORN

Autumn and Winter

The coming of autumn is a time of change, of slowing, of approaching dormancy. In the Eastern Deciduous Forest, leaves change color and drop, perhaps the most dramatic characteristic of autumn, but many more events are occurring in nature as well. In the northern Boreal Forest winter-hardening, the filling of the coniferous trees with resin that acts as antifreeze, prepares the spruces and firs for winter's cold. In southern forests, where there is a mixture of deciduous and evergreen species, the season preceding winter is not as extremely felt, but the many fruit-producing species of trees and shrubs and the large flocks of birds visiting them are the field marks of autumn.

The flush of newly fallen leaves covering most eastern forests in fall signals the commencement of another cycle of decomposition, to be followed by the recycling of the many chemicals necessary for new leaves to grow in the spring. The forest litter layer, which fills anew with dead leaves, is the habitat of a remarkable community of decomposer bacteria, fungi, and various animals, ranging from tiny scorpionlike creatures to "large" beetles and earthworms. The richly diverse fungal community is never more evident than in autumn, when various reproductive structures such as colorful mushrooms, puffballs, and brackets seem to sprout overnight (see p. 418).

Flocks of migrant passerine birds, many searching for fruits, briefly visit fields and forests, joined by such other migrants as Monarch butterflies and Red Bats, all moving southward. Some mammals, including the Eastern Chipmunk, Woodchuck, several bat species, and Black Bear, are preparing to enter various degrees of deep sleep that will carry them peacefully through the winter. Others, like White-footed Mice, Gray and Fox squirrels, and White-tailed Deer, will remain active throughout the winter.

As winter arrives, fields and forests may seem almost devoid of activity, but such a perception is mistaken. A woodlot will suddenly be visited by mixed flocks of birds, each species foraging in a particular manner, some probing bark, some gleaning tiny overwintering insects or insect egg cases from tree branches. Flocks of sparrows feed on seeds of goldenrod and other old field weeds. Certain northern birds may become extraordinarily abundant in winter, so much so that they are called *irruptive species* (see p. 445), because their numbers vary so much from one winter to the next. Even some mammals, such as Meadow Voles, vary dramatically in abundance from year to year.

Within the most snow-covered woodland there are animals to be seen. Several remarkable insects, such as tiny gray springtails, may become incredibly abundant over the surface of newly fallen snow.

SOIL NATURAL HISTORY

Soil is vital. Like sunlight and water, it provides raw materials necessary for green plants to produce food that sustains the entire ecological community. Calcium, potassium, nitrogen, phosphorus, magnesium, sodium, and other elements essential to living systems are taken up from soil by plant root systems. These elements are *recycled,* having been extracted from dead vegetation, animal carcasses, and animal wastes that accumulate as organic litter. A complex community of bacteria, fungi, protozoans, and many different kinds of animals dwell within the soil and these myriad creatures facilitate the process of *decomposition,* followed by *biogeochemical cycling,* the movement of atoms between the soil and the plants. Autumn is an ideal time to observe the recycling drama from its beginning, and to become acquainted with its many actors.

The Litter Layer

Across eastern North America, autumn is a time when leaves drop. Though fallen leaves are present throughout the year, fall is the season of greatest input and accumulation, except in forests consisting solely or principally of evergreen species. (Evergreens drop needles or leaves, a few at a time, throughout the year and most temperate evergreen forests have very well-developed litter layers.) Observe the carpet of dried leaves, twigs, and other plant debris that covers the forest floor. This is called the *litter layer,* and it is here that recycling begins. Carefully examine leaves, beginning with the uppermost ones and working downward through the plant litter. The upper leaves of the litter are dry, mostly intact, and (usually) not too heavily damaged by herbivorous insects. A brief survey of a dozen leaves or so should demonstrate to you that most newly fallen leaves are easily recognizable as to tree or shrub species and are, indeed, largely undamaged when they drop. Newly fallen dead leaves contain a great deal of potential food energy, which is tapped by the many inhabitants of the soil.

As you probe deeper, sifting through the litter layer, the leaves become increasingly moist and flimsier. Many have tiny holes in the blade and along the edges. These holes, made by minute insects and mites (see following section), allow fungi and bacteria to easily enter the interior of the fallen leaf and represent the initial stages of decomposition. Leaves deeper in the litter become increasingly skeletonized, some with just the petiole and veins remaining but the softer tissues gone. Decomposing leaves become darker in color, some virtually black, and often feel slimy, due to a coating of microorganisms. Look closely, preferably with the aid of a magnifying glass or hand lens, and you will often see thin, white fungal networks crisscrossing the leaves. Deeper still, the leaves become broken up, fragmentized, and basically unrecognizable. Here the litter layer is dark, gritty, and moist. This dark layer of decomposing organic material is the site of *humus formation,* the point of transition between the complex organic matter of formerly living systems, and simple more elemental inorganic material, the raw materials needed by living systems. Below the rich, dark humus layer is the soil itself, the temporary repository for the atoms of decomposed plants and animals.

Humus Formation and Decomposition

As you sift through the leaf litter from topmost, newly fallen leaves to humus to soil you are witnessing a succession through time. Each leaf is destined to be converted to humus and further broken down to simple chemicals by hosts of bacteria, fungi, protozoans, and many kinds of animals. The process by which dead vegetation, dead animals, and animal waste products are converted from complex carbohydrates and proteins to basic atoms is called *decomposition.* Without decomposition, all atoms would remain "locked up" in animal or plant tissue and recycling could not occur. Just as life on earth depends on the sun for energy, so too does it depend on the presence of decomposers to carry out recycling.

Types of Humus

OBSERVATIONS: Humus varies from forest to forest. There are generally two basic types of humus, one called *mor,* the other called *mull.* Many forests have a mixture of the two types. The best way to identify mor or mull humus is by forest type and the kinds of animals that you find in the litter (see following section on soil-dwelling animals).

Mor humus is characteristic of forests with thick litter layers, such as coniferous forests. Many sandy soils support a mor humus. Decomposing material is in a distinct surface layer, usually dark in color. Mor humus is acidic and low in calcium. Most decomposers are fungi, including many fungal species that live mutualistically with tree roots (see fungi, below). Bacteria, especially those that are involved in capturing nitrogen from the atmosphere, are in short supply. Animals of a mor humus are the tiny springtails and mites (see p. 433), often in extreme abundance. Earthworms are sparse in mor-humus soils.

Mull humus is typical of deciduous forests with few if any conifers. Forests with an abundance of basswood, aspen, alder, or elm, or old fields with Red Cedar are notable for the formation of mull humus. Mull is an alkaline or chemically neutral humus, and hence is not acidic. Calcium levels are high and the soil tends to be a rich, well-mixed loam. Fungi are present but there is also a rich flora of bacteria, especially those that capture nitrogen from the atmosphere. Many earthworms can be found, in addition to slugs, millipedes, and many insect larvae.

EXPLANATION: Both leaf chemistry and amount of rainfall influence decomposition. Rain water contributes both hydrogen and oxygen to the soil and the more hydrogen there is, the more acidic (generally) the soil. Hydrogen atoms from rainfall replace, or exchange, with atoms such as calcium and potassium that are electrostatically attached to particles of clay. The calcium and other minerals removed by rainfall are washed or "leached" into lower soil layers. As hydrogen accumulates, the soil becomes increasingly acidic; thus acidity is strongly influenced by rainfall. Leaf decomposition, especially of conifer needles, also adds to the acidity of soils. The more acidic a soil is, the more slowly decomposition will occur, since acidity retards both the growth and activity of microorganisms. Humus type is therefore the product of both the physical environment and chemistry of leaves.

The terms *mor* and *mull* may be easily confused. Remember, soils with a mor humus are "mor" acidic than soils with a mull-type humus. Mull that over.

THE DECOMPOSERS

Dead organisms are used as a source of energy by myriad bacteria, fungi, protozoans, and soil-dwelling animals. You will not be able to directly observe bacteria or protozoans—they are much too small and can only be seen with the aid of a microscope magnifying them from 100 to 1500 times. Many of the soil-dwelling animals are also so tiny that they are best observed with a low-power microscope, though a good hand lens or powerful magnifying glass will suffice. Soil animals are best observed by extracting them using a simple device called a Berlese funnel, described on p. 423 (see Fig. 55).

Fungi are largely microscopic, but their thicker strands can be observed with the unaided eye as a mesh network of white fibers growing on decaying logs and leaves. Autumn is when many fungi reproduce and the colorful though ephemeral mushrooms, puffballs, and brackets reveal the presence of the subterranean fungal network.

Fungi (Pl. 53)

Fall is a great season for the mushroom fancier. After an autumn rainfall forests abound with mushrooms, puffballs, and brackets of many shapes and colors. Mushrooms, which are actually reproductive (fruiting) structures, can be observed throughout spring and summer, but peak in abundance and diversity in late summer and throughout autumn. A walk through a freshly moistened woodlot should reveal many species. This is not a book on mushroom identification (see *A Field Guide to the Mushrooms*) and you must bear in mind that edible mushrooms and poisonous mushrooms can look confusingly similar.

Never eat a mushroom unless you are 100% certain that it is nonpoisonous and safe to eat.

Fungi lack chlorophyll and so cannot photosynthesize. They must take in complex chemicals, the carbohydrates, fats, and proteins, just as animals do. Because of this need, all fungi, like all animals, depend on being able to obtain organic material. Fungi grow right upon or into their food sources, penetrating leaves and other food sources with thin strands called *hyphae.* A typical forest-dwelling fungus exists as a dense, irregularly shaped underground network of hyphae called a *mycelium.* Mycelial networks penetrate soil, dead leaves, branches, logs, animal carcasses, feces, and often living root systems. Some, like the notorious Athlete's Foot fungus, penetrate naturalists. Hyphae secrete digestive enzymes that dissolve the food source, permitting the mycelium to take up simple food molecules.

You will only be able to see the largest of the mycelial fibers unless you have a microscope at your disposal. Mycelia appear as a soft white mesh on the undersides of decomposing leaves and inside moist rotting logs. The mycelium is the "main body" of each fungus species and it may live for many hundreds of years, regenerating and replacing broken or dead hyphae. Periodically, mycelia generate their reproductive structures, most of which are readily visible as mushrooms above ground (truffles are an exception—they remain subterranean).

Fungi reproduce by making billions of extremely tiny dust-sized *spores*. Once liberated, spores depend on being blown by wind, washed by rain, or accidentally transported by an animal. The few spores fortunate enough to land in a suitable location begin a new mycelium. Only a tiny percentage of spores survive to disperse the fungal species to a new location.

Types of Fungal Reproductive Structures

1. GILLED MUSHROOMS are the "typical" mushrooms. Many are whitish but many others are highly colorful (deep crimson, scarlet, golden-orange, or purple). Many are quite poisonous. Spores are shed from the gilled undersides of the caps. There are numerous families of fungi with gills and all tend to grow from the ground or decomposing branches and stumps. These fungi are important forest and field decomposers.

2. FLESHY PORE MUSHROOMS (Family Boletaceae) differ from gilled mushrooms because the spores are inside numerous small pores on the underside of the cap. Most tube mushrooms, also called boletes, are yellowish tan and the stems are covered with coarse, grayish brown scales. They often turn reddish and then black when the cap is cut. You will notice that tube mushrooms tend to grow near tree trunks. These fungi grow in intimate relationship with the root systems of many tree species (see mycorrhizal fungi, p. 422).

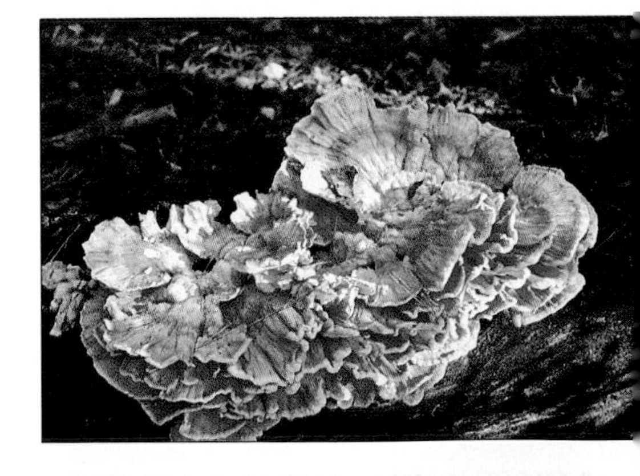

Turkeytail Bracket Fungus. Most of this fungus is inconspicuous, hidden in the tree itself. This is merely the spore-producing structure.

3. STINKHORNS (Family Phallaceae) are sharply pointed fungi, usually with pink or red stems and dark caps. They grow from a case resembling an egg. The cap has a strong odor of rotting meat, which attracts flies and carrion beetles that inadvertently help disperse the spores. Stinkhorns are common among wood chips and decaying litter.

4. BRACKET OR PORE FUNGI (Family Polyporaceae) grow attached to decomposing logs and tree trunks and may be quite hardened (woody) compared with most mushrooms, which are moist and soft. Brackets, because of their woody character, tend to last longer than other mushrooms. Colors range from browns and whites to bright orange, red, and deep mahogany. As in the tube mushrooms, spores are inside tiny pores on the underside of the bracket. Brackets sometimes occur singly, but are most often clustered, sometimes in bracket "bouquets."

5. PUFFBALLS (Family Lycoperdaceae) are distinctive round balls, usually brownish or tan, which "explode" when struck, emitting a tiny dust cloud containing over a trillion spores. The Collared Earthstar is a common species found in deciduous woodlands. It releases spores during heavy rainfall.

6. CUP FUNGI (Family Pezizales) resemble small inverted mushrooms. As the name implies, the cup edge turns upward, not downward, though some, like the morels, have wrinkled or deeply ridged caps rather than smooth, upturned caps. Many cup fungi are brightly colored—scarlet, orange, yellow, or turquoise. Spores are in the upper inside part of the cup or within the ridges on the outer surface.

7. CORAL FUNGI (Family Clavariaceae) are well named—most resemble clumps of coral. They grow as branched clumps with spores produced on the branches. Most coral fungi grow on decomposing logs and stumps but some inhabit living root systems. Typical colors are ivory and light brown.

8. CHANTERELLES (Family Cantharellaceae) look like mushrooms with wavy, upwardly turned edges. The underside is deeply ridged, and spores are produced within the ridges.

9. JELLY FUNGI (Family Tremellales) are colorful (orange and red) and appear shiny and slimy, though they are usually dry. Most are irregularly shaped, resembling a mass of jelly, but some grow in ribbonlike patterns. Spores are produced on the surface following moistening by rain. They grow on branches and logs.

Fungal and Bacterial Decomposition

Consider the vast number of leaves that fall inside a forest in autumn. The total weight, or biomass, of these leaves represents the result of the previous growing season, having been produced from early spring through autumn. Though many, if not most, leaves just fallen from a tree will show some signs of damage by herbivorous insects, it is easy to see that the leaves are generally intact. What this means is that the vast majority of the energy captured by the forest plants and stored within the leaves never enters herbivores. Instead, it drops with the leaf from the tree, to enter the *decomposer food chain*, at the base of which are the bacteria and fungi. In excess of 90% of the dry matter made by forest plants during the growing season goes directly to the decomposers.

Bacteria and fungi thrive in warm, moist conditions, hence there is little decomposition in the cold of winter except in the warmer southern climes. Decomposition is most intensive during growing season. As this year's new leaves are made in the forest canopy, last year's are decomposed in the litter and soil (though leaves from some species of trees routinely take up to 3 years to decompose—see below).

Leaves are not easy to decompose. They are covered by a waxy outer layer and are made up in large part of very big molecules, cellulose and lignin, which are hard to break down, even for fungi and bacteria. In addition, leaves contain various defense compounds such as tannins and phenolics. These too are hard to decompose and must leach out, or "leak," from the leaves before effective decomposition occurs. In areas where tannins are abundant in the leaves, streams will be dark brown, a "tannin tea."

Fungi and bacteria enter leaves when soil animals chew tiny holes in the leaf tissue. As soil animals, fungi, and bacteria digest the softer, simpler compounds in the leaf, the leaf skeletonizes, leaving the thicker, tougher veins temporarily intact. Decomposition accelerates as leaves are chewed by insects and mites (see below) because these animals effectively increase the surface area where fungi and bacteria colonize.

The rate at which a leaf decomposes is strongly influenced by the ratio of carbon to nitrogen in the leaf. In general, carbon reflects difficult to decompose chemicals such as cellulose, while nitrogen is representative of easier to digest chemicals such as proteins. Thus, the lower the carbon to nitrogen (C/N) ratio, the easier and quicker the decomposition. An animal carcass, which (aside from the mineral-laden bones) is very easy to decompose, has a C/N ratio of about 3–5:1. Organic humus, an intermediate product of the decomposition process, has a C/N ratio of 20:1.

Tree species vary in C/N ratio. The lowest values are found in such species as alders (15:1), Sugar Maple (20:1), hornbeams (23:1), and elms (28:1). These species typically require only a year to a year and a half for complete decomposition. Mid-range C/N ratios occur in species such as basswoods (37:1) and some oaks (47:1). These species require up to 2½ years for a leaf to decompose. High C/N ratios are typical of some maples other than Sugar Maple (52:1), some oaks (50–60:1), aspens (63:1), beech (51:1), pines (66:1), and larch (113:1). These species can require more than 3 years for a leaf or needle to completely decompose.

The combination of a short growing season, high levels of phenolics and/or tannins, and an unfavorable C/N ratio act to slow decomposition and thus gradually thicken the litter layer, especially in Northern Hardwood forests, pine forests, and Boreal Forests, making the litter dense and carpetlike in these woodlands.

Mycorrhizal Fungi

Though most fungi live on dead organic material, some are parasitic, living inside root systems of living trees. However, most root fungi are not parasitic but mutualistic. In other words, the fungi and trees are interdependent and each gains from the relationship (see lichens, p. 206). Many of the fungi that produce fleshy pored mushrooms, as well as a few of the gilled mushroom species, are *mycorrhizae,* a general term for fungi that live mutualistically in tree roots. The mycelial strands of these fungi grow into the roots and take some of the sugary compounds produced by the tree during photosynthesis. However, mycorrhizal fungi benefit the tree because they take in minerals from the soil, which are then used by the tree. Many trees depend heavily on mycorrhizae and may not be successful without them. Mycorrhizae are very common near conifers and many hardwoods such as oaks.

Slime Molds

Slime molds are unique protozoan colonies that closely resemble fungi in appearance. They are most often observed on decomposing logs, appearing as a mass of bright yellow slime superficially resembling a fungus. The mass is actually an amoeba-like organism, called a *plasmodium,* which eventually makes spores. The spores are contained in tiny *sporangia,* which grow from tiny stalks on the body of the plasmodium. The developing plasmodium grows by feeding on bacteria and decomposed organic material.

Litter and Soil Animals

OBSERVATIONS: Many different animals inhabit the soil, ranging from mouse-sized moles to tiny insects. Some, like the diminutive mites and springtails (described below), occur in immense numbers. The soil-animal community depends on fallen debris from the forest canopy. Dried leaves and twigs, diseased stumps, decomposing logs, and animal remains make up the food base for protozoa, roundworms, earthworms, potworms, millipedes, isopods, mites, springtails, and many others. These are fed upon by predators—false-scorpions, spiders, centipedes, beetles, predatory mites, and moles.

Seeing litter and soil animals requires patience and careful observation. In the field, the best method is to pick carefully through the litter layer, having a hand lens or magnifying glass at the ready. Many of the larger soil- and litter-dwellers are active foragers and can be spotted among the decomposing leaves. Another excellent way of finding these creatures is to carefully inspect a decomposing log. Peel away loose bark, pick through the soft, moist decaying wood, and you will find carpenter ants and other ants, millipedes, centipedes, slugs, isopods, spiders, and maybe a Red-backed Salamander or Brown Snake. The log may even host a termite colony.

The most effective way to see the tiny creatures of the soil, the springtails, mites, roundworms, false-scorpions, and smaller insect larvae, is with a device called a *Berlese funnel,* which is easy to make (see Fig. 55). A Berlese funnel is merely an ordinary fun-

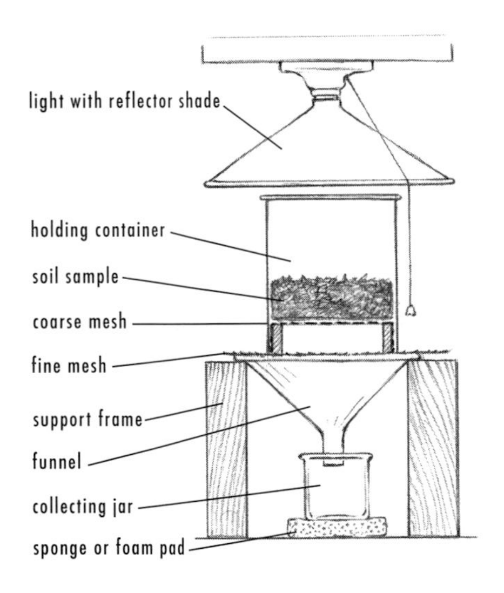

Fig. 55. Berlese funnel.

nel with a wide opening covered by wire mesh, topped with a layer of burlap or other porous cloth. The funnel is placed beneath a low-wattage light bulb (refrigerator bulbs work well for this—too much heat will kill the animals). A soil sample is put on the burlap mesh and heated by the light. Beneath the thin neck of the funnel is a collection jar, filled with either water or alcohol. The gentle heat from the light bulb causes the soil to dry out, forcing the tiny soil animals ever downward, away from the drying. Soon they reach the bottom of the soil sample, crawl through the burlap cloth and slide down the neck of the funnel, dropping into the collecting jar. This remarkably simple procedure is very effective for seeing many of the most interesting soil inhabitants. Though they are tiny, they will be readily visible in the collection jar and can be observed under a hand lens or magnifying glass. A low-power microscope is ideal.

There is a remarkable diversity of animal life beneath your feet as you tread over the forest litter. For practicality in observing these creatures, it is convenient to divide them according to size.

1. *Animals large enough to be seen well without a hand lens.* These include the slugs, snails, earthworms, isopods, millipedes, centipedes, beetle adults and larvae, ants, various spiders, moles, and shrews.

2. *Animals too small to be seen well without a hand lens.* These include the many kinds of mites, the springtails, the false-scorpions, the roundworms, the potworms, and fly larvae. These animals are best observed after extraction through a Berlese funnel.

3. *Animals too small to be seen without a powerful microscope.* These animals are mentioned here to complete the account, but you will not be able to observe them without strong optical aid. These include rotifers and protozoans such as various amoebas and ciliates. They will not be treated further in this guide.

Larger Animals of Litter and Soil

Slugs and Snails

Slugs and snails are members of the phylum Mollusca, class Gastropoda. Mollusks are basically aquatic creatures; land snails and

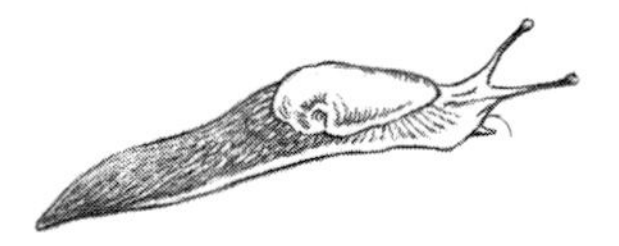

Fig. 56. Slug.

slugs must remain moist or perish. In the damp soil and litter they find satisfactory conditions. Slugs are shell-less and very slimy, leaving slime trails as they glide about on their nocturnal foraging excursions. Slugs and snails feed on decomposing leaf litter and animal carcasses, and some slug species eat fungi and living vegetation (becoming serious pests to gardeners). Snails are most common in areas with high calcium soil. Slugs are much more cosmopolitan, occurring in all soil types.

Earthworms

Earthworms are without question the best-known soil animals. Members of the phylum Annelida, the segmented worms, their segmentation is readily visible and their musculature permits them to both contract and lengthen, allowing them to burrow effectively. Charles Darwin, in his last major work, demonstrated how effectively earthworms loosen and aerate soil. In 1881, Darwin estimated that English pastures contained approximately 23,000 earthworms per acre. Earthworms are most numerous in soils that are either neutral or slightly alkaline. They are much less common in acid soils. They favor soils with high levels of calcium, where they can be as abundant as 300–400 per square meter. However, most forest populations rarely exceed 100 per square meter. Often called "night-crawlers," earthworms forage in the cool of evening, coming to the surface of their burrows to feed on decaying leaves. They have definite preferences among leaves, prefering those with high sugar and high nitrogen levels (low C/N ratio). Alder, birch, Sugar Maple, ash, sycamore, and elm leaves are preferred; beech and oak leaves and pine, spruce, and larch needles are least appetizing. Earthworms are hermaphrodites, meaning that each animal has both male and female reproductive organs. This condition is quite adaptive to creatures living in the dark soil, forging random burrows. It ensures that when, by chance, two animals meet, they can mate. Earthworms lay eggs in tiny cocoons. There are many different earthworm species and it is not easy to separate them.

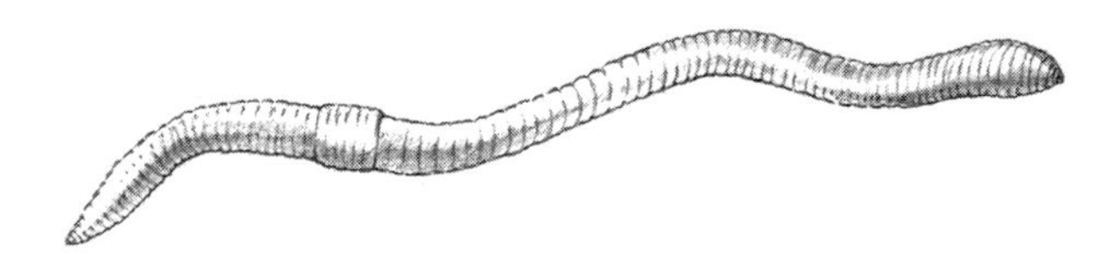

Fig. 57. Earthworm.

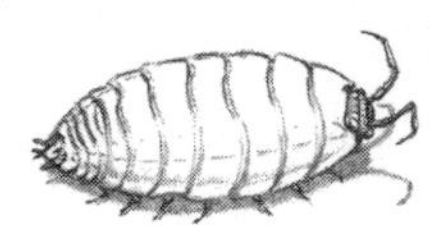

Fig. 58. Isopod (wood louse).

Isopods

Called "wood lice" or "pillbugs," isopods are members of the huge phylum Arthropoda and are in the subphylum Crustacea, which includes the crabs, lobsters, and shrimps. Though they live on land, isopods require very moist conditions because, like other crustaceans, they breathe by gills. Consequently, isopods remain under cover of litter, decaying logs, and other detritus, and forage at night, especially during and after rainfall. Look for them under loose bark in decaying logs. Their thick, segmented exoskeleton is covered by a water-resistant cuticle, which helps keep them from drying out. They resemble tiny gray tanks, ambling along on 14 pairs of short legs. A conspicuous pair of antennae protrudes from their tiny heads. When threatened, many species curl up into a ball. There are hundreds of isopod species and isopods are found in all soil types, whether field or forest. They feed on decaying vegetation, carrion, and animal feces.

Millipedes

Millipedes are segmented and covered by an exoskeleton and cuticle, and thus look somewhat like very long, thin isopods. They are also arthropods but are in their own class, the Diplopoda, a name referring to their 2 pairs of legs per body segment. They are quite susceptible to drying and remain within the litter and under logs. Different species occur in different soil types but virtually all soils contain at least some species. Millipedes feed on dead leaves and decaying wood. When threatened, they curl up into a tight ball with their legs tucked within the curl.

Fig. 59. Centipede, compared with a millipede.

CENTIPEDES

Centipedes are also arthropods, in class Chilopoda. Though they superficially resemble millipedes, they are quite different because they are mostly predators (some species do occasionally feed on dead plant material). They actively seek and capture insects such as springtails, flies, and aphids, as well as spiders, millipedes, slugs, roundworms, and even other centipedes. Their segmented bodies have 1 pair of legs per segment and their legs are long, permitting them to move rapidly. The legs on the first body segment past the head are modified into poisonous claws, with which they capture prey. Most species have long antennae.

Compare a centipede, a millipede, and an isopod. Notice that they have much in common regarding body form, a result of their adaptations to the same type of habitat, but note also how the centipede is specialized as a predator by virtue of its long legs and poisonous claws.

BEETLE ADULTS

Beetles are insects, also in the Arthropoda. Insecta is the largest class of any animals and, within the Insecta, the beetles comprise the largest order (Coleoptera). Some entomologists estimate that there are 300,000 species in the order Coleoptera. By compari

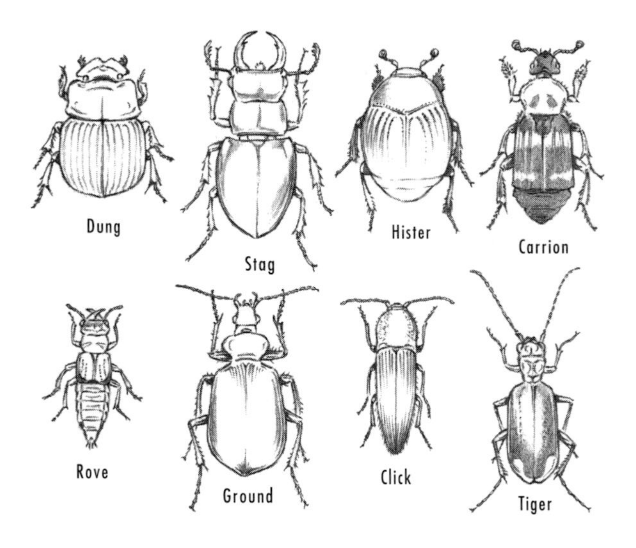

Fig. 60. Beetles.

son, there are an estimated 8700 bird species on earth. The most characteristic feature of adult beetles is the elytra, or hardened forewings, which are often metallic in color and give the insect a shiny, armored appearance. Though many beetle families have ground- and soil-inhabiting species, the following families are most common in soil and litter. These are:

SCARAB BEETLES (Scarabaeidae)
STAG BEETLES (Lucanidae)
HISTER BEETLES (Histeridae)
CARRION BEETLES (Silphidae)
ROVE BEETLES (Staphylinidae)
GROUND BEETLES (Carabidae)
CLICK BEETLES (Elateridae)
TIGER BEETLES (Cicindelidae)

Search for beetles in the litter, under rocks, and in decaying logs. Many beetles can be attracted to lights at night. In this guide, only brief general descriptions will be included about these families. For additional information, consult *A Field Guide to the Beetles.*

SCARAB BEETLES include such familiar insects as june bugs and Japanese Beetles. They are rounded and often metallic-looking and shiny. Adults do not actually live in the soil but larvae do (see below). Many adult scarabs live on the ground, including the dung beetles and tumblebugs. These curious insects are noted for their habit of shaping feces into a dung ball and rolling it into a burrow they have excavated. Larvae use the ball of dung as a food source. Tumblebugs are very rounded and black in color.

STAG BEETLES are among the largest insects of the forest floor. Both males and females are large (up to 2 in.), black or brown. Males have very long jaws, resembling antlers. Though often seen on the forest floor, they feed on leaves and bark. Stag beetles are easily attracted to lights. *Be careful*—the jaws are strong and these insects can give a good pinch.

Charles Darwin wondered why male stag beetles have long, antlerlike jaws and hypothesized that these structures aided males in competition against each other. Males are larger than females, a fact that suggested to Darwin that the large size was an additional advantage to males in competition. He called this theory Sexual Selection, because the characteristics of the males were largely determined by how they helped them secure females. Recent research has shown that male stag beetles do indeed engage in combat with other males, with the successful male dislodging the loser from the tree branch.

HISTER BEETLES are among the smaller ground-dwelling beetles. They are rounded, black, greenish, or bronze, and generally shiny.

They are found around carrion or feces, where they prey on other animals attracted to the organic material. Like insect versions of the turtle, hister beetles retract their legs and antennae when disturbed, remaining motionless until the danger passes.

CARRION BEETLES are very well named. These large, flattened, and often colorful insects are attracted to dead animals, which they often bury. To watch several of these lumbering beetles patiently burying a deceased chipmunk is to witness a remarkable feat of nature. Both adults and larvae feed on the carcass. Most species of carrion beetles are black with large spots of red or orange. They are active on the forest floor at night.

ROVE BEETLES are easy to identify by their peculiar habit of scurrying across the forest floor with an upraised abdomen. Their elytra, or thickened forewings, are very short and thus these beetles have more conspicuous abdomens than other beetles. Rove beetles are predatory and can bite vigorously. Many species frequent carrion, where they prey on other insects.

GROUND BEETLES, as the name implies, inhabit litter, logs, and are under rocks in wooded areas. Most species are glossy black, though some are iridescent green and blue. Like tiger beetles (see below), they have long legs, conspicuous eyes, and are fast moving. Most are nocturnal, pursuing insect prey. One group of ground beetles, the bombardiers, has a very unique defense system. They emit a toxic liquid from their anal glands that actually "puffs" like smoke, permitting the beetle to escape. Other ground beetles are known to emit strong odors when disturbed.

CLICK BEETLES are named for their habit of snapping their bodies (emitting an audible *click*) when turned upside down. This action usually returns the insect to right-side up. They are elongate beetles, with long legs, and they frequent the forest litter. One common species, the **EASTERN EYED CLICK BEETLE,** has very conspicuous black eyespots on its thorax.

TIGER BEETLES are long-legged, large-eyed beetles that usually have a metallic sheen. Their long legs make them fast runners and they are active in the daytime, scurrying about on the ground. They mostly frequent sandy dry soils, in which they burrow, though some species occur in moist woodlands. Adults are voracious predators, with large jaws used to capture insects.

BEETLE LARVAE

Beetles, like butterflies and moths, undergo a complete metamorphosis, and larval beetles, called *grubs,* are common inhabitants of the soil and litter. They superficially resemble small, pale segmented worms, but a closer look will usually reveal 3 pairs of legs on the thorax, a prominent abdomen, and a pair of short antennae. Note, however, that some beetle grubs have vestigial legs and some lack legs entirely (see the wireworms, below). Many species are predatory and a look at the insect through a hand lens will enable you to see its tiny jaws.

Grubs vary depending upon which family the grub belongs to. Scarab beetle grubs are very thick and juicy, and are actually eaten as food in some tropical countries. The common alien Japanese Beetle is a scarab; birds such as European Starlings (also aliens) probe soil in search of its grubs. Scarab grubs feed on decaying vegetation and live roots, and can be pest species to gardeners. Grubs in the rove beetle family are robust but not as thick as scarabs, and tend to be more bristly. Most are carnivores. Ground beetle grubs are flattened and elongate, and can move quickly. They are also carnivores. Hister beetle grubs are thickened and have extremely short legs and antennae. They prey on grubs of other beetles as well as on fly larvae. Click beetle grubs are called wireworms, and appear more wormlike than beetle-like. They lack legs entirely and resemble fly larvae (see below) more than beetle larvae. They also have extremely short antennae and bristly bodies. Wireworms are predators, especially on other beetle grubs. Many also feed on live plant material and are considered pest insects. The tiger beetles have long-legged, predatory grubs that live in burrows in sandy soils. They have conspicuous dorsal hooks on the raised upper surface of their fifth abdominal segment.

Ants

Ants are familiar to anyone who has ever taken a picnic lunch into the field. They are members of the order Hymenoptera, along with the bees and wasps. Like many hymenopterans, ants are primarily social insects. They form large colonies organized by caste. There is a single queen, who is cared for by thousands of workers, all of which are genetic sisters. The tasks of the workers are to gather food, care for larvae, and attend the queen. There may also be soldiers in the colony, large ants with formidable jaws that defend the colony or capture other ants, used as "slaves." Ants communicate through an elaborate biochemistry. They emit external hormones, called pheromones, which effectively signal a trail to food. Periodically, reproduction takes place when winged males and females leave the colony *en masse* to mate. Fertilized females begin new colonies, while males simply perish.

There are many species of ants in fields and forests and they occupy several ecological roles. Some are predators, some scavengers, some nectar-feeders. When ants take decomposing vegetation below ground into their burrows, they help recycle this ma-

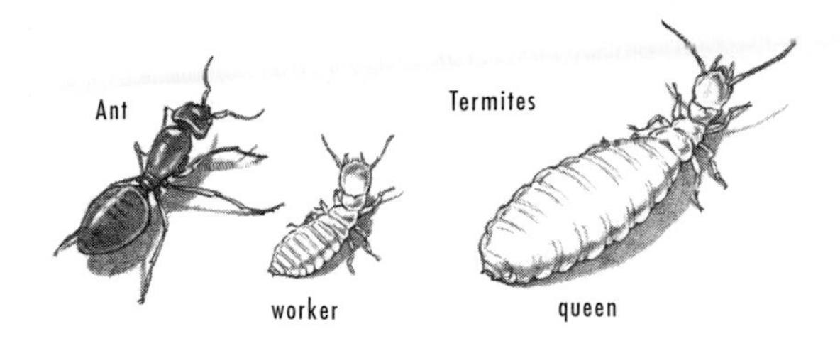

Fig. 61 . Ant, compared with a termite worker and termite queen.

terial directly to the soil. Their burrowing activities also contribute to aeration of the soil.

Two common species inhabiting forests and surrounding areas (including houses) are the **BLACK CARPENTER ANT** and the **LITTLE BLACK ANT.** The Carpenter Ant is large and brownish black; it tunnels inside wood, including houses, as well as dead stumps and trees. Tunneling is done to provide space for the colony, and the ants eat insects and sugary substances. Carpenter Ants can bite but do not sting. Little Black Ants are notorious for gaining entry into kitchens and pantries. The colonies are in subterranean nests, but the workers forage above ground and have no reluctance to enter houses.

RED ANTS, identified by their color, are also common subterranean woodland inhabitants. Some species are known for their habit of "milking" aphids. The ants stroke the aphids and cause them to evacuate sweet "honeydew," which the ants relish as a food source.

The reddish, large-jawed **FIRE ANT** is the most irritating ant in our area, presently confined to the Southeast. These ants can both sting and bite, and their colonies are to be avoided.

TERMITES

Termites, members of the insect order Isoptera, are among the only animals capable of eating and digesting wood. They feed on decaying trees, fallen logs, and, unfortunately, houses. Termites, like ants, are social insects, and live in huge subterranean colonies where a single very large queen provides all of the eggs. The queen, eggs, and larvae are attended by thousands of pale whitish workers. Termites, especially the long-winged, blackish reproductive males and females, are frequently confused with ants, but termites are somewhat more flattened and lack the sharply constricted abdomen so characteristic of ants. The common eastern termite species, *Reticulitermes flavipes,* ranges from southern states north to Massachusetts. Termites, as a group, are much more abundant in the tropics than in the temperate zone.

Termites survive through a unique association with certain protozoa that inhabit their guts. These single-celled protozoa, most of which have numerous long, whiplike flagella (giving the name "flagellates" to the group), actually digest the cellulose in the wood eaten by the termite. The termite uses some of the products of protozoan digestion and thus both protozoans and termites benefit—yet another example of mutualism. If a termite is purged of its protozoans, it will continue to eat wood, but will not be able to digest it and will starve.

Spiders

Spiders are active predators that live in fields, on the forest floor, and foliage. Walk through a dew-laden field in early morning and you will be impressed by the abundance of spider webs made visible by dew cover. Like insects, spiders are members of the Arthropoda, but are in the class Arachnida, along with the mites, ticks, Harvestmen (Daddy-longlegs), and scorpions. Arachnids lack antennae, have simple eyes (usually 8) rather than compound eyes, and have 4 pairs of walking legs. Spiders have powerful jaws connected with poison glands and immobilize their prey with the paralytic effect of the poison. Once the prey is immobilized, the spider sucks out the soft inner tissue leaving an empty exoskeleton where once there was an insect. The dried corpses of former meals are readily visible in most spider webs.

The most common ground-dwelling spiders in our area are the wolf spiders of the family Lycosidae. These animals do not construct webs; they make burrows and forage in the litter, mostly at night. Eyeshine from wolf spiders is often easy to pick up by sweeping the forest floor at night with a strong-beamed flashlight. Wolf spiders are long-legged, brownish, mottled spiders, very well camouflaged as they scurry over leaves. Though their bite is irritating, they are not dangerous, unlike the famed Black Widow.

Harvestmen

Also called Daddy-longlegs, Harvestmen superficially resemble spiders. They are arachnids, but they belong to their own order, the Opiliones, and they lack the powerful jaws and poison glands of spiders. They are generally brownish and are most easily recognized by their 4 pairs of extremely long legs. If disturbed, they will vigorously wave the second pair at their tormentor. Active at night and abundant, Harvestmen feed on small insects, plant juices,

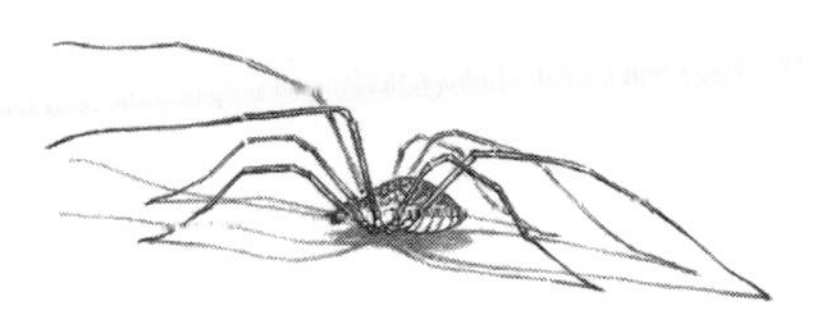

Fig. 62. Harvestman (Daddy-longlegs).

and decaying vegetation. Harvestmen tend to aggregate, so you may encounter many at a time.

Animals Extracted with a Berlese Funnel

Springtails

Abundant in the soil are the tiny insects of the order Collembola, the springtails. These diminutive gray, brown, and whitish animals occur in densities ranging from 500 to 50,000 per square meter. When a soil sample is put through a Berlese funnel (see above) springtails are usually among the most numerous of the groups extracted. Lacking wings and having simple wormlike bodies, short legs, and short, clublike antennae, springtails are considered among the most primitive of insects. They are named for a unique abdominal structure, not possessed by all springtails, which permits them to "spring" or jump suddenly.

Springtails are very important components of the decomposer food web. They feed on fungi, decaying plant material, and some live plant material. Some species are carnivorous. Springtails ingest bacteria, which may be an important food source for them, but they may also utilize bacteria, much as termites depend upon flagellate protozoans (see above). Certain bacteria may digest cellulose inside the gut of the springtail, making some of this digested material available to the insect, and hence a mutualistic relationship between insect and bacteria.

Springtails occur in soils of fields, meadows, and forests. However, they are most abundant in acidic soils with mor humus (see p. 417). Berlese extractions of soils from conifer forests usually abound with springtails. Some springtails (called "snow fleas") form dense aggregations on the surface of snow.

Soil Mites

Mites, like spiders, are arachnids, eight-legged little creatures, some so tiny that they are only visible under a microscope. Fortunately, most soil mites, though diminutive, can be observed with a magnifying glass or hand lens. Like springtails, mites are in the litter and soil in immense numbers. One abundant group of soil mites, the beetle mites, or oribatids, are rounded and glossy, often tan or deep brown in color. These mites suggest tiny turtles. Beetle mites, along with springtails (see above), are usually the most common soil animals extracted in a Berlese funnel. Beetle mites eat mostly fungi, and can significantly reduce fungal populations. They also feed on decomposing plant material and roundworms (see below). In addition to beetle mites, there are predatory mites called prostigmatids and metastigmatids. These animals look like tiny spiders. Their legs are much longer than those of beetle mites. They feast upon beetle mites, springtails, roundworms, and each other, including eggs and larvae.

Soil mites are present in all soil types but are most common in acidic soils topped with a mor-type humus (see p. 417).

False-scorpions

False-scorpions (also called pseudoscorpions) resemble tiny versions of the well-known scorpion, except that they lack the slender abdomen tipped with a sharply pointed poison gland. False-scorpions do have poison glands, but they are associated with the 2 large, lobsterlike claws, or pedipalps. False-scorpions are predators, feeding on mites, large springtails, and potworms (see below). Their predatory success depends on their ability to detect odors and sense vibrations, as they completely lack eyes and ears. Considering that they dwell in the dark soil, it is not unusual that sense organs other than those of sight and sound would develop in these creatures. They are not highly abundant and are probably most common in soils with a mor-type humus. A good Berlese funnel extraction of a mor-soil sample should yield a false-scorpion or two among the dozens of springtails and mites.

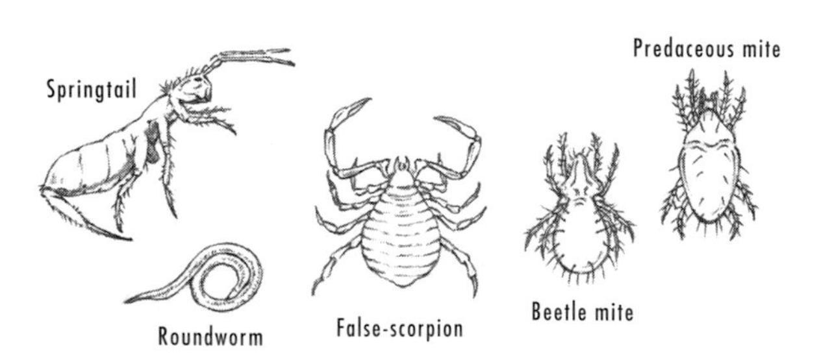

Fig. 63. Tiny soil animals.

Roundworms

Roundworms, or nematodes, are among the best-known parasites in the world. Such afflictions as trichinosis and hookworm are caused by parasitic roundworms. However, most roundworms are tiny, free-living soil-dwellers. They are easily differentiated from earthworms and potworms because they lack segmentation. They thrash their bodies back and forth—unlike segmented worms, they are unable to expand and contract. Mostly pale whitish, they are ubiquitous in all soil types as well as fresh water. Some soils have been estimated to contain nearly 20 million roundworms per square meter! They feed on fungi, bacteria, and decaying vegetation, though some are herbivores, feeding on plant roots. Roundworms are not numerous in Berlese extractions. Most are probably killed by the heat before they drop out of the soil.

Some fungi have turned an incredible evolutionary trick on roundworms. The fungal mycelium develops tiny "nooses," which tighten and trap foraging roundworms. The fungus then digests the roundworm instead of the other way around. In some fungal species, noose formation is stimulated by the presence of roundworms, possibly because the fungus detects a chemical secreted by the worms.

Potworms

The potworms, or Enchytraeidae, are tiny white versions of earthworms. Rarely exceeding 5 cm (2 in.), potworms seem to replace earthworms as the dominant segmented worm type in acidic mor-type soils, where they may number from 1000–100,000 per square meter. They feed on decomposing vegetation and fungi.

Fly Larvae

Flies are in the insect order Diptera. Adult flies are attracted to feces and carrion, where they lay eggs, and larval flies, sometimes called maggots, are common soil animals. Fly larvae are generally white or pale brown. They have distinct segmentation, lack legs, and may be bristly. Depending on species, they eat decomposing leaves, carrion, or feces. Some are parasitic on earthworms, mollusks, and other insects.

The Litter-Soil Food Web

The vast amount of organic material represented by fallen leaves, needles, twigs, branches, and tree trunks, as well as by animal carcasses and feces, supports the abundant soil animals. Often called *cryptozoa,* because they live essentially below ground, these animals inhabit the damp, dark recesses of the soil. They play a key ecological role in the process of biogeochemical cycling because fallen organic material makes up the base of the litter-soil food web.

At the base of the litter-soil food chain are the bacteria and fungi. They get up to 80–90% of the energy available in the fallen debris. The remainder of the plant and animal material, as well as the bacteria and fungi themselves, make up the food base for the soil animals. Many mites, springtails, ants, termites, roundworms, and protozoans feed primarily on bacteria and fungi. Others, such as earthworms, potworms, isopods, millipedes, and some springtails and mites feed directly on the organic litter. There are many generalists, animals that feed on both the fungal-bacterial component and the dead organic material of the soil. These include many roundworms, beetle mites, springtails, and fly and beetle larvae. Also, because of the close proximity of living roots to soil, it is not surprising that many soil-dwellers, such as slugs, beetle larvae, and roundworms, have become root-feeders. Such an abundance of animals provides ample food for predators, such as spiders, predatory mites, false-scorpions, centipedes, predatory roundworms, and beetles. The forest is as alive underground as it is in the canopy.

PLANTS IN FALL

Leaf Color Changes

OBSERVATIONS: Perhaps the most charming characteristic of Eastern Deciduous Forests is the succession of fall colors highlighting the

Oak Leaves. Changes in leaf color result from decomposition of chlorophyll, unmasking underlying pigments (as seen here), as well as from the chemical formation of new pigments.

termination of the growing season. Slowly at first, and then rapidly, green leaves change color. Streaks or blotches of orange-red, yellow, or tan appear, eventually covering the entire leaf. Green gives way to yellow, orange, scarlet, deep red, or shades of brown and tan. Note the progression of color among the trees in your area. Which turn color first, and which last? What colors characterize each species? In the North you will find proportionately more deciduous species and thus observe more leaf color changes. In the South, you will notice a larger percentage of evergreen species. Below are some of the most common species, and their respective fall colors.

OAKS

WHITE OAK—Orange-brown, sometimes quite reddish. Many leaves often remain on the tree throughout the winter, a characteristic of many oaks.

RED OAK (northern and southern) and **BLACK OAK**—Reddish brown; quite variable, often dull brown.

SCARLET OAK—Bright scarlet to deep red.

CHESTNUT OAK—Brownish, with orange tinge.

BEAR OAK—Yellowish brown.

BLACKJACK OAK, PIN OAK, TURKEY OAK—Reddish, with orange tinge.

POST OAK—Brown.

VIRGINIA LIVE OAK—Green! This is an evergreen oak.

HICKORIES

All hickories have bright yellow leaves in autumn.

MAPLES

RED MAPLE—Red, often very bright. Related to soil acidity. The more highly acidic the soil, the deeper the red.

SUGAR MAPLE—Bright orange-red, becoming yellow.

STRIPED MAPLE—Yellow.

BOX-ELDER—Yellow, occasionally reddish.

SILVER MAPLE—Pale yellow.

BIRCHES

Birch leaves turn shades of yellow, from bright (Yellow Birch) to relatively pale (Gray and Paper birches).

BLACK TUPELO (BLACK GUM) — Dark red, very deep.
AMERICAN BEECH — Light yellow, becoming increasingly brownish tan.
QUAKING ASPEN — Yellow, sometimes pale, sometimes deep.
BIGTOOTH ASPEN — Orange-yellow, becoming pale yellow.
SWEETGUM — Quite orange-red, but becoming yellow.
TULIPTREE — Bright yellow.
SASSAFRAS — Reddish, becoming yellow.
FLOWERING DOGWOOD — Deep red.
AMERICAN CHESTNUT — Brownish yellow.
WHITE ASH — Maroon, dark reddish green.
SUMACS — Orange, becoming bright red.
PIN AND WILD CHERRIES — Reddish, becoming bright yellow.
EASTERN COTTONWOOD — Yellow.
EASTERN SYCAMORE — Brown.
TAMARACK (AMERICAN LARCH) — Bright yellow needles.

EXPLANATION: The factor signaling the beginning of foliage color change is temperature. Times of peak color vary from year to year in relation to temperature trends. Early cool weather brings early fall colors. The combination of decreasing day length and cooler nights triggers activation of cells in the *abscission layer* between the leaf and stem. These cells eventually cut off the leaf from the stem, essentially by depriving it of water and minerals.

The pigment chlorophyll, which makes the leaf green, is not the only pigment present. Other pigments are masked by chlorophyll's abundance, but when chlorophyll breaks down, these accessory pigments are revealed. Pigments called carotenoids and xanthophylls give the yellow and brownish colors of many leaves. In addition, as chlorophyll and other pigments break down, they become brownish.

Some fall colors are byproducts. The reds in sumacs, Sugar Maple, and other species are caused by anthocyanin, a pigment produced by leaves with high sugar content. Once the abscission layer is formed, the leaf can no longer transport sugar to the stem, and the sugar in the leaf is gradually converted to anthocyanin. The production of anthocyanin is the reason why so many species turn reddish initially, but later become yellow. Anthocyanin is briefly produced, only to decompose and reveal other pigment. Abnormally dry weather can significantly reduce fall color intensity because the parched leaves do not produce enough sugar to make anthocyanin.

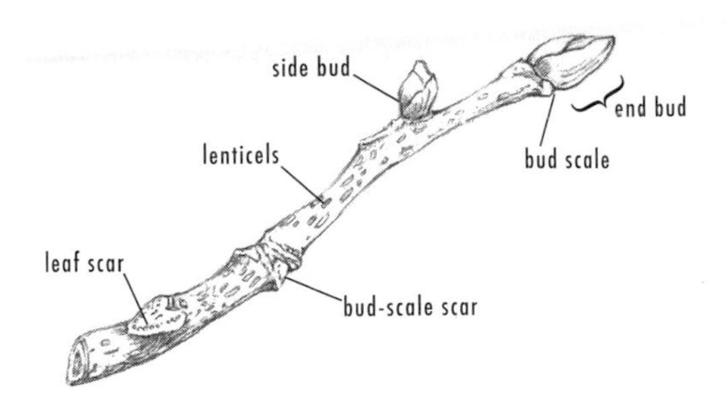

Fig. 64. Parts of a twig.

Anatomy of a Twig

After the leaves drop from the trees, next year's growth is contained in buds at the branch tips. A close look at a twig or branch will help you to understand the basics of tree growth. The tip of the branch contains the *end bud,* covered by *bud scales.* End buds vary in appearance and can be used to identify tree species during winter, when no foliage is present. The bud remains dormant throughout the winter, awaiting the warming temperatures and longer days of spring before opening. All growth of the branch occurs just below the area of the bud. Below the bud along the twig are *leaf scars,* marking where leaves have dropped off. Just above a leaf scar may be a *side bud,* usually considerably smaller than the end bud. A thickened concentric ring located some distance below the end bud is the *bud-scale scar,* marking the end bud of the previous year. You can measure the annual amount of growth by noting the distance from the bud-scale scar to the end bud. Tiny dots on the twig are really minute openings called *lenticels.* These function for gas exchange between the tree and the atmosphere. The tree takes in oxygen and emits carbon dioxide, just as we do.

Tree Trunks and Growth Rings

OBSERVATIONS: Sawed tree trunks or logs reveal much about the growth pattern of trees. Looking closely at the cross-section of a tree trunk with a hand lens should reveal growth rings, as well as two basic kinds of wood, *heartwood* and *sapwood* (see Fig. 65). The inner area, usually darkest in color, is called heartwood and the outer, thinner area is called sapwood. The bark rests against the sapwood. Growth rings can be seen in both heartwood and sap-

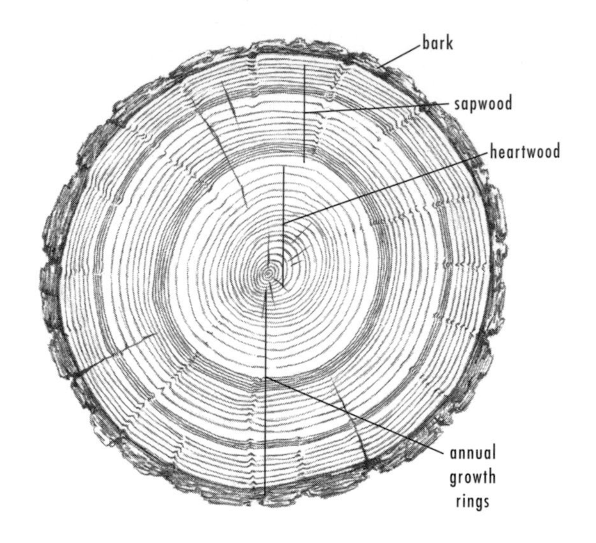

Fig. 65. Cross-section of tree trunk.

wood. A close look at the growth rings will show that they are not all equal. Some rings are considerably wider than others, and some may be tinged blackish. Try and count the growth rings (this requires a steady eye and patience). The number of growth rings is the age of the tree.

EXPLANATION: Only the sapwood is living tissue. The heartwood makes up the skeletal support system of the tree, but is no longer alive. The outermost part of the sapwood, called the cambium layer, is the actual growing part of the tree. It is the cambium layer that adds the annual growth ring. Growth rings are the result of seasonality. The tree grows only during the warm period of the year and is dormant throughout winter. Beginning in spring and continuing through the growing season, trees add tissue, producing the growth ring. Growth rings vary in width because growing seasons vary. Drought years result in narrow growth rings because the tree was stressed and did not have much energy to put into growth. Favorable seasons result in wider growth rings. Occasional blackish growth rings reveal that fire burned the bark and scorched the cambium layer.

OBSERVATIONS: Many species of trees in the temperate zone have "mast" years, when their seed production is extraordinarily high. Among the most common masting species are White Pine and other conifers such as firs and spruces, the various oaks and hickories, and American Beech. Other species also tend to have periodic bumper-crop seed years as well. You can easily observe masting by noting the relative numbers of conifer cones, acorns, beech nuts, etc., from year to year. Masting is conspicuous. Several years of low to moderate seed production will be followed by a year in which huge numbers are produced. Masting is synchronous over large areas so that almost *all* White Pines, not just a few, are masting. You will also see that mast-consuming species, such as squirrels and jays, are much in evidence during masting years. In years of poor seed crop, numbers of such species are very depressed. Seed-dependent birds tend to migrate in low seed years.

EXPLANATION: Why do trees go several years with little or no seed production and then suddenly "flood the market" with their reproductive products? Masting is probably an adaptation to aid the trees in escaping the potential ravages of seed predators. By alternating between occasional bumper crops and more usual poor crops, trees conserve energy, enabling them in a given year to produce more seeds than all seed predators combined could ever hope to eat. Though jays and squirrels eat and cache many acorns during a mast year, many still remain on the forest floor to sprout into oak seedlings. Masting in White Pines is responsible for the rather uniform age of many White Pines growing together in a field. Seeds first blow into the field during a mast year.

The **PERIODICAL CICADAS** are the insect equivalent of masting trees. These cicadas, which are sometimes mistakenly called "locusts,"

Fig. 66. Periodical cicada.

spend either 13 or 17 years as grubs underground before the adults emerge synchronously. The temporary abundance of cicadas assures that many will survive to breed, since there is no way that predators could consume so many in such a brief time. The 13- or 17-year period is an adaptation related very directly to reproductive success. Both numbers, 13 and 17, are prime numbers, not divisible except by themselves. This characteristic assures that predators with shorter life cycles (virtually all predators!) will not evolve to synchronize their life cycles with the cicadas.

Mayflies, which are aquatic as larvae, tend to emerge as adults at nearly the same time, creating clouds of swarming mayflies around summer ponds. Because they live very short adult lives, this synchrony aids mayflies both in finding mates and, as masting does, in releasing simultaneously far more insects than predators could possibly eat. Trees, unlike cicada or mayfly populations, are not nearly as predictable as to when masting will occur, probably because climatic factors that serve as masting triggers may vary considerably over several years.

BIRDS: MIXED FORAGING FLOCKS (PL. 51)

OBSERVATIONS: Beginning in midsummer, as nesting season approaches its end, and continuing throughout fall and winter into early spring, mixed foraging flocks of birds patrol forests and fields. Forest flocks will typically contain mostly insectivorous species such as chickadees, nuthatches, and woodpeckers, while field flocks are comprised mostly of seed-eaters, such as sparrows and juncos. Mixed foraging flocks, especially in woodlands, are active and nomadic, moving together through the habitat as each bird searches for food. A forest will appear empty of birds in winter, only to suddenly have the trees swarming with vocalizing chickadees, titmice, and other species. Once you encounter a mixed foraging flock, try to identify and note the abundance of each species, a task that is often not easy, as the birds move about quickly. You might try "pishing" by making a thin *psshh, psshh* sound, or perhaps kissing the back of your hand. These odd sounds, which may mimic distress calls, often attract foraging birds, and some may come very close to you.

In spring and fall mixed flocks usually contain both migrant and resident species. Migrating warblers, such as Yellow-rumped Warblers, American Redstarts, Magnolia Warblers, and Black-and-white Warblers, and many other species join mixed foraging flocks of resident species.

In woodland mixed flocks there is typically a "nucleus species," usually a member of the titmouse family, that tends to be both the most numerous and most vocal. The nucleus species seems to set the pace at which the flock moves. Insectivorous mixed flocks move through forests at a fairly quick pace. You will have to walk briskly to keep up with them. Sparrow flocks in fields also have nucleus species, often Tree Sparrows or Northern Juncos, and field flocks can, occasionally, be made up of but a single species. Field flocks are more sedentary, often remaining in the same field for days. The difference in mobility between forest and field flocks is because insect food is more widely scattered and far less abundant than weed seeds. Insectivorous birds must constantly be searching for food that is in low abundance, and thus must cover a lot of ground to achieve success. Seed-eaters often locate a rich field, abundant with goldenrod, ragweed, and other seed sources, and remain to exploit the food source until it is depleted.

In insectivorous mixed flocks, notice *how* each species feeds. Chickadees, such as the Black-capped and Carolina chickadees, tend to feed on the outer branches, often high in the tree. The Tufted Titmouse tends to search the thicker inner branches and often picks among the leaves on the ground. In a mixed flock containing both Carolina Chickadees and Tufted Titmice, both species may feed on the ground, but titmice do this more frequently than chickadees. Warblers, like chickadees, tend to search the canopy and outer branches, but one species, the Black-and-white Warbler, feeds almost the way a nuthatch does, methodically probing the thick branches and trunk. Kinglets are very active and often hover momentarily as they snatch insects from the outermost branches. Forest mixed-species flocks also contain several bark-feeders. Woodpeckers drill into the wood, extracting insects with their long, barbed tongues. Nuthatches poke beneath bark and grab insects with their slightly upturned bills. The Brown Creeper methodically spirals up a tree trunk snaring tiny spiders and insects that it encounters with its thin, downcurved, forceps-like bill (see Pl. 51).

Seed-eaters also accompany woodland mixed foraging flocks. Northern Juncos, White-throated Sparrows, Rufous-sided Towhees, and Fox Sparrows hunt among the leaf litter.

A typical mixed flock in the Northern Hardwood or Boreal Forest contains different bird species from those of mixed flocks in the southern and central states, as the following lists show.

BIRDS OF MIXED FLOCKS IN NORTHERN FORESTS

Black-capped Chickadee
Boreal Chickadee (boreal region only)

White-breasted and/or Red-breasted Nuthatch
Downy Woodpecker (occasionally also the Hairy)
Brown Creeper
Golden-crowned Kinglet
Yellow-rumped Warbler (southern areas)
Northern Junco
White-throated Sparrow

BIRDS OF MIXED FLOCKS IN SOUTHERN STATES

Carolina Chickadee
Tufted Titmouse
White-breasted Nuthatch
Brown-headed Nuthatch (pines only)
Downy and Red-bellied woodpeckers
Yellow-throated Warbler
Yellow-rumped Warbler
Pine Warbler

EXPLANATION: Why do birds forage in mixed-species flocks? Food can be very scarce in winter and it would seem that foraging in a group is counterproductive, in that it could intensify competition. Wouldn't an individual be more assured of success by foraging quietly alone? Flocking is a form of dynamic clumping of individuals. The fact that birds, not only in our area, but also in tropical forests (where mixed flocks can have several dozen species), forage in mixed-species flocks, suggests that there must be some strong advantage to each individual that remains with the group.

One obvious potential advantage of flock membership is added protection from predators (see pp. 295, 307). Many pairs of eyes and ears are superior to one. Predator detection is much facilitated by flocking, and a predator has a more difficult time concentrating on a single subject amongst the cacaphony of an excited, dynamic flock. Flock membership affords significantly increased awareness of predators, and is thus a great advantage to each individual member. Each bird can forage for a longer time period because other birds are also on watch for possible predators, making foraging more efficient. In the tropics, certain species act as "sentinels," warning the others through loud and specific call notes, when a predator is detected. Titmice and chickadees probably act as sentinels in our area. They become very noisy upon detecting a Screech Owl, weasel, or other potential predator. Only when a predator actually makes an attack do these birds become quiet. Should a Sharp-shinned Hawk make a sweep at a foraging flock, the birds will remain quiet and out of sight in the underbrush for many minutes after the attack.

Efficient feeding may be facilitated by flock membership. Moving with a flock systematically through a habitat may ensure that each section of habitat is thoroughly exploited to each bird's ultimate advantage. Flock membership may facilitate the discovery of food sources (again, many pairs of eyes are superior to one). Note how quickly nuthatches and Downy Woodpeckers visit suet feeders after chickadees discover them.

Why do insectivorous species tend to specialize, feeding in a certain manner? Feeding specialization is evident both in behavior and in anatomy. Some birds are active gleaners, taking insects from dried leaves, twigs, etc. Others forage on the ground among the leaf litter, some kicking up the dried leaves. Still others tend to remain on the tree trunk. Some have thin, forceps-like bills; others have thick, seed-crushing bills; and still others have strong, chisel-like bills for drilling into bark. Feeding specialization reflects the fact that food is distributed in many kinds of packages in nature. One method of feeding or one bill type is not sufficient to successfully exploit all food types. Specialization for feeding may even be evident in the entire anatomy of the bird. A woodpecker and creeper are each totally adapted to life on vertical tree trunks. Each has a stiff tail that acts as a prop, holding the bird up against the trunk. Feeding specialization is, therefore, an evolutionary response to the diversity of potential food sources. Just as tools are molded to do specific tasks, so bird behaviors and bill characteristics enable their owners to do certain tasks very well. Insects and spiders are well camouflaged and well protected under leaves and bark. Birds have to work to find and catch them. A bird's behavior and bill are the tools of its trade.

Feeding specialization may also significantly reduce competition among species. Several species of warblers in northern coniferous forests feed in the same spruce tree at the same time (see p. 65). They do not, however, feed in exactly the same place in exactly the same way. One species will tend to feed on the topmost branches, one will concentrate on inner branches, one will feed low on the tree, one will hover, picking caterpillars off the outermost branches. The same is true of mixed foraging flocks in general. By specializing, each species "captures" a certain range of food resources for itself, exploiting this source better than any other species.

Irruptions of Bird Species (Pl. 52)

OBSERVATIONS: During fall and winter there are occasional large-scale movements of certain bird species into northern, central, and, occasionally, southern states. These dramatic mass movements, called *irruptions,* are unusual both because they involve large numbers of birds and because, unlike migration, they are not generally predictable. A given year may or may not witness the invasion of irruptive species. You can look in vain for crossbills, siskins, redpolls, and Snowy Owls for many years, only to be inundated by them without warning during a given winter. There is no local indication that an irruption will occur. The events setting off the mass movement occur far from the area where the birds eventually arrive. Irruptions involve bird species that nest in the northern states and/or Canada, which "erupt" from their nesting ranges and "irrupt" into more southern latitudes. Two general categories of irruptive species exist — seed-eaters and raptors.

IRRUPTIVE SEED-EATING SPECIES

PINE GROSBEAK — berries, ash and conifer seeds

EVENING GROSBEAK — conifer seeds, Box-elder and ash seeds, and sunflower seeds at bird feeders

PURPLE FINCH — a generalist, feeding on many different seeds

RED AND WHITE-WINGED CROSSBILLS — conifer seeds

PINE SISKIN — birch and alder seeds

COMMON REDPOLL — birch and weed seeds

RED-BREASTED NUTHATCH — pine and spruce seeds

BLACK-CAPPED CHICKADEE — generalist seed-feeder; also feeds on arthropods. Erupts only in northern part of its range.

BOREAL CHICKADEE — conifer seeds; also arthropods

BOHEMIAN WAXWING — berries, especially Mountain-ash (Rowanberry)

IRRUPTIVE RAPTOR SPECIES

SNOWY OWL — lemmings, voles, hares

GREAT GRAY OWL — lemmings, voles, hares

NORTHERN HAWK-OWL — lemmings, voles, birds

NORTHERN GOSHAWK — birds, hares, lemmings, voles

ROUGH-LEGGED HAWK — hares, voles, lemmings

NORTHERN SHRIKE — a passerine, not a raptor, but feeds on mice, small birds

The appearance of irruptive species is called a *flight year,* and flight years vary in degree of irruptiveness. In some years, a few Snowy Owls may invade the mid-central states, a poor irruption. Other years may bring many owls. Swirling flocks of Common Redpolls, often numbering hundreds of individuals, winter in weedy fields and birch clumps during a good flight year. Crossbills are particularly sporadic, often being absent for many years, only to invade in large numbers during a good flight year. Irruptive species generally move from north to south but may also move west to east. The Evening Grosbeak was originally not an eastern species, but moved eastward during flight years. Its range is now firmly established in the East.

EXPLANATION: Irruptions of bird species are thought to be caused by periodic unpredictable food shortages in the breeding ranges of these species. Seed-eating species may irrupt in years following the cessation of *masting* (see p. 441). Many young are produced when seeds abound during masting, producing an overpopulation. When seed crops drop precipitously (in a *crash*), seed-dependent species such as crossbills, Pine Grosbeaks, and Pine Siskins are forced southward. Irruptive raptors such as the Snowy Owl, Great Gray Owl, and Rough-legged Hawk are dependent on lemming populations, which are highly *cyclic* (see below). The appearance of large numbers of individuals of these species signals a crash in the Arctic lemming population. Not all individuals of the irruptive species leave the nesting areas, however. Irruptive flocks tend to be comprised predominantly of young birds. Of adults, females seem to outnumber males, though data are not well established on this point.

Winter Ranges of the Northern Junco

OBSERVATIONS: The Northern Junco (Pl. 6) is a common winter resident in weed fields and woodland undergrowth throughout eastern North America. If you look carefully at junco flocks in your area, you will see that some birds are dark slate gray, others a warmer brownish gray. The dark slaty birds are males, the brownish gray ones females. Juvenile males are also brownish and can be confused with females, but full adult males are usually readily identifiable by their dark gray hoods. As you survey junco flocks, note the proportion of adult males to females. You can do this with feeder flocks or flocks you encounter in fields and woodlands. If you live in the northern states, your junco flocks should contain mostly adult males. If you are in the central states, the flocks will be divided about evenly between males and females. Southern junco flocks have up to 70% females. There is, therefore, a trend toward increasing proportions of females and juvenile males in the winter range of the Northern Junco, going from north to south. This trend also is seen in other species, including the Northern Mockingbird, Tree Sparrow, Song Sparrow, and Mourning Dove, but none of these species is sufficiently sexually dimorphic (see Chapter 6) to permit you to accurately sex the birds in the field.

EXPLANATION: Why do male juncos winter farther north than females? Or, why do females winter farther south than males? Could the differing winter ranges of the two sexes represent adaptations? Though it is not possible to see this in the field, males are a bit larger than females. The average wing length for males is 81 mm (approx. 4¼ in.) and for females is 76.5 mm (approx. 3 in.). Males average 1.22 grams more in weight than females (average junco weight is 22 grams). One characteristic noted for many warm-blooded (endothermic) animals is that individuals are larger in the northern, colder regions of the species' range. Large body size seems adaptive in cold regions. This trend applies to animals from Grizzly Bears to Robins. It is such a general trend that it is called Bergmann's Law. Could the larger body size of male juncos be related to their more northern wintering range?

The advantage of large body size in colder climates is that it maximizes the heat-producing part of the animal, its volume, relative to the heat-losing part, its surface area. Animals with a bulky "furnace" are better able to produce the heat needed to survive winter's cold. Small animals are more prone to lose heat. The easiest way to understand this is to consider a block of ice versus an equal volume of ice cubes. Which will melt first, the block or the cubes? The answer is, of course, the cubes. A small ice cube exposes much surface area to melting relative to its volume, which retains the cold. An ice cube melts quickly. A block of ice minimizes surface area to volume and thus melts far more slowly. The larger volume of junco males means that at freezing temperature (32 degrees F), the average male junco should be able to fast 4% longer than the average female, a time period amounting to about 1.6 hours. A heavy male might be able to stand fasting for nearly 11 hours longer than a light female, quite a difference. These differences may seem trivial to humans, but they are not trivial to birds. Snow, ice, cold, and blustery winds all prevent birds from feeding. To be able to stand an additional hour or so without food can be a huge advantage in surviving the rigors of a northern winter. Males, because they are larger, are probably better able to survive the northern winter than females.

But, why don't males "make it easy on themselves" and go south with the females? There are two possible reasons why males stay far to the north. First, migration is both risky and costly in terms of energy. Migrating birds are constantly exposed to unfamiliar terrain and possible predation, and having to feed intensively to obtain fuel to continue their journeys. A brief migration for males may be the safest and certainly the "cheapest" kind of migration. Conversely, males may be the reason females winter farther south. Juncos are very pugnacious birds. Watch them interacting at a bird feeder and see how frequently one bird will displace another. Females may be just as "willing" to remain in northern areas but are forced out by larger, more dominant males. Secondly, males compete among themselves for breeding territories. Those that winter farthest north are closest to the breeding range and have the best chance of getting back quickly in the early spring and securing territories. Being able to return to the breeding grounds quickly may give some males a very essential benefit in reproduction.

Winter Adaptations of Birds

Birds are active throughout cold winter days. A small passerine like a Black-capped or Carolina Chickadee has a normal body temperature of 107 degrees F, a temperature many degrees above normal air temperature. On a day when the mercury is zero or below, chickadees and other species forage without apparent discomfort.

A thick covering of body feathers provides birds with effective insulation. Birds often fluff their body feathers on cold days. This behavior helps trap warm air from their own bodies. By moving its body feathers, fluffing them when cold and smoothing the feathers when warm, a bird helps control its degree of insulation, retaining or losing body heat as the situation demands. A chickadee has approximately 2000 body feathers and has muscular control over their movement. Raising feathers to trap heat and lowering feathers to cool down are the means by which birds adjust to cope with temperature stress. Birds also remain generally dry when it rains or snows because they coat their body and flight feathers with oil secreted by a gland at the upper base of the tail. This oil is very water-repellent and permits the bird's body to remain dry while the bird is feeding during a pelting rainstorm.

Birds feed on suet and visit bird feeders more frequently in cold weather and during times of snow cover. Cold stresses the bird's physiology, forcing it to produce added heat to remain warm. Suet is very high in readily digestible calories and birds such as the tiny kinglets and overwintering warblers can survive extreme cold by eating suet. Sunflower seed is also oily, full of fats, and provides much energy to birds. Both suet and sunflower seed help birds produce sufficient oil to keep feathers dry. When the ground is covered with snow, ground-feeding species are forced to feed on only those weeds protruding above the snow. Ice storms pose very serious situations for birds because branches are covered with hard ice, preventing foraging birds from reaching their food sources.

Birds are normally active from dawn until dusk, and spend most of their winter days foraging. During winter, birds spend a far greater proportion of time in search of food than they do during the summer. However, birds may remain relatively inactive on very cold winter days, especially during inclement weather. The energy lost foraging for scarce food is too costly; it is in the bird's physiological best interest to remain inactive and protected, tucked in a dense shrub or cedar clump, preserving as much of its body heat as possible until conditions moderate.

Many birds roost in tree cavities in winter. Species such as Brown Creeper, White-breasted Nuthatch, Winter Wren, and Eastern Bluebird often roost together, crowded densely into a bird box or tree cavity. Other species, such as the Eastern Screech Owl, Downy and Hairy woodpeckers, and Black-capped and Carolina chickadees, roost singly. In a confined cavity, the bird's body heat is trapped and the bird is further protected from possible wind chill. Birds are required to use far less energy to remain warm when protected inside cavities.

Finches, sparrows, crows, jays, and doves roost in dense clumps of cedars or other conifers. The thick cover of evergreen branches helps reduce wind chill, providing a warmer surrounding for birds. Both cavities and evergreen roosts also provide an added measure of protection against nocturnal predators.

Blackbirds, such as Red-winged Blackbirds, Common Grackles, and Brown-headed Cowbirds, roost in immense flocks in southern states. These aggregations afford a large measure of protection from predation (see flocking, p. 295).

MAMMALS: VOLE CYCLES

OBSERVATIONS: Meadow voles periodically undergo major population increases, followed by precipitous decreases in cyclic fashion. Though the cycles are somewhat irregular, they tend to occur approximately every 2–4 years, a cycle similar to that of lemmings in the Arctic. Often, vole cycles are generally synchronous over large areas. You will see voles during the peak of their population cycle much more easily than at other times. These husky, dark brown rodents live in dense, grassy fields and their runways are very apparent when they are abundant. Occasionally, voles are so abundant that the grass crop is obviously depleted. A walk through such a field often reveals scurrying voles underfoot. Another indicator of vole abundance, particularly in winter, is the abundance of raptors (American Kestrel, Red-tailed Hawk, Rough-legged Hawk) and diurnal owls (Snowy, Great Gray, and Hawk owls) around vole fields. Mammalian predators, including feral (stray) cats, are also more abundant during peak vole years.

EXPLANATION: Why do voles have population cycles? Several other animal populations have cycles. In the Arctic, both Collared Lemmings (close relatives of voles) and Arctic Hares undergo population cycles, but of differing durations. The Arctic Hare cycle is well known because the abundance of the hares influences that of the Lynx, their major predator. Lemmings cycle at approximate 4-year intervals, but hares cycle at 10-year intervals. Lemming predators include Snowy Owls, jaegers, and Arctic Foxes. Snowy Owl clutch size is influenced by lemming abundance. Lemming-rich years are those in which the owls lay the most eggs and fledge the most young. Ruffed Grouse and Snowshoe Hares, neither of which lives in the Arctic, also cycle in approximately 10-year intervals. Thus far no single causative factor has been shown to be responsible for the cycles.

Many factors have been suggested as causes for the cycles. One obvious cause could be that voles deplete their food source as they reproduce and thus their population eventually crashes for lack of food. A more sophisticated suggestion is that the rodents eat the most nutritious foods when their populations are lowest, but high populations force them to rely more on plants with toxic compounds in their leaves, thus poisoning the voles and causing a reduction in their population. So far, there is little evidence to suggest that either idea is accurate. Voles can be very abundant without seriously depleting their grassy cover, though some reduction can often be noticed.

Another suggestion is that voles become more aggressive as their population increases and that the increased level of aggression interferes with their reproductive success, thus reducing the population. Male voles tend to disperse more during the increase phase of the cycle, and voles with wounds, presumably caused by fighting, have been observed during high-population times. However, behavioral studies on vole "psyches," as affected by population density, have thus far failed to conclusively show that behavioral changes are the root cause of the cycle.

Both weather factors and predators affect vole populations, especially when populations are declining. Predators, attracted by vole abundance, remain during vole decline, placing even more pressure toward reducing vole population size. But no evidence exists to show that either weather effects or predators initiate the cycle.

Physiological changes in the immunological systems of the voles, perhaps induced by the presence of microparasites, have also been suggested as a factor responsible for the cyclic populations. As the population increases, voles are increasingly stressed, and crowding causes a parasite epidemic that reduces vole numbers. No conclusive study exists to support this idea either.

The phenomenon of vole cycles is real. The explanation is thus far elusive. Try to observe a vole cycle and consider for yourself its possible causes.

MIXED FORAGING FLOCKS (P. 442)

The plate shows a mixed flock foraging in a woodland; the birds shown could be found throughout most of our area. Species composition of a flock will vary geographically and with the season.

Branch Foragers

BLACK-CAPPED CHICKADEE

Probes and gleans for small arthropods. Often hangs upside-down, and often feeds on upper and outer branches. See Pl. 7.

TUFTED TITMOUSE

A close relative of the Chickadee, but slightly larger. Also probes and gleans but tends to favor lower branches, closer to trunk, and will often forage among the ground litter. See Pl. 14.

GOLDEN-CROWNED KINGLET

Smaller than a Chickadee; olive, with 2 wing bars. Male has an orange crown, female a yellow crown. Very active feeder, often hovering momentarily to pick an insect from an outer branch. Favors conifers.

YELLOW-RUMPED WARBLER

Yellow rump; yellow on head, forewing. Joins mixed foraging flocks when migrating. Feeds on arthropods and forages anywhere from high in the tree to ground level.

Bark Foragers

BLACK-AND-WHITE WARBLER

Streaked black and white, like an avian zebra. Forages for arthropods on bark, creeping methodically over trunk and large limbs. May also forage in canopy.

BROWN CREEPER

A little brown bird with a stiff woodpeckerlike tail and a slender decurved bill. Spirals up tree trunks, probing for arthropods on bark.

WHITE-BREASTED NUTHATCH

"Upside-down bird"—forages by climbing head-first down tree trunks, probing the bark for arthropods. Black cap, blue back. Bill turned slightly upward. Call a nasal *yank, yank.*

DOWNY WOODPECKER

The smallest of our woodpeckers. Small, straight bill and black spots on white outer tail feathers; otherwise, like Hairy Woodpecker (see Pl. 39). Hitches itself up trunks and branches, drilling for arthropods. May forage on outer branches also.

male
YELLOW-RUMPED
WARBLER
male
GOLDEN-
CROWNED
KINGLET
BLACK-CAPPED
CHICKADEE
male
DOWNY
WOODPECKER
BLACK-AND-WHITE
WARBLER
male
WHITE BREASTED
NUTHATCH
TUFTED
TITMOUSE
BROWN
CREEPER
DOWNY
WOODPECKER
WHITE-BREASTED
NUTHATCH
BROWN
CREEPER

IRRUPTIVE BIRD SPECIES — PERIODIC WINTER INVADERS (P. 445)

EVENING GROSBEAK

A chunky, 8-in. finch with a large white bill. Males golden, with white wing patches and a yellow stripe above eye; females grayish yellow. Flocks are attracted to ash trees and bird feeders with sunflower seed. Call a pleasant two-note warble.

PINE GROSBEAK

A robin-sized, long-tailed finch with a thick black bill. Adult males very rosy; juvenile males and females grayish, with an orange-yellow rump. Often very tame. Flocks feed on crabapples, ash and conifer seeds. Rarely comes to bird feeders. Call note a thin, whistled *chew, chew, chew.*

GREAT GRAY OWL

A very large (up to 33 in.) gray owl with large gray facial disks and yellow eyes. No ear tufts. Roosts in conifers; feeds on rodents and rabbits in open fields, often during the day. Very easy to approach.

COMMON REDPOLL

Goldfinch-sized. Streaked with brown; red patch on forehead. Males have a rosy breast. Flocks, often containing over 100 birds, occur in old fields and feed on ragweed and other seeds. Attracted to thistle feeders. Call note, given on the wing, a dry chatter.

PINE SISKIN

Resembles a Redpoll, but lacks any red and is more darkly streaked. Yellow wing bars; yellow on tail. Bill sharply pointed. Flocks favor old fields, where they feed on thistle and other seeds. Attracted to bird feeders. Call note a strident, dry *chee-up.*

RED CROSSBILL

A chunky, 6-in. finch with a crossed bill, visible at close range. Males brick red; females greenish yellow, bright on rump. Flocks favor conifers. White-winged Crossbill (see Pl. 4) is also an irruptive species. Both species are often tame.

SNOWY OWL

A large (up to 27 in.) white owl with blazing yellow eyes. No ear tufts. Males almost pure white; females streaked with brown. Frequents open areas, shorelines. Diurnal.

EVENING
GROSBEAK
male
PINE GROSBEAK
male
GREAT
GRAY OWL
PINE SISKIN
male
COMMON
REDPOLL
male
RED
CROSSBILL
male
SNOWY OWL

SELECTED FUNGI (P. 418)

Each species shown represents a major group of fungi.

CUP FUNGI (**Pezizales**)
Small, mushroomlike, but cup upturned. Cup fungi grow on decaying bark and logs and are generally colorful. This species is **Blue-green Cup** *(Chlorosplenium aeruginosum)*.

BOLETES or TUBE MUSHROOMS (**Boletaceae**)
Boletes have small, tubelike openings (pores) instead of gills under the cap. These fungi range in color from orange-red to yellow or brown. Shown is **Bitter Bolete** *(Tylopilus felleus)*.

AMANITAS (**Amanitaceae**)
A ***highly poisonous*** group, containing such species as **Destroying Angel** *(Amanita virosa)* and **Deathcap** *(A. phalloides)*. Color ranges from pure white (Destroying Angel) to gray, pale green, yellow, brown, or orange-red. Shown is **Fly Agaric** *(A. muscaria)*.

BRACKET or PORE FUNGI (**Polyporaceae**)
Grow horizontally from logs and trunks. Some species are large (up to 20 in.) and colorful. Most are at least partly woody. Underside of bracket lined with rounded pores, sometimes very tiny. Shown is **Sulphur Shelf** *(Laetiporus sulphureus)*.

EARTHSTARS (**Geastraceae**)
Starlike shape, with rays radiating from a puffball-like center. Closely related to puffballs (Lycoperdaceae). Spores "puff" from the center sac. Found in sandy soils, hardwood forests. **Collared Earthstar** *(Geastrum triplex)* is shown.

CHANTERELLES (**Cantharellaceae**)
Cup with wavy edges; underside with large, thick-ridged gills that run down the stalk. Colors quite variable from one species to the next. **Scaly Chanterelle** *(Gomphus floccosus)* is ***poisonous.***

CORAL FUNGI (**Clavariaceae**)
Ephemeral clusters of pale orange or white fungi that resemble fingers of coral. These fungi lack gills. They grow from moist, decaying logs or directly from the forest floor. A few species are mycorrhizal (see text). **Crown Coral** *(Clavicorona pyxidata)* shown.

STINKHORNS (**Phallaceae**)
Highly ephemeral. No cup; erect stalk with a sharply pointed, foul-smelling tip that attracts flies, carrion beetles. Stinkhorns grow from soil or decaying wood. **Dog Stinkhorn** *(Mutinus caninus)* is shown.

JELLY FUNGI (**Tremellales**)
Slimy-looking but hard to the touch. Some bright orange-red or yellow. Grow on decaying logs, bark. **Witches' Butter** *(Tremella mesenterica)* is shown.

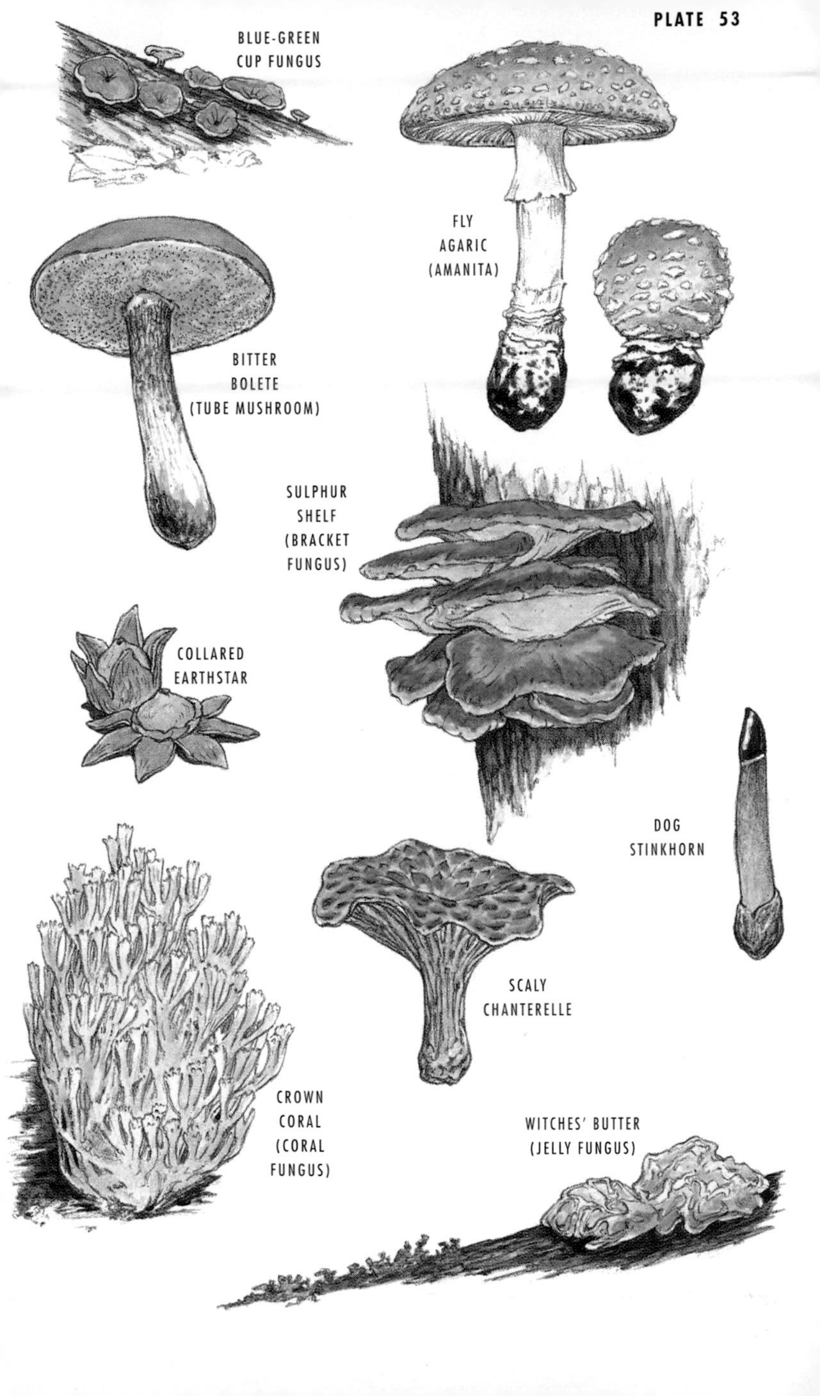
BLUE-GREEN
CUP FUNGUS
FLY
AGARIC
(AMANITA)
BITTER
BOLETE
(TUBE MUSHROOM)
SULPHUR
SHELF
(BRACKET
FUNGUS)
COLLARED
EARTHSTAR
DOG
STINKHORN
SCALY
CHANTERELLE
CROWN
CORAL
(CORAL
FUNGUS)
WITCHES' BUTTER
(JELLY FUNGUS)

INDEX

INDEX

Numbers in *italics* refer to pages where drawings appear; references in **boldface** type are to illustrations on the plates at the end of each chapter.

THE PETERSON SERIES®

PETERSON FIELD GUIDES®

BIRDS

ADVANCED BIRDING (39) North America 53376-7
BIRDS OF BRITAIN AND EUROPE (8) 66922-7
BIRDS OF TEXAS (13) Texas and adjacent states 92138-4
BIRDS OF THE WEST INDIES (18) 67669-X
EASTERN BIRDS (1) Eastern and central North America 91176-1
EASTERN BIRDS' NESTS (21) U.S. east of Mississippi River 48366-2
HAWKS (35) North America 44112-9

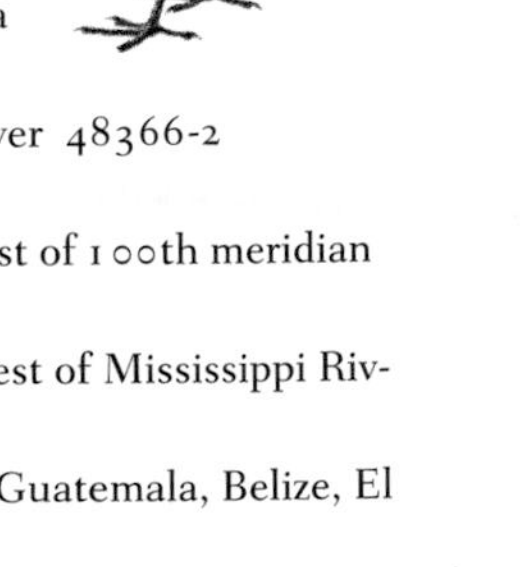

WESTERN BIRDS (2) North America west of 100th meridian and north of Mexico 91173-7
WESTERN BIRDS' NESTS (25) U.S. west of Mississippi River 47863-4
MEXICAN BIRDS (20) Mexico, Guatemala, Belize, El Salvador 48354-9
WARBLERS (49) North America 78321-6

FISH

PACIFIC COAST FISHES (28) Gulf of Alaska to Baja California 33188-9
ATLANTIC COAST FISHES (32) North American Atlantic coast 39198-9
FRESHWATER FISHES (42) North America north of Mexico 91091-9

INSECTS

INSECTS (19) North America north of Mexico 91170-2
BEETLES (29) North America 91089-7
EASTERN BUTTERFLIES (4) Eastern and central North America 90453-6
WESTERN BUTTERFLIES (33) U.S. and Canada west of 100th meridian, part of northern Mexico 41654-X
EASTERN MOTHS North America east of 100th meridian 36100-1

MAMMALS

MAMMALS (5) North America north of Mexico 91098-6
ANIMAL TRACKS (9) North America 91094-3

ECOLOGY

EASTERN FORESTS (37) Eastern North America 9289-5
CALIFORNIA AND PACIFIC NORTHWEST FORESTS (50) 92896-6
ROCKY MOUNTAIN AND SOUTHWEST FORESTS (51) 92897-4
VENOMOUS ANIMALS AND POISONOUS PLANTS (46) North America north of Mexico 35292-4

PLANTS

EDIBLE WILD PLANTS (23) Eastern and central North America 31870-X
EASTERN TREES (11) North America east of 100th meridian 90455-2
FERNS (10) Northeastern and central North America, British Isles and Western Europe 19431-8
MEDICINAL PLANTS (40) Eastern and central North America 92066-3
MUSHROOMS (34) North America 91090-0
PACIFIC STATES WILDFLOWERS (22) Washington, Oregon, California, and adjacent areas 91095-1
ROCKY MOUNTAIN WILDFLOWERS (14) Northern Arizona and New Mexico to British Columbia 18324-3
TREES AND SHRUBS (11A) Northeastern and north-central U.S. and southeastern and south-central Canada 35370-X
WESTERN TREES (44) Western U.S. and Canada 90454-4
WILDFLOWERS OF NORTHEASTERN AND NORTH-CENTRAL NORTH AMERICA (17) 91172-9
SOUTHWEST AND TEXAS WILDFLOWERS (31) 36640-2

EARTH AND SKY

GEOLOGY (48) Eastern North America 66326-1
ROCKS AND MINERALS (7) North America 91096-X
STARS AND PLANETS (15) 91099-4
ATMOSPHERE (26) 33033-5

REPTILES AND AMPHIBIANS

EASTERN REPTILES AND AMPHIBIANS (12) Eastern and central North America 90452-8
WESTERN REPTILES AND AMPHIBIANS (16) Western North America, including Baja California 38253-X

SEASHORE

SHELLS OF THE ATLANTIC (3) Atlantic and Gulf coasts and the West Indies 69779-4
PACIFIC COAST SHELLS (6) North American Pacific coast, including Hawaii and the Gulf of California 18322-7
ATLANTIC SEASHORE (24) Bay of Fundy to Cape Hatteras 31828-9
CORAL REEFS (27) Caribbean and Florida 46939-2
SOUTHEAST AND CARIBBEAN SEASHORES (36) Cape Hatteras to the Gulf Coast, Florida, and the Caribbean 46811-6

PETERSON FIRST GUIDES®

ASTRONOMY 93542-3
BIRDS 90666-0
BUTTERFLIES AND MOTHS 90665-2
CATERPILLARS 91184-2
CLOUDS AND WEATHER 90993-6
DINOSAURS 52440-7
FISHES 91179-6
INSECTS 90664-4
MAMMALS 91181-8
REPTILES AND AMPHIBIANS 62232-8
ROCKS AND MINERALS 93543-1
SEASHORES 91180-X
SHELLS 91182-6
SOLAR SYSTEM 52451-2
TREES 91183-4
URBAN WILDLIFE 93544-X
WILDFLOWERS 90667-9
FORESTS 71760-4

PETERSON FIELD GUIDE COLORING BOOKS

BIRDS 32521-8
BUTTERFLIES 34675-4
DESERTS 67086-1
DINOSAURS 49323-4
ENDANGERED WILDLIFE 57324-6
FISHES 44095-5
FORESTS 34676-2
INSECTS 67088-8
MAMMALS 44091-2
REPTILES 37704-8
SEASHORES 49324-2
SHELLS 37703-X
TROPICAL FORESTS 57321-1
WILDFLOWERS 32522-6

PETERSON NATURAL HISTORY COMPANIONS

LIVES OF NORTH AMERICAN BIRDS 77017-3

AUDIO AND VIDEO

EASTERN BIRDING BY EAR
cassettes 50087-7
CD 71258-0

WESTERN BIRDING BY EAR
cassettes 52811-9
CD 71257-2

EASTERN BIRD SONGS, Revised
cassettes 53150-0
CD 50257-8

WESTERN BIRD SONGS, Revised
cassettes 51746-X
CD 51745-1

BACKYARD BIRDSONG
cassettes 58416-7
CD 71256-4

MORE BIRDING BY EAR
cassettes 71260-2
CD 71259-9

WATCHING BIRDS
Beta 34418-2
VHS 34417-4

PETERSON'S MULTIMEDIA GUIDES: NORTH AMERICAN BIRDS
(CD-ROM for Windows) 73056-2

PETERSON FLASHGUIDES™

ATLANTIC COASTAL BIRDS 79286-X
PACIFIC COASTAL BIRDS 79287-8
EASTERN TRAILSIDE BIRDS 79288-6
WESTERN TRAILSIDE BIRDS 79289-4
HAWKS 79291-6
BACKYARD BIRDS 79290-8
TREES 82998-4
MUSHROOMS 82999-2
ANIMAL TRACKS 82997-6
BUTTERFLIES 82996-8
ROADSIDE WILDFLOWERS 82995-X
BIRDS OF THE MIDWEST 86733-9
WATERFOWL 86734-7
FRESHWATER FISHES 86713-4

WORLD WIDE WEB: http://www.petersononline.com

PETERSON FIELD GUIDES can be purchased at your local bookstore or by calling our toll-free number, (800) 225-3362.

When referring to title by corresponding ISBN number, preface with 0-395.